International Library of Ethics, Law, and the New Medicine

Volume 66

The book series International Library of Ethics, Law and the New Medicine comprises volumes with an international and interdisciplinary focus. The aim of the Series is to publish books on foundational issues in (bio) ethics, law, international health care and medicine. The 28 volumes that have already appeared in this series address aspects of aging, mental health, AIDS, preventive medicine, bioethics and many other current topics. This Series was conceived against the background of increasing globalization and interdependency of the world's cultures and governments, with mutual influencing occurring throughout the world in all fields, most surely in health care and its delivery. By means of this Series we aim to contribute and cooperate to meet the challenge of our time: how to aim human technology to good human ends, how to deal with changed values in the areas of religion, society, culture and the self-definition of human persons, and how to formulate a new way of thinking, a new ethic. We welcome book proposals representing the broad interest of the interdisciplinary and international focus of the series. We especially welcome proposals that address aspects of 'new medicine', meaning advances in research and clinical health care, with an emphasis on those interventions and alterations that force us to re-examine foundational issues.

More information about this series at http://www.springer.com/series/6224

Rebecca A. Greenberg · Aviva M. Goldberg
David Rodríguez-Arias
Editors

Ethical Issues in Pediatric Organ Transplantation

Springer

Editors
Rebecca A. Greenberg
The Hospital for Sick Children
Toronto, ON
Canada

and

Department of Paediatrics,
Faculty of Medicine
University of Toronto
Toronto, ON
Canada

Aviva M. Goldberg
Department of Pediatrics
Winnipeg Regional Health Authority
Winnipeg, MB
Canada

and

Department of Pediatrics and Child Health,
Faculty of Health Sciences
University of Manitoba
Winnipeg, MB
Canada

David Rodríguez-Arias
Department of Philosophy I
University of Granada
Granada
Spain

ISSN 1567-8008 ISSN 2351-955X (electronic)
International Library of Ethics, Law, and the New Medicine
ISBN 978-3-319-29183-3 ISBN 978-3-319-29185-7 (eBook)
DOI 10.1007/978-3-319-29185-7

Library of Congress Control Number: 2016934416

Printed on acid-free paper

This Springer imprint is published by Springer Nature
The registered company is Springer International Publishing AG Switzerland

Acknowledgments

This book has been a labor, mostly of love. It has reinvigorated our passion for ethics and transplantation and allowed us to connect with many of the brightest in the field to make this book the best it can be. We are indebted to our authors, who wrote innovative, well-researched, and passionate chapters and met our deadlines even when they came up against holidays and vacations. Our sincere appreciation to our senior colleagues whose support and guidance we relied on throughout this process; in particular we want to thank Randi Zlotnik Shaul, Lainie Ross, and Richard Fine for their honest and straightforward advice. We would like to give much thanks to David Heti, our copy-editor, for his excellent insights and meticulous attention to detail. Our thanks to the publishers, who recognized the need for a book specifically addressing ethical issues in pediatric transplantation and allowed us to gather experts in the field to contribute. To our universities, hospitals, and organizations—thank you for encouraging us and giving us the time and space to write, edit, and discuss. To our friends and families—thank you for putting up with our late nights, deadline stress, and the generalized grouchiness that writing and editing seems to involve—this book could not have happened without your patience and support.

Rebecca A. Greenberg thanks those who have supported and mentored her work in bioethics, in particular: Michael Gordon, Sue MacRae, Pat Murphy, Barbara Secker, George Webster, and Linda Wright. She also thanks all of the colleagues, trainees, students, patients, and families she has worked with for providing inspiration, in particular: Jonathan Hellmann, Parviz Manji, and Randi Zlotnik Shaul. Her greatest thanks and love are extended to her family. To her parents, who have provided relentless encouragement and inspiration. To her husband, Marshall, for his ongoing support, friendship, and love, and whose humor and high-level perspective propelled her forward.

Aviva M. Goldberg thanks the mentors who inspired her to pursue pediatrics, nephrology, and ethics, the students who keep asking questions, and the patients who inspire and remind her of what is really important. Special teachers, mentors, and role models along the way include Julian Midgley, Patricia Birk, Tom

Blydt-Hansen, Joel Frader, Craig Langman, Rick Cohn, Kayhan Parsi, Kathryn Montgomery, Katie Watson, Ian Mitchell, Peter Nickerson, and David Rush. To her parents Marilyn and Norman Goldberg, and darling boys Sam and Isaac: you are my beginning, my middle, and my end. Finally to her husband Greg: my best friend, best editor and best shirt—you make my heart happy.

David Rodríguez-Arias would like to remember in this acknowledgment James Lovette (RIP), the first successful child heart transplantee in history and one of the longest survivors of a second transplant. He thanks the two donors and the many transplant professionals who made it possible for him to meet James and to become his bioethics classmate and friend in Paris and Cleveland. I owe more than what I know on transplantation ethics to my mentors Stuart Youngner and Linda Wright and am deeply grateful for their encouragement. A special thank you to David Paredes, a transplant coordinator in Spain whose intellectual curiosity and open mind give him the ability to listen to ethical challenges without necessarily hearing criticism. I'm also grateful to Iván Ortega, a Ph.D. student from whom I learn more than what I teach. Finally, thanks to Carissa Véliz, the smartest philosopher I know and the trigger of my happiness.

Contents

Editors and Contributors

About the Editors

Rebecca A. Greenberg is a Bioethicist at The Hospital for Sick Children. She is Assistant Professor in the Department of Paediatrics, a Project Investigator at SickKids' Research Institute, and a member of the Joint Centre for Bioethics, at the University of Toronto. Rebecca has a Bachelor of Arts (psychology) from the University of Manitoba, and a Bachelor of Science in Nursing and Ph.D. (Bioethics) from the University of Toronto. She has completed an ethics internship at Baycrest Centre and a postdoctorate fellowship in Clinical and Organizational Ethics at the University of Toronto Joint Centre for Bioethics. Prior to working as a Bioethicist, Rebecca worked as a nurse in the area of general medicine. She is the co-chair of the Canadian Society of Transplantation Ethics Committee. Her main research interests include transplant ethics, ethics education, and priority setting.

Aviva M. Goldberg is Section Head of Pediatric Nephrology in the Department of Pediatrics and Child Health in Winnipeg, Manitoba Canada and Associate Professor, Department of Pediatrics and Child Health in the Faculty of Health Sciences at the University of Manitoba. She is co-director of the Professionalism program in Undergraduate Medical Education and Ethics Medical Advisor to the Winnipeg Regional Health Authority and the Manitoba Provincial Health Ethics Network. She is past chair of the Canadian Society of Transplantation (CST) Ethics Committee, the chair of the pediatric group of the CST as well as a newly elected CST board member. She completed her pediatric nephrology fellowship and ethics fellowships at McGaw Medical Centre of Northwestern University Feinberg School of Medicine and Children's Memorial Hospital (now Lurie Children's Hospital of Chicago). She received her master's degree in Bioethics from Loyola University Chicago. She is interested in transplant ethics, pediatric ethics, and in improving medical education in ethics and professionalism.

David Rodríguez-Arias is a Ramón y Cajal researcher at the Philosophy I Department of the University of Granada (Spain), and belongs to the ESOT-ELPAT

Working-Group on Public Issues. In December 2008, he simultaneously obtained a Ph.D. in Moral Philosophy at the Philosophy Department of the University of Salamanca, and a Ph.D. in Medical Ethics at the School of Medicine of the Université Paris-Descartes. His Ph.D. dissertation on death determination and organ transplantation won a national bioethics prize by the Víctor Grífols i Lucas Foundation. David has been a visiting fellow at the Hastings Center, Case Western Reserve University Department of Bioethics, and the University of Toronto Joint Center for Bioethics. His research is mainly devoted to organ transplantation ethics and global health ethics. More information is available at www. davidrodriguez-arias@weebly.com.

Contributors

Sandra Amaral is a pediatric nephrologist and co-director of the Pediatric Kidney Transplant Program at the Children's Hospital of Philadelphia (CHOP) as well as a clinical epidemiologist and Assistant Professor of Pediatrics and Epidemiology at the University of Pennsylvania. Dr. Amaral's research and clinical work focuses on understanding barriers and facilitators to optimal and equitable pediatric kidney transplant access and outcomes. She is also a public health advocate who has voluntarily served on the Kidney and Pediatric Committees of the United Network for Organ Sharing to promote the unique needs of children in the national allocation system in the United States.

Dr. Gregory Borschel graduated from the Johns Hopkins University School of Medicine in 1997 and completed a residency in Plastic and Reconstructive Surgery at the University of Michigan in 2005. He completed a two-year postdoctoral research fellowship in neuromuscular tissue engineering at the University of Michigan. He completed a fellowship in Pediatric Plastic and Reconstructive Surgery at the Hospital for Sick Children in Toronto, Canada and began his surgical career at Washington University in St. Louis in 2006.

In 2009 Dr. Borschel joined The Hospital for Sick Children and University of Toronto and has investigated nerve regeneration both clinically and in his research laboratory. This work has resulted in new strategies to treat nerve injuries. His clinical work focuses on pediatric facial paralysis, nerve injury, hand surgery, and microvascular reconstruction. He is Associate Professor of Surgery with a cross-appointment in the University of Toronto Institute of Biomaterials and Biomedical Engineering.

The laboratory focuses on ways to enhance peripheral nerve regeneration using a variety of approaches, including cytokines and growth factors, electrical stimulation, novel surgical manipulations, synthetic small molecules, and nerve allografts.

He has published a textbook on plastic surgery, 53 peer reviewed publications, and 35 book chapters. Dr. Borschel is an officer in the American Society for Peripheral Nerve and has been active in the Plastic Surgery Research Council.

He is the recipient of the American Association of Plastic Surgeons Academic Scholar Award. He was also awarded the C. James Carrico Faculty Research Award by the American College of Surgeons. He is the Principal Investigator on two federal operating grants administered by the Canadian Institutes for Health Research.

Daniel Z. Buchman is a bioethicist at the University Health Network in Toronto, Ontario, and provides ethics services to the Multi-Organ Transplant Program. Daniel is a Member of the Joint Centre for Bioethics and is the co-Director for the course Empirical Approaches in Bioethics for the MHSc in Bioethics at the University of Toronto. He was previously an Ethics Fellow at Providence Healthcare in Vancouver, British Columbia and the Senior Ethics Fellow at the Toronto Central Community Care Access Centre.

Jennifer A. Chandler joined the University of Ottawa (Canada) Faculty of Law in 2002, after practicing law in Canada with a national law firm and serving as a law clerk to the Hon. Mr. Justice John Sopinka of the Supreme Court of Canada. She is Associate Professor at the Faculty of Law, where she teaches "Mental Health Law and Neuroethics" as well as "Medical-Legal Issues" and "Tort Law." She has also taught a graduate level course called "Technoprudence," which addresses the philosophy of law and technology. During her sabbatical in 2010–2011 she was a Visiting Associate Professor at Centre for Biomedical Ethics at the National University of Singapore.

Jennifer Chandler researches and writes about the legal and ethical aspects of advances in biomedical science and technology, with particular interest in neuroethics, organ donation, and regenerative medicine. Recently, she has written on the legal implications of advances in neurotherapies and neuroimaging technologies, regulatory policy related to medical practices such as organ donation and transplantation, and the ethics and law of scientific inquiry.

She was also the recipient in 2012 of the James Kreppner Fellowship from Canadian Blood Services, which provides support for her work in relation to organ donation and transplantation. She is a co-leader of the law and ethics sub-group within the recently announced Canadian National Transplant Research Program, and a member of the ethics working group of the Trillium Gift of Life Network.

Dr. Sonny Dhanani is a pediatric intensivist at the Children's Hospital of Eastern Ontario (CHEO) in Ottawa and Associate Professor (Pediatrics) at the University of Ottawa. He is the chief medical officer—donation for Trillium Gift of Life, Ontario's organ procurement organization. He is a co-lead for the training program of the Canadian National Transplant Research Program and is leading national research pertaining to practices and standards for determining death after cardiac arrest for the purposes of donation.

Genieve Dhani works in the operating room at the Toronto General Hospital, Toronto, Canada. She has 17 years' nursing experience, 11 of those being in the operating room. Her specialty focus is in hepato-pancreato-biliary surgery and liver, kidney, and pancreas transplantation. Her interests are in clinical research and ethical issues surrounding liver transplantation. She has revised and maintained clinical support tools to facilitate learning needs to staff related to transplantation and was a keynote speaker on Living Related Liver Transplants at the American Nurses Association in Denver Colorado.

Dr. Douglas S. Diekema is Professor of Pediatrics at the University of Washington School of Medicine with adjunct appointments in the Departments of Bioethics and Humanities and Internal Medicine in the School of Medicine and the Department of Health Services in the School of Public Health. He is also an attending physician in the emergency department at Seattle Children's Hospital and serves as Director of Education for the Treuman Katz Center for Pediatric Bioethics at Seattle Children's Hospital. He has been a member of the Seattle Children's Hospital ethics committee since 1991, has served as an ethics consultant since 1993, has been chairperson of the institutional review board since 2000, and founded the Center for Pediatric Bioethics at Seattle Children's in 2004. He is past-Chair of the Committee on Bioethics of the American Academy of Pediatrics and is a member of the Ethics Committee of the American Board of Pediatrics. Dr. Diekema is the author of numerous scholarly publications in clinical and research ethics and an editor of Clinical Ethics in Pediatrics: A Case-based Textbook. He was honored by the American Academy of Pediatrics as the 2014 recipient of the William G. Bartholome Award for Ethical Excellence.

Anne I. Dipchand is Professor of Paediatrics at the University of Toronto and the head of the Heart Transplant Program at The Hospital for Sick Children (SickKids) in Toronto. Her clinical practice focuses on heart failure and transplantation. Dr. Dipchand is actively involved in clinical research, especially registry-based analyses, and is the Past-President of the Pediatric Heart Transplant Study (PHTS)—an international study group. She is currently the Associate Director—Paediatrics for the International Society of Heart and Lung Transplantation (ISHLT) Registry and sits on the Executive Committee of the ISHLT Registry for Mechanically Assisted Circulatory Support (IMACS). Nationally, she is on the Executive of the Canadian Cardiac Transplant Network, the Past-Chair of the Canadian Society of Transplantation (CST) Paediatric Committee, and the Past-Chair of the CST Education Committee. Internationally, she is the Secretary-Treasurer of the International Paediatric Transplant Association (IPTA), Past-Chair of the AST Pediatric Community of Practice, and Past-Chair of the IPTA Education Committee. Dr. Dipchand is the Founding President and CEO of the Pediatric Heart Transplant Study Foundation, a not-for-profit foundation dedicated to advancing the science and treatment of children with endstage heart failure listed for or following a heart transplant. From a community perspective, Dr. Dipchand is a member of the Board of Directors of the David Foster Foundation, an organization committed to helping families of children who undergo a solid organ transplant and to increasing organ donor awareness in North America. She also actively spearheads opportunities for children and families of children with organ transplantation including educational symposia and written educational materials.

Sabrina F. Derrington received her medical degree from the University of California Davis School of Medicine and completed training in pediatrics and pediatric critical care at Children's Hospital Los Angeles. She currently practices pediatric critical care and palliative care at Ann and Robert H. Lurie Children's Hospital of Chicago, and holds an appointment as assistant professor of pediatrics at Northwestern University Feinberg School of Medicine. Dr. Derrington's research interests focus on the ethics of prognostication and medical decision-making. She enjoys teaching ethical and palliative care topics, particularly those that relate to the pediatric intensive care unit.

Dr. Annie Fecteau is a pediatric surgeon who specializes in pediatric gastrointestinal transplantation. After her training in pediatric surgery at McGill University, she went on to complete further years of training in liver transplantation in Cincinnati and Paris. In 1998 Dr. Fecteau joined the University of Toronto Liver Transplant team as the only pediatric surgeon. In addition to a full pediatric surgery practice, she manages with the help of a multidisciplinary team, the pre- and post-transplant care of children with end-stage liver disease and intestinal failure.

Dr. Fecteau also has an interest in ethics and has completed a Master's degree in Bioethics at the Joint Center for Bioethics at the University of Toronto. She is the ethics coordinator for the Division of Pediatric Surgery and is actively involved in teaching ethics at all levels of training. Dr. Fecteau is a member of many hospital, national and international committees related to ethics.

Chris Feudtner is the Steven D. Handler Endowed Chair of Medical Ethics and the Director of the Department of Medical Ethics at The Children's Hospital of Philadelphia and Professor of Pediatrics, Medical Ethics and Health Policy at the Perelman School of Medicine of the University of Pennsylvania. His career has been devoted to improving the lives of children with complex chronic conditions and their families. He has published extensively in pediatric palliative care, epidemiology, health service use and quality, child outcomes, and medical ethics.

Dr. Richard Fine serves as a Pediatric Nephrologist at Stony Brook Children's. Previously, he was Professor and Chairman of Pediatrics (1991–2005), Dean of the School of Medicine (2005–2010), and a Distinguished Service Professor of Pediatrics (2006). Dr. Fine earned his medical degree at Temple University Medical School in 1962 and completed his residency at Boston City Hospital, Jr. where he served as Assistant Resident from 1963 to 1964 and at the Children's Hospital of Los Angeles where he served as Chief Resident in 1965. He is licensed in Massachusetts, Pennsylvania, California, Nevada, and New York and is certified by the American Board of Pediatrics and the sub-board Pediatric Nephrology.

Dr. Fine's career has been focused on education and innovation and is a current member of several national and international committees which focus on improvements in research and transplant technology. Dr. Fine is credited for being an editor for major medical journals including Pediatric Nephrology, Transplantation, and the International Journal of Pediatrics. Throughout his career, his research has been focused on renal disease, clinical dialysis and dialysis therapy, and pediatric solid organ transplantation. He has worked as a consultant for several prominent pharmaceutical companies and research laboratories for over thirty years in addition to his work most prominently with the Children's Hospital of Los Angeles and Stony Brook Children's. His honors include Stony Brook University President's Award for Excellence In Team Achievement (2008), the Pediatric Nephrology Seminar's Jose Strauss Visiting Professorship Award (2010), and the Children's Memorial Hospital in Warsaw, Poland 'Medal of Friend' (2010), The Henry L. Barnett Award for Outstanding Teaching and Clinical care for Children with Kidney disease of the American Academy of Pediatrics, Section of Nephrology (2003), The Henry P. Laughlin Alumnus of the Year Award from Temple University School of Medicine (2007), The Founders Award from the American Society of Pediatric Nephrology (2007), and The Ernest Hodge Award of American Society of Transplantation.

Jennifer Flynn is Assistant Professor of Bioethics in the Faculty of Medicine at Memorial University. She is cross-appointed to Memorial's Department of Philosophy. Her Ph.D. is in Philosophy from the University of Virginia, and her Bachelor of Arts (Honours) and her Masters of Arts are from Queen's University. While at Virginia, she was a Social Sciences and Humanities Research Council of Canada doctoral fellow, and a Dissertation Year Fellow of the University of Virginia's College of Arts and Sciences. Prof. Flynn has completed the fellowship program in clinical and organizational ethics at the University of Toronto's Joint Centre for Bioethics, and was Visiting Fellow at Western University's Rotman Institute of Philosophy.

At Memorial, Prof. Flynn teaches bioethics to undergraduate and postgraduate medical students, teaches within Memorial's Masters of Health Ethics program, and teaches undergraduate and graduate philosophy students. Along with her Memorial colleagues in bioethics, Prof. Flynn does some clinical ethics work for the local health region and the province of Newfoundland and Labrador more generally. Her research interests are moral philosophy, the foundations of bioethics, and within applied ethics, issues concerning reproduction and pediatrics.

Marie-Chantal Fortin is a nephrologist with the kidney transplant team of the Centre hospitalier de l'Université de Montréal (CHUM), a researcher at the research center of the CHUM and Associate Professor at the Université de Montréal. She completed a Ph.D. in bioethics at the Université de Montréal in 2008. She holds a CIHR operating grant to look at the ethical issues related to the recruitment of compatible pair in kidney exchange programs. She is also a member of the Ethical, Economic, Legal and Social Issues Platform of the Canadian National Transplant Research Program. Finally, she holds a KRESCENT New Investigator award and she is a research scholar of the Fonds de la recherche en santé du Québec.

Joel E. Frader is the A Todd Davis Professor of Academic General Pediatrics and Professor of Medical Humanities and Bioethics at Northwestern University's Feinberg School of Medicine and Ann & Robert H. Lurie Children's Hospital of Chicago. He also serves as the Medical Director of the Bridges Pediatric Palliative Care Program at Lurie Children's. Dr. Frader received a B.A. from Columbia University (1970), his MD from Tufts (1974), and an M.A. in Sociology from the University of Pennsylvania (1980) where he was a Robert Wood Johnson Clinical Scholar. He is active in and has served in leadership positions for national organizations concerned with pediatrics and with bioethics. His research interests focus on ethical issues involving children in the health care system. His published research includes work on ethics and psychosocial issues in the pediatric intensive care unit; end-of-life decision making in pediatrics; ethical problems associated with innovation in medicine; organ transplantation; children with disorders of sex development and gender identity; and conscientious objection by health care professionals. His major clinical focus is in palliative and hospice care for children.

Michael A. Freeman is Assistant Professor of Pediatrics in the Division of Pediatric Nephrology and Hypertension at Penn State Hershey Children's Hospital and Penn State College of Medicine. In addition to his clinical service, he has also served as an ethics consultant, ethics educator, and as a member of the hospital ethics committee. He graduated with a medical degree with distinction in biomedical ethics from Albany Medical College and is currently a Master's Candidate in Bioethics at the Center for Bioethics and Health Law and the Dietrich School of Arts and Sciences at the University of Pittsburgh. He has also served as a member of the executive committee of the Pediatric Community of Practice of the American Society of Transplantation. His current clinical and scholarly research focuses on racial disparities in kidney transplantation and on the ethical implications of emerging genetic predictors of transplant outcomes.

Dr. David Grant is Professor of Surgery at the University of Toronto where he is the Surgical Director, Transplantation, University Health Network and a staff surgeon at SickKids. He has published more than 250 papers and book chapters. Research interests include: ethical issues in transplantation; small bowel transplantation, living donor liver transplantation, transplantation tolerance, and xenotransplantation. Dr. Grant performed the first liver-small bowel transplant in 1989. He established and directs the international Intestine Transplant Registry. He is the Current Treasurer of the Canadian Transplant Society, past Chair of the Canadian Liver Transplant Section of the Canadian Transplant Society; and the past President of the International Intestine Transplant Society.

Vanessa Gruben is Associate Professor at the Faculty of Law, University of Ottawa. Professor Gruben teaches in the areas of property law, family law, and health law and policy. Professor Gruben's research focuses on health law including various public health law issues, organ donation, and the legal regulation of various aspects of assisted human reproduction including contractual disputes over frozen embryos, privacy and access to information, gamete donor anonymity, and the constitutionality of the *Assisted Human Reproduction Act*.

Dr. Mark D. Hanson is Associate Professor, Department of Psychiatry, University of Toronto and a staff psychiatrist at The Hospital for Sick Children. Dr. Hanson is Child and Adolescent Psychiatrist with certification from the Royal College of Physicians and Surgeons of Canada. Dr. Hanson's primary clinical interest is the provision of psychiatric consultation for children and adolescents with chronic medical and surgical conditions. For several years he provided psychiatric consultation for children and adolescents on The Hospital for Sick Children Burn Unit. Dr. Hanson is associate dean of admissions and student finances, Undergraduate Medicine, Faculty of Medicine, University of Toronto, Toronto, Ontario, Canada. Dr. Hanson's primary academic interest is medical education scholarship. He has presented and published extensively on educational topics relating to Child and Adolescent Psychiatry and medical school admission.

Charlotte H. Harrison is the Hospital Ethicist and Director of the Office of Ethics at Boston Children's Hospital, where she also co-chairs the Ethics Advisory Committee. She is a Faculty Associate of the Harvard Center for Bioethics. While at Children's she has led clinical ethics consultations and educational programs for hospital staff, directed the hospital's literature and medicine program, and served as a leader for organizational ethics task forces and working groups, including a hospital-wide Task Force on Donation after Cardiac Death. Before entering the field of bioethics, she graduated from Harvard Law School and practiced law for fifteen years, both at private firms and at Massachusetts General Hospital. She has taught public health ethics at Tufts University. Her current interests include organizational ethics, transplantation ethics, and the ethics of professional collaboration.

Jonathan Hellmann is Professor of Paediatrics in the Department of Medicine at the University of Toronto, and Medical Bioethics Associate in the Department of Bioethics at The Hospital for Sick Children. He is a medical graduate of Witwatersrand University in Johannesburg, South Africa. He completed a fellowship in Neonatology at the Hershey Medical Centre in Pennsylvania and moved to The Hospital for Sick Children in 1983. He was the Clinical Director of the Neonatal Intensive Care Unit for 17 years and has a long standing interest in ethical issues in the newborn. He completed an MHSc in Bioethics in 2006 and is on the teaching faculty of the Joint Centre for Bioethics at the University of Toronto.

Dr. Vicky Lee Ng received her MD from the University of Toronto in 1993. She then completed her Paediatrics residency at The Hospital for Sick Children, including a year as Associate Chief Resident, prior to going to the Children's Hospital Medical Center in Cincinnati, Ohio, where she completed a three-year clinical and research fellowship in Paediatric Gastroenterology, Hepatology and Liver Transplantation.

As Associate Professor with the University of Toronto, Staff Gastroenterologist and Medical Director of the Liver Transplant program at SickKids, Dr. Ng's clinical research interests focus on improving outcomes of children following transplantation, including quality of life after pediatric liver transplantation. She is the co-Creator of the PeLTQL (Pediatric Liver Transplant Quality of Life), a disease-specific health-related quality of life tool for children who have undergone life-saving liver transplantation, with funding by an operating grant from the Canadian Institutes of Health Research (CIHR).

Pamela Nathanson is the Policy and Program Coordinator for the Department of Medical Ethics at The Children's Hospital of Philadelphia. After beginning her career in public health, she received her Master of Bioethics from the University of Pennsylvania. She recently co-wrote the hospital's current guidelines for staff on the use of social media in the context of therapeutic relationships.

Ivan Ortega-Deballon is a specialist in law (LL.B.) and in medical ethics. He is also an Emergency Nurse Practitioner (NP) and works in emergency nursing care in the Helicopter Emergency Medical Service in Madrid, Spain where he is responsible for the care of potential organ donors during transport.

He teaches at several universities in Madrid and he is associate professor at the Faculty of Medicine and Health Sciences of the University of Alcala de Henares, where he has done a Masters in University Teaching. He belongs to the Teaching Innovation Group of this University, recently honored after his works in High Fidelity clinical simulation in the learning process (Advanced Life Support training and Resuscitation).

His doctoral thesis project has been accepted (Ph.D. candidate) on the topic of the relationship between programs of donation after circulatory death (DCD) and international high-quality resuscitation research projects. He is involved in several projects that seek for the public access to Automated External Defibrillation (AED) looking for increasing the survival outcomes rates of out-of-hospital cardiac arrest. He has been participating in uncontrolled DCD since 2000, involved in the care of more than 60 potential donors as part of his work in the HEMS of Madrid which is currently the most active uncontrolled DCD program worldwide.

Dr. Erin Paquette is currently an Instructor in Pediatrics at Northwestern University and Attending Physician in the Division of Critical Care Medicine at Lurie Children's Hospital of Chicago. She completed her graduate training in medicine, law, and ethics at the University of Pennsylvania, followed by pediatrics residency at the University of Chicago and pediatric critical care fellowship at Boston Children's Hospital. Dr. Paquette is a physician-lawyer-bioethicist with academic interests focused at the intersection of law, medicine, ethics, and public policy. Her previous work has included examining approaches to treatment refusals in pediatrics, competency of pediatric residents in providing end-of-life care, confidentiality at end-of-life care, and informed consent in the research and clinical settings.

Lisa S. Parker a philosopher, is Associate Professor of Human Genetics and Behavioral and Community Health Sciences in the Graduate School of Public Health at the University of Pittsburgh. She serves as Director of Graduate Education for the Center for Bioethics and Health Law, where she is Director of the interdisciplinary Master of Arts in Bioethics program in The Dietrich School of Arts and Sciences. With over 40 peer-reviewed articles and over 20 book chapters, Dr. Parker has published extensively on ethical concerns related to the design and conduct of research, particularly genetic research and mental health research, as well as confidentiality and informed consent in both research and clinical care. With a national working group of the Association of Schools of Public Health, Dr. Parker edited *Ethics and Public Health: Model Curriculum* http://www.asph.org/document.cfm?page=782 (Association of Schools of Public Health 2003). She is co-author of the second edition of *Informed Consent: Legal Theory and Clinical Practice* (Oxford 2001) and is co-editor of *Mutating Concepts, Evolving Disciplines: Genetics, Medicine, and Society* in the Philosophy and Medicine Series published by Kluwer Academic Publishers in 2002.

Dr. Parker helped to develop the University of Pittsburgh protocol for living-related lung lobe donation/transplantation. Her paper (with Pittsburgh colleagues) won the 1990 American Federation for Clinical Research Nellie Westerman Prize in Ethics for its analysis of allocation and informed consent issues in left ventricular assist device research. Her recent research has focused on clinical genetics and genetic research, particularly the ethical management of incidental findings in both research and clinical care. In 2012, she was appointed to the Genomics and Society Working Group of the National Advisory Council for Human Genome Research for the NHGRI, where she has served on the study section of the ELSI Program (the Ethical, Legal, and Social Implications Program) and on two national working groups examining the management of incidental findings and research results, as well as on the Expert Scientific Panel of the Electronic Medical Records and Genomics (eMERGE) Network.

Dr. Lainie Friedman Ross is the Carolyn and Matthew Bucksbaum Professor of Clinical Medical Ethics, Professor of Pediatrics, Medicine, Surgery and The College and Associate Director, MacLean Center for Clinical Medical Ethics at the University of Chicago. Dr. Ross did her undergraduate studies at Princeton University where she majored in the Woodrow Wilson School of International and Public Affairs, graduating with an AB in 1982. She graduated from the University of Pennsylvania School of Medicine in 1986 and did her pediatric training at the Children's Hospital of Philadelphia and Columbia University. She then went to Yale University to study philosophy and earned an M.Phil in 1992 and a Ph.D. in 1996. She has written more than 100 research articles on ethical and policy issues in organ transplantation, genetic testing, pediatrics, and human subjects protections. Her most recent book, Transplantation Ethics, 2nd edition, is co-authored with Robert M. Veatch (Georgetown University Press, 2015). Currently, Dr. Ross serves on the National Institutes of Health Recombinant DNA Advisory Committee. Dr. Ross is a 2013 recipient of a Robert Wood Johnson Investigator Award in Health

Policy with Dr. Dick Thistlethwaite. They are writing a book on living donor organ transplantation. Dr. Ross was also awarded the 2014 John Simon Guggenheim Memorial Foundation Fellowship.

Maja Segedi is a Hepatopancreatobiliary and Liver Transplant Surgeon at the University of British Columbia, Vancouver, Canada. She trained as a surgical fellow at the University of Toronto and holds a Masters of Public Health degree from Harvard School of Public Health. In 2012 she completed the MacLean Centre Summer Course in Clinical Ethics at the University of Chicago. Her recent work is in the area of expanding indications for living donor liver transplantation, and the accuracy of preoperative anatomic mapping for living donor hepatectomy. Her interests are in the ethics of transplantation, living organ donation, and surgical innovation.

Dr. Randi Zlotnik Shaul is Director of the Bioethics Department at The Hospital for Sick Children. Randi is Associate Professor in the Department of Pediatrics at the University of Toronto, has an appointment in the Department of Surgery and is a member of the University of Toronto's Joint Center for Bioethics. Randi has a degree in political science from McGill University, a law degree from Osgoode Hall, and both a Master of Laws degree and a Ph.D. in Bioethics from the University of Toronto. Before graduate school Randi practiced law at the Ontario Ministry of Health. She has published on healthcare issues that interface law and ethics—including: priority setting, accountability, innovation, and research ethics. Randi's current research interests bioethics pedagogy for diverse trainee populations, and the tensions and synergies between models of patient and family centered care.

Sam D. Shemie is the Bertram Loeb Chair in Organ Donation at the University of Ottawa, practicing Paediatric Critical Care physician and director of Extracorporeal Life Support program at the Montreal Children's Hospital, and Professor of Pediatrics, McGill University. Dr. Shemie's area of specialty is organ replacement in critical illness. His recent research interests have included the development and implementation of national ICU-based organ donation strategies. He is the former chair of the Donation Committee of the Canadian Council for Donation and Transplantation. In April 2003, he chaired a Canadian Forum entitled "From Severe Brain Injury to Neurological Determination of Death" which has developed new medical standards for brain death determination and organ donation in Canada for all age groups. In February 2004, he chaired a Canadian Forum entitled "Medical Management to Optimize Donor Organ Potential" which has developed national consensus guidelines to optimize organ donor function for the purposes of transplantation. In February 2005, he chaired a Canadian Forum on 'Donation after Cardiocirculatory Death'. In December 2006, he was appointed as the Bertram Loeb Chair in Organ and Tissue Donation in the Faculty of Arts at the University of Ottawa. The mandate of the Loeb chair is to provide research leadership in current and emerging issues related to organ donation, with an emphasis on collaborative interdisciplinary research (ethical, philosophical, religious, cultural, legal, technological and biomedical). In August 2008, he was appointed Medical Director (Donation), Organs and Tissues, for the Canadian Blood Services.

Jodi M. Smith is Associate Professor of Pediatrics and Adjunct Associate Professor of Epidemiology at the University of Washington. Dr. Smith is the Medical Director of the Kidney Transplantation Program and the Pediatric Clinical Research Center at Seattle Children's Hospital and a recognized leader in pediatric kidney transplantation policy and research.

J. Richard Thistlethwaite the senior member of the University of Chicago Transplant Surgery faculty, has been performing transplants for more than 20 years. In that time, he has performed more than a thousand kidney transplants, 300 liver transplants, and approximately 180 pancreas transplants.

Dr. Thistlethwaite performed Illinois' first pancreas transplant and was also a member of the team that performed the first liver transplant in Illinois. One of his patients is the longest surviving recipient of a living-donor liver transplant. The University of Chicago Transplant Surgery section developed this operation. Some of Dr. Thistlethwaite's other patients include the youngest recipient of a deceased donor liver transplant, and the second youngest recipient of a deceased donor kidney transplant worldwide.

Dr. Thistlethwaite is a past president of the International Pediatric Transplant Association, the Illinois Transplant Society, and the Medical Staff at the University of Chicago. He has also served as board chairman for the Regional Organ Bank of Illinois, a governor of the American College of Surgeons, and a councilor of both the American Society of Transplant Surgeons and the United Network of Organ Sharing. Dr. Thistlethwaite's Ph.D. is in immunology, and he has more than 200 published research papers on transplantation, immunology, and anti-rejection medications

Christopher Tomlinson is a neonatologist at the Hospital for Sick Children, Toronto, and Assistant Professor at the University of Toronto. His undergraduate medical degree is from Glasgow University, Scotland and completed neonatal fellowships in both Toronto and The West of Scotland. He is lead neonatologist for organ donation at SickKids and sits on the Organ and Tissue Donation Committee which has a focus on exploring the issues around organ donation from newborn infants.

Dr. Aaron Wightman is a pediatrician completing fellowships in pediatric nephrology and pediatric bioethics at the University of Washington, Seattle Children's Hospital, and the Treuman Katz Center for Pediatric Bioethics.

Linda Wright is Director of Bioethics at the University Health Network (UHN), Toronto, Assistant Professor in the Department of Surgery, Faculty of Medicine, and a member of the Joint Centre for Bioethics at the University of Toronto. Linda provides clinical and organizational ethics consultation and teaching, has research expertise in organ transplantation ethics, and writes on current issues in transplantation ethics. Her research focuses mainly on ethical issues of living organ donors. In addition to working on national and international projects in donation and transplantation, Linda is ethics section editor of the journal Progress in Transplantation.

Ronald M. Zuker received his MD degree from the University of Toronto in 1969. After a rotating internship and period of family practice, he entered the McGill University Training Program in Surgery, completing the FRCSC in Plastic Surgery in 1976, through the Gallie Program at the University of Toronto. After a Fellowship in Microvascular and Pediatric Plastic Surgery, he joined the staff at The Hospital for Sick Children in 1978, obtaining the rank of Full Professor in 1994. Dr. Zuker also served as Chief of the Division of Plastic Surgery at SickKids from 1986 to 2002.

As a Pediatric Plastic and Reconstructive Surgeon, Dr. Zuker's main interest is in facial reanimation, cleft lip and palate and giant congenital melanocytic nevi. He is the Past Chairman of The American Academy Pediatrics Section of Plastic Surgery, Past President of The American Association of Pediatric Plastic Surgeons, and Past Present of The American Society of Reconstructive Microsurgery. He has been instrumental in structuring the Vascularized Composite Allotransplantation Team at The University of Toronto.

Introduction

With continuing medical advances, clinicians, scholars, and policy makers continue to grapple with complex ethical questions, and pediatric organ transplantation is especially fraught. Obtaining and allocating organs raise ethical challenges regarding how to conciliate the interests of donors with those of recipients, family members, clinicians, and the society at large. To a great extent, these challenges are an unavoidable by-product of the pervasive problem of organ shortage. The increasing societal demand for organs created by higher rates of end organ disease, and the medical advances that make transplant possible for more complicated patients, create an ever-increasing demand for organs. With the purpose of increasing the organ supply, transplant professionals and policy makers first relied solely on living donors, then progressively turned to donation after brain-death and to donation after circulatory death; they have designed campaigns and considered incentives to increase the donor pool; they have explored different models of consent for organ donation and have attempted different allocation models in order to increase efficiency. There are ongoing attempts to reduce the demand for organs by improving the detection, management, and prevention of those diseases that lead to end-stage organ failure. Despite these innovations, there is still a broad chasm between organ supply and demand.

The public places a significant trust in transplant professionals and healthcare organizations. However, this trust comes with the corresponding responsibility of protecting donors and recipients from unacceptable risks, of being transparent with families, and, more generally, to maintain fairness in the transplant system. Further, available organs are generally considered to be societal resources for which physicians are tasked with being responsible stewards. Ethical dilemmas for transplant professionals and organizations can be particularly challenging as they must balance their, at times competing, duties to the public at large and their fiduciary duty to the individual patient.

Bioethics has brought a helpful lens for transplant professionals and policy makers to grapple with the vexing ethical issues in transplantation. A rich body of bioethical academic reflection and clinical practice has materialized in the relatively recent bioethical subdiscipline of *organ transplantation ethics*. Organ transplantation ethics attempts to identify conflicting values at stake and to consistently provide reasonable ways to address moral conflicts related to organ transplantation.

For the past 50 years, the moral debates in this field have mostly been informed by cases in adult transplantation. Ethical issues in pediatric transplantation, which accounts for about 10 % of all transplants,[1] have not been extensively and systematically discussed. Pediatric transplant medicine involves unique circumstances that require a tailored ethical approach. As a result, there is a growing interest in the field of pediatric transplant ethics. This is the first book that addresses pediatric transplant ethics as its sole subject. It offers a theoretical and practical overview of the specific ethical issues as well as the legal issues they raise.

This book delves further into the uniqueness of the circumstances and questions raised in pediatric transplantation. First, while pediatric recipients receive priority for allocation in many deceased donation systems, some organs require size matching (like heart transplants), so have a more limited donor supply than adult candidates. Second, there is a relative paucity of evidence-based therapies in this age group. The literature specifically describing donor and recipients' outcomes is more limited both in sample size and sample diversity than in adults. This limitation creates a greater margin of uncertainty and difficulty in assessments of benefit and risk. Third, pediatric transplantation not only occurs in a context characterized by higher levels of risk and uncertainty, but also within additional complexities related to the parent–child relationship and the associated emotions which can include hope, fear, and guilt, all of which may influence the timing and nature of organ donation discussions with families and may require very strong trust relationships between families and clinicians. Fourth, children's dependence on caretakers creates unique issues. For instance, when a child is the potential recipient of a living donation from a parent, the parents may be evaluated simultaneously for both their psychosocial readiness to care for their child post-transplant and their readiness to serve as living donors. Whereas the evaluation of potential adult recipients does not rely on the ability of others to look after them post transplant. Fifth, complexities related to the consent process in pediatrics can also complicate the decision-making process that should ideally be—but rarely is—driven by the expressed preferences of the minors directly involved. Lack of information regarding any prior expressed wishes to donate, in the case of potential pediatric donors, and developing capacity in the case of both donors and recipients may raise important difficulties for substitute decision-makers who try their best to foster children's interests. Even when their preferences are known, pediatric living donors and recipients may not have developed the level of mental capacity required to provide appropriate consent to

[1] According to UNOS, between 1988 and 2012, patients less than 18 year old accounted for 7.5 % of all recipients. http://www.transplantliving.org/?module=data&trend=transplants-age-year.

the transplant process. Sixth, as a result of the above, children are generally considered a vulnerable population for which the standard of 'best interest' is commonly appealed to guide decision-making. However, the 'best interest' standard can be challenging to apply in practice when subjective views differ around what is best for the child. Seventh, children have differential conditions that require special consideration for candidacy to become organ recipients and it is these considerations that are often referenced for why children are often prioritized. They are in a developmental phase that is crucial for their lives: life-threatening organ failure, waiting for and then eventually receiving a transplanted organ can have a profound, lasting physical, psychological, and emotional impact on a child or adolescent and on his or her parent(s) and family. Prioritization of children can be somewhat controversial when adolescents, specifically, are often a group with high rates of risk taking and non-compliant behaviors. Eight, the physiology of children is different from that of adults in ways that are ethically relevant for transplantation procedures. For example, standards for death determination in adults may not be appropriate for newborn donors.

This book is intended to offer a theoretical and practical overview of the specific ethical issues in pediatric organ transplantation as well as the practical and legal issues they raise. We invited an international group of experts to write chapters that identified and explored complex ethical questions pertaining to pediatric organ transplantation. The authors shared rich insights, analyses, and reflections from their views as ethicists, physicians, nurses, lawyers, trainees, academics, and philosophers.

The book is structured into five parts:

Part I—Living Donation

- In Chap. 1, Lainie Friedman Ross and J. Richard Thistlethwaite, Jr. discuss whether it is ever ethically acceptable to accept minors as living organ donors. While they acknowledge the situations in which minors as donors may be requested (e.g., a teenager donating to his parent) they offer arguments to accept these donors only in very limited circumstances, and only after careful analysis.
- In Chap. 2, Maja Segedi, Genieve Dhani, Vicky Lee Ng, and David Grant explore ethical aspects of living liver donation in the case of pediatric acute liver failure. In these situations, parents may have just a few days to decide whether to be living liver donors for their critically ill children, without the "cooling off" period suggested in early discussion of living liver donation. The authors explore the arguments for and against asking parents to make such high pressure decisions, and conclude that, in certain situations, allowing parents to serve as living liver donors can be ethically justified.
- In Chap. 3, Michael A. Freeman and Lisa S. Parker explore the situation of misattributed paternity that is uncovered during the transplant evaluation. While the transplant evaluation is not designed to detect misattributed parentage, there are test results in the evaluation that can raise the possibility (e.g., a father who

shares no human leukocyte antigen (HLA) types with his child). They discuss arguments for and against the disclosure of this information to the pediatric recipient and his parents, and provide a model for policy in this rare but important clinical situation.

Part II—Deceased Donation

- In Chap. 4, Jennifer A. Chandler and Vanessa Gruben examine the ethics of using persuasive techniques to encourage families to consent to organ donation by their deceased loved children. They review the literature on the psychological effects of donation on families, consider when and how the treatment of families in the requesting and donation process might put their future psychological well-being at increased risk, and also consider lawsuits brought against healthcare personnel. They conclude that in some scenarios persuasive techniques are ethically justified.
- In Chap. 5, Sam D. Shemie, Ivan Ortega-Deballon, and Sonny Dhanani describe how the advances in life-sustaining technologies and transplantation achieved during the past five decades have both helped to understand and complicate the biology of the dying process and death. They overview current controversies in death determination for donors who are declared dead according to both neurologic and circulatory criteria. Death is understood as a single phenomenon based on the permanent cessation of brain function, with two mechanisms to reach that point: permanent absence of circulation or subsequent to a catastrophic brain injury. They conclude that there is a need to enhance pediatric-specific research in regard to the biology of death.
- In Chap. 6, Erin Paquette, and Joel Frader provide an informative overview of contemporary ethical controversies related to controlled donation after circulatory determination of death (cDCDD), with a focus on issues specific to pediatrics, including debates about the dead donor rule, informed consent, ante-mortem interventions intended to preserve the organs, conflicts of interest, perspectives from parents and health professionals, and organizational issues. Although ethical concerns in cDCDD are not unique to pediatrics, they conclude that the unique role of parents as decision-makers heightens reactions in this special population. Honesty with families is for these authors the only legitimate way to advance in these debates.
- In Chap. 7, Christopher Tomlinson and Jonathan Hellmann explore the clinical uncertainties and the ethical challenges created by the possibility of organ donation from deceased newborns. Through two case studies of neonatal patients who were considered as potential organ donors, this chapter shows, with great level of detail, the difficulties of a decision-making process where some compromises on the idea of newborns' best interests might be required in order to optimize organ retrieval and graft function in the recipient.
- In Chap. 8, Charlotte H. Harrison describes the efforts of one institution in developing a protocol for controlled donation after circulatory determination of

death within a pediatric hospital community in which there were deep and longstanding divisions about the ethics of pediatric cDCDD. Her chapter narrates the experience of a Task Force created at the Boston Children's Hospital in dealing with questions about the compatibility of such cDCDD programs with the overall mission of the hospital.

Part III—Emerging Challenges

- In Chap. 9, Annie Fecteau reviews the ethical issues relevant to small bowel transplantation and short bowel syndrome which often leave clinicians with some moral distress. This chapter succinctly reviews the relevant medical literature before discussing the ethical issues that include the ethics of emerging technology, best interest principle, accessibility to care, and the societal cost of small bowel transplantation.
- In Chap. 10, Randi Zlotnik Shaul, Gregory Borschel, Jennifer Flynn, Mark D. Hanson, Linda Wright, and Ronald M. Zuker explore the ethical implications of Vascularized Composite Allotransplantation (VCA) in pediatrics. They review ethics issues related to establishing acceptable risk–benefit ratios for recipients, informed consent, and considerations for establishing pediatric programs. The chapter focuses on the state of the science and international experience with VCA; the balancing of risks and benefits; issues of informed consent; and establishing pediatric VCA programs which includes donor and recipient selection, resource allocation, and accountability pathways for surgical innovation.
- In Chap. 11, Sandra Amaral, Pamela Nathanson, and Chris Feudtner review the history of social media and how social media works as a means of persuasion, particularly in healthcare cases. They provide an overview of social media crises, and present an ethical framework to inform and assist policy makers, hospital centers, and healthcare providers in responding to public outcry, while managing legal constraints to maintain patient confidentiality and ethical constraints to promote equity and utility in transplant access. In conclusion, they address preventive strategies to mitigate both patient harm and public misperceptions.

Part IV—Allocating Organs

- In Chap. 12, Aaron Wightman, Jodi Smith, and Douglas S. Diekema discuss whether children with neurodevelopmental disabilities should be considered as candidates for solid organ transplant. After highlighting the differences in policy that exist for different organs and reviewing studies on attitudes and outcomes of these children, they discuss how arguments based on the ideas of "equality of opportunity," "utility," "quality of life," and "social value" intervene in the rationale for or against the consideration of these infants as organ recipients.

While they acknowledge that the issue of consideration of children with intellectual disabilities as candidates for organ transplantation remains contentious, the authors offer empirical data and ethical reasons for the equal consideration of these children as potential organ recipients. They also offer some recommendations intended to better justify the allocation of organs while avoiding unjustifiable discrimination against this population.
- In Chap. 13, Marie Chantal Fortin and Rebecca A. Greenberg discuss the situation of transplantation for foreign national children. Taking an international lens, they explore the existing policy and legislation toward this issue throughout the world and offer arguments for and against the practice of offering transplant to children who are not residents or citizens of the country in which the transplant is performed. Ultimately, they advocate for transparent national policies that will ensure that all children will be treated equally in this thorny situation.

Part V—Recipients Issues

- In Chap. 14, Daniel Z. Buchman and Anne I. Dipchand use pediatric heart transplantation as a model for discussing risk and uncertainty with families of very sick children. They advocate a commitment to epistemic humility and strong trust building to achieve the goal of truly family centered care for this population.
- In Chap. 15, Sabrina F. Derrington, Aviva M. Goldberg, and Joel E. Frader consider psychosocial aspects of pediatric organ transplantation for transplant candidates, organ recipients, and their families—including living related donors. They outline an ethical approach to considering several ethical issues including how mental health issues, substance use, and assessment of available psychosocial support structures ought to play into decisions about organ transplant candidacy in pediatrics, and if psychological sequelae of transplantation for children and adolescents should affect medical decision-making.
- In Chap. 16, Aviva M. Goldberg and Richard Fine explore the ethical issues that arise in the care of the adolescent transplant recipient. They discuss the ethical dilemmas that arise in four clinical situations: (1) the adolescent who refuses transplant, (2) the non-adherent teen who is being considered for transplant, (3) contraception for female transplant recipients, and (4) transition from pediatric to adult care.

We feel privileged to have worked with such an esteemed group of authors. It is our hope that this book will spark further dialogue and debate about the important and challenging ethical issues in pediatric transplantation.

Part I
Living Donation

Chapter 1
Children as Living Donors

Lainie Friedman Ross and J. Richard Thistlethwaite Jr.

Abstract Within three years of the first successful kidney transplant in the United States between adult identical twins (1954), physicians were in court getting permission to perform kidney transplants between minor identical twins (1957). In the 1970s and 1980s, at least two dozen additional cases involving donation from non-identical twin siblings as well as siblings with cognitive disabilities underwent judicial review. With some notable exceptions, the courts authorized the donations, usually on the grounds that it was in the best interest of both the donor and the recipient. There are data that show that children have also served as donors for other solid organs. In this chapter we examine the ethics of involving minors as living solid organ donors. Our position is that ethically children can serve as living solid organ donors only in very limited circumstances.

Keywords Donor risks · Ethics · Identical twins · Living donor · Minor · Parental decision-making · Policy

Abbreviations

A2ALL	Adult to adult living liver transplantation cohort study
AAP	American Academy of Pediatrics
AST	American Society of Transplantation
BCKA	Branched chain keto acid dehydrogenase
MSUD	Maple syrup urine disease
OPTN	Organ Procurement and Transplantation Network
UNOS	United Network of Organ Sharing
US	United States

L.F. Ross (✉) · J.R. Thistlethwaite Jr.
University of Chicago, Chicago, IL, USA
e-mail: lross@uchicago.edu

J.R. Thistlethwaite Jr.
e-mail: jthistle@surgery.bsd.uchicago.edu

R.A. Greenberg et al. (eds.), *Ethical Issues in Pediatric Organ Transplantation*, International Library of Ethics, Law, and the New Medicine 66,
DOI 10.1007/978-3-319-29185-7_1

1.1 Introduction

One of the first successful kidney transplants in the United States (US) occurred between adult identical twins in 1954 [1]. Three years later, in 1957, three transplants were performed between identical twins aged 14, 14, and 19 years [2–4]. The choice of identical twins was not random because immunosuppression was in its infancy. Because the age of majority in the US at the time was 21 years old, all three cases involved minor donors and underwent judicial review. In all three cases, the potential donor, potential recipient, parents, and surgeons all wanted the transplant to take place. Despite the unanimous consensus, the hospital went to court for permission to do the transplant in order to release the hospital from legal liability [5]. In each case, the court found that the transplant was in both minors' best interests. A number of other cases were approved by the courts in various US jurisdictions over the next two decades including, in 1972, the approval by a Connecticut court of a donation by a seven year old kidney donor to her identical twin sister [6].

Data about living kidney donors are incomplete. In the US, The United Network for Organ Sharing (UNOS) only began to collect social security numbers of living donors in 1994, making it difficult to track donors prior to that time [7]. UNOS only began to collect limited follow up data in 2000 [7], so it is unknown how many minors have participated as living donors and how they are doing long-term. Two recent publications provide data that living kidney donors may be at increased risk of developing kidney disease over the long term [8, 9], highlighting the need for long-term data collection from pediatric kidney donors. A recent long-term study of minors found that minors are not at increased risk of shortened survival, hypertension, diabetes or proteinuria than young adults who served as living donors [10]. However, 92.3 % of the 39 minor donors who donated at the University of Minnesota were white, whereas the literature show that young minority donors may be at greater risk [11, 12]. The Minneapolis researchers have data for 31.8 ± 8 years but clearly longer follow-up is still needed given that these donors are expected to live more than fifty additional years [10].

What data do exist? In 2002, Delmonico and Harmon reported that there had been at least 60 living pediatric kidney donors in the US between 1987 and 2000 [13]. There is also a case report and follow-up from Italy of donation of small bowel by a 13-year-old to his identical twin [14, 15]. In 2011, a 17 year old Canadian donated a liver lobe to his mother [16]. In the US, the UNOS/Organ Procurement and Transplantation Network (OPTN) website reports that 16 children have served as living liver donors (four under the age of 10 and 12 between the ages of 11 and 17) as well as two living heart donors (both under the age of 10) since 1988 [17]. Although no additional information is given, the hearts must represent domino donor transplants (discussed below), and this may or may not be true of some of the living liver donors as well.

Children then do serve as living donors. The question is: should they? The professional community is divided. Thys et al. [18] identified 39 guidelines and position papers on living kidney donation by minors and found that 27 (69 %)

endorsed an absolute prohibition. The review included a number of international consensus statements that say that being below the age of majority should be an absolute contraindication to donation [19–21]. In contrast, the Consensus statement on the Live Organ Donor sponsored by the National Kidney Foundation, and the American Societies of Transplantation, of Transplant Surgeons, and of Nephrology argued that minor age was a relative not an absolute contraindication [22], as did the American Academy of Pediatrics (AAP) eight years later [5]. The Live Donor Consensus Group delineated four conditions that must be met for a child to ethically serve as a living donor which the AAP adopted with minimal modification and to which it added a fifth (to be discussed further below).

In this chapter we will examine the ethical issues raised when children are considered for live solid organ donation, focused mainly on kidney and liver lobe donation. Part 1 examines the issue from a utilitarian risk-benefit calculation. Part 2 examines what special safeguards exist and whether they are adequate. Part 3 evaluates the issue from a principle-based approach, and Part 4 considers three special circumstances. Our position is that children can ethically serve as living solid organ donors only in the very limited circumstances delineated by the AAP statement (for which we were the lead authors).

1.2 Is It Ethical for Children to Serve as Living Donors?

1.2.1 Benefits and Risks

To determine whether it is ethical for children to serve as living donors, one can do a utilitarian calculation of the benefits and risks and then decide whether donation is in a child's best interest. Clearly a unilateral nephrectomy or liver lobe resection is not in a child's medical best interest, and so it would be hard to justify if an alternative source of organs existed. There are short term perioperative risks including the risk of death. There are also potential long-term risks of unilateral nephrectomy that are just being understood now. Liver lobe resection for live donor transplantation has been performed since 1989 [23–25] without reported long-term risks to-date.

A utilitarian calculation must consider both the clinical risks as well as the psychological risks and benefits to the potential donor. Psychological benefits to the donor include: (1) greater self-esteem; (2) positive feedback from family and friends; (3) helping a loved one; (4) relief from the burden of a seriously ill family member (while the child-donor may not be the caregiver, relieving the parent of this burden means that the parent has more time and energy for the child-donor); and (5) improved quality time for intrafamilial activities [5, 26, 27]. The benefits then include both self-regarding interests (e.g., self-esteem) and other-regarding interests (e.g., helping a loved one) [27]. These benefits may even accrue if the transplant fails as the child and parents know that they did everything possible to save the recipient's life [5, 28, 29].

There are also psychological risks to the donor. First, donors may feel neglected because post-donation, the focus returns to the recipient who is considered ill while the donor is considered healthy [30]. Some living donors express dissatisfaction by the scar and it may cause a change in body image and lower self-esteem [5, 31]. The Adult to Adult Living Liver Transplantation Cohort Study (A2ALL) found approximately 4 % of adult living liver donors experienced severe mental health issues following donation [32]. The University of Minnesota found that 4 % of adult living kidney donors found the experience "extremely stressful" and another 8 % found it "stressful" [33]. Finally, if the donor is a child, the donation may prevent the child from playing some contact sports [34] or joining the military [35]. And if the transplant fails, the donor may feel guilty and experience significant mental health problems [36–38]. There are cases of adult donors committing suicide after a transplant fails [39].

To justify living donation, the potential benefits to the donor must outweigh the risks. Schoeman argues that the focus on donor benefit and risk is too individualistic and that one should also weigh in the benefits and risks to the potential recipient and the family at large [40]. In an intrafamilial donation, the benefits are obvious: the recipient gets a healthy living graft and the family is relieved of the burden of a chronically ill child. The risks to the recipient include a feeling of indebtedness and guilt if the donor experiences complications [5, 41]. Unless the recipient is an identical twin, the recipient now has the risks of immunosuppression, but that is a risk the recipient has regardless of the donor's identity. If one includes the risks and benefits of the recipient in the utilitarian calculation, a positive calculation is almost certain because the organ transplant is life-prolonging for the recipient.

The problem with expanding the risk: benefit calculation to include both donor and recipient risks and benefits is the concern that the pediatric donor could be used as a means for the well-being of others. Under that logic, parents can choose to sacrifice one family member if the benefits outweigh the harm. In ethical terms, such a calculation would fail to respect the donor as a person with his or her own ends [26, 42]. This is not to say that the risk: benefit calculation must only include donor self-regarding interests, but only that recipient health benefit should only be counted to the extent that it is a benefit to the donor. Donors can have other-regarding interests that should be included in the calculation, but in the end, the risks to the donor are only justified if they are outweighed by donor benefit evaluated from the perspective of the donor. And these benefits are only speculative at best.

But in pediatrics, even if the benefits outweigh the risks, one must ask whether this is sufficient. In general, parental decision making is held to a "best interest" standard [43–48]. And that was the standard used by most of the US courts that reviewed the sibling organ and tissue donor cases [2–4, 49]. In fact, when courts did not authorize the donation, it was on the grounds that the donation did not serve the child's (or the incompetent's) best interest [50, 51]. The problem is that an "appeal to the best interests of the nonconsenting donor child is morally specious at best," [52 at p. 162] even if one defines best interest broadly to include a child's best interest all things considered [53]. Rather, it is more apt to conclude, like Buchanan and Brock, that "the best interest principle is to serve only as a regulative ideal, not as a strict and literal requirement, because parents' obligations toward their other children as well as their own legitimate self-interests can conflict with doing what maximizes the child's well-being, and sometimes may take precedence over it" [46 at p. 236]. This is the current approach by many bioethicists [26, 27, 48, 54–57]. And, in fact, in at least one case involving bone marrow donation by a minor sibling, the court took the approach that its job was not to determine if the donation was in the child's best interest, but only to determine if the parents considered all of the risks and benefits. Satisfied that the parents had, they approved the donation [58].

Even if we do not require that parents act in their child's best interest, how much leeway should we give them? Ross argues that within an intimate family, parents have the authority to balance the needs of interests of the various family members, provided that they do not sacrifice the needs of any one particular family member [26]. Dwyer and Vig similarly argue that "a transplantation is justified when there is a proper fit between the relationship and the risks and benefits" [59 at p. 11]. With stringent limitations, that is the position adopted in the AAP guidelines [5].

1.2.2 Informed Consent

Even if donor benefits outweigh donor risks, one is still left with the question of who can consent for a child to participate as a living donor. In the US, we assume that children lack decision making capacity and so it is their parent (or legal guardian) who consents on their behalf.

Can parents legally authorize a non-medically indicated procedure on one child to benefit his sibling? This was one of the questions asked of the courts in the early days of kidney transplant, and most courts found that parents did have that authority when the children were siblings being raised in a family setting. However, when the potential donor was developmentally delayed and institutionalized, the courts were less likely to say donation was in his or her best interest [50, 51] unless the family could show a significant relationship between the potential donor and the recipient [60].

Although there are no federal regulations surrounding surrogate consent in organ transplantation, one can look to the research setting for an analogous situation. In the sphere of research, parents have authority to consent for a research procedure on a healthy child that offers no medical benefit to the child only if the procedure entails, at most, minimal risk [61]. A unilateral nephrectomy or the removal of a liver lobe entails more than minimal risk, but organ transplantation is different from research in that the beneficiary is not society at large, but an emotionally-related recipient. Ethicists have argued that this difference gives parents greater leeway [26, 42, 54, 57].

While parental permission is necessary for a child to be a living donor, it is still not clear that it is or ought to be sufficient. Additional safeguards are needed.

1.3 What Additional Safeguards Are Needed?

The Live Donor Consensus Group outlined four conditions that would need to be met in order for a minor to ethically serve as a live organ donor: "(1) when the potential donor and recipient are both highly likely to benefit (as in the case of identical twins); (2) when the surgical risk for the donor is extremely low; (3) when all other opportunities for transplantation has been exhausted, no potential adult living donor is available, and timely and/or effective transplantation from a cadaver donor is unlikely; and (4) when the minor freely agrees to donate without coercion (established by the independent donor advocate)" [22 at p. 2924]. The AAP endorsed the idea that age was a relative contraindication, adopted the four criteria delineated by the Live Organ Donor Consensus Group although it removed the parenthetical example in the first condition about identical twins (discussed below). The AAP also added a fifth condition: (5) when the emotional and psychological risks to the donor are minimized [5].

The first condition requires that both donor and recipient are highly likely to benefit. The AAP argues that this condition will most likely be met if the child donor is restricted to intrafamilial donations where the psychological benefit to the child donor of a healthy recipient is likely to be significant on a daily basis. This condition restricts children from donating to neighbors, family acquaintances or even distant relatives where long-term psychological benefit is more speculative. It also restricts donation in low-success, last ditch efforts, even if it is being proposed to save a dying sibling, in order to reduce the risk of psychological harm if the transplant fails [5]. It would permit a minor to donate to a sibling or parent assuming that all of the other conditions are also met.

The second condition requires that the surgical risk to the potential donor is extremely low. While neither the Consensus group nor the AAP quantified the acceptable level of risk, the short-term risks of kidney donation include a .03 risk of mortality, a 3–6 % risk of major morbidity (Clavien[1] grade 3 or higher) and ~22 % risk of minor morbidity (Clavien grade 1 or 2) [63–65].

The mortality risk of living liver donation is approximately 0.17–0.4 % [21, 66]. The Vancouver Forum on the Care of the Living Organ Donor estimated 6000–7000 live donor liver resections and a 0.4–0.6 % risk of catastrophic complication including 14 deaths, two donors requiring liver transplantation and one donor being left in a persistent vegetative state [21]. The risks of major and minor morbidity depends on whether it is a right or left lobe resection. If the recipient is a very young child, the living liver donor may only need to donate a left lateral segment which has much lower risk than full lobe donations [67, 68], although at least two living donor

[1]The Clavien system was designed as a classification of surgical complications. See [62]. Clavien grade 1 requires no treatment other than antiemetic, antipyretic, diuretic, analgesic, electrolytes or physiotherapy; Grade 2 requires medical therapy beyond grade 1; Grade 3 requires radiological or surgical intervention; Grade 4 signifies at least one organ failure, a life threatening condition, or an intensive care unit admission; and Grade 5 equals death.

deaths from left lateral donations have been reported [69–71]. Again, the exact rate of complications is difficult to know because there is no formal registry. Because of the scarcity of living liver donations by minors, there are no data about its safety unless one extrapolates from the adult literature. With the adoption of the Clavien method to quantify risks of different medical procedures, living liver donor morbidity is calculated at approximately 40 % in the US with 0.3–1.8 % of the patients experiencing Clavien grade 3 or 4 complications, depending on era and center [72, 73]. Data from Japan has somewhat better results [74, 75], in part, because of greater experience with living donors given the dearth of deceased donor organs.

Although regeneration does occur in the remnant donor liver and the transplanted graft, it does not regenerate to 100 % of the pre-donation mass [76]. Consequently, the regenerative capacity and long-term health consequences of such a resection are not fully known.

Given the significant perioperative risks of solid organ donation, one can argue that these risks are too great for any minor. At most, minors should be allowed to donate a kidney, but clearly not a liver lobe, or even left lateral segment, even though long term risks may be low because of the liver's regenerative capacity.

The third condition requires that alternatives have been exhausted such that the child is a donor of last resort. The AAP stated more specifically than the Live Donor Consensus Group that this means that a child should not serve as a donor unless all adult candidates have been evaluated and found unable to donate and the potential recipient is likely to die while waiting for a deceased donor organ to become available. Current UNOS policy gives priority to children for all kidneys from donors younger than 35 years [77]. The new UNOS policy that was implemented in late 2014 continues to give priority access to children for the best deceased donor organs (kidneys that are judged to have a kidney donor profile index less than 35 %) [78]. As such, children are likely to get a high quality deceased donor graft quickly which reduces the need for a sibling donor.

Novel living donor options also challenge whether the third condition can be met. Historically, potential adult donors were excluded from donation based on medical conditions (e.g., hypertension and/or diabetes) or psychosocial conditions (e.g., psychological instability; substance problems). Although some institutions are now willing to accept some less healthy donors [79], to a large extent, these exclusions still hold. However, potential adult donors were also excluded based on ABO-incompatibility and recipient donor-specific anti-HLA antibodies, but desensitization procedures for both ABO- and HLA-incompatibility are becoming more accepted [80]. Antibody incompatible adult donors can also participate in paired exchanges [where Donor A who cannot donate directly to an intended recipient (Recipient A) can donate to a third-party in end stage renal disease (Recipient B) whose intended donor (Donor B) donates to recipient A] [81]. Alternatively they can participate in a kidney transplant chain [where Donor B who cannot donate directly to an intended recipient (Recipient B) can donate to a third party (Recipient C) and his/her intended recipient gets a kidney from another donor (Donor A) who may be a member of a donor-recipient pair or a nondirected donor] [82]. By last resort, we mean that potential adult donors should not only be excluded from being eligible for direct donation but also from paired or chain donations before permitting a minor to serve as a living solid organ donor [83].

Condition four requires that the minor give his or her assent and that this is affirmed by a living donor advocate. This condition includes two very important safeguards. The first is the need for the child's assent. This is not meant to be a passive assent but an affirmative agreement to be a donor. Although the Consensus group did not provide an age limit, the AAP suggested that a firm lower age limit was 11 years old but encouraged a case by case assessment and not to use any age as a bright line given the wide variability in cognitive ability and social maturity of minors [5].

UNOS currently mandates living donor advocates or living donor advocate teams for all living organ donors. The purpose is to ensure that the donor understands the risks and benefits and is donating voluntarily, free of coercion. The donor advocate is not meant to usurp parents. A child should not begin the transplant process unless his or her parents are on board. The donor advocate should "ensure that the degree of emotional intimacy can justify the risks from the perspective of the minor donor, that there are no alternative donors who are adults, and that dialysis is not a realistic possibility for the recipient as a bridge to deceased donor transplantation" [5 at p. 457].

The fifth condition is that the emotional and psychological risks to the donor are minimized. This condition stems from the bone marrow transplant literature which has data to show that donors often feel neglected post-donation. The living donor advocate can help prepare the donor to know what to expect, and to educate the family about the needs of the donor both pre- and post-donation [5].

1.4 Principle-Based Arguments

An alternative approach to evaluate whether it is ethical for a child to serve as an organ donor is a deontologic or principle-based approach. This requires that minors are always treated with respect as ends in themselves, and not solely as a means. The deontologist does not have to deny that the donation treats the donor, regardless of age, as a means, but rather, he/she must only show that the child is being treated simultaneously as an ends. To the extent that the donor actively assents to the procedure, he or she is being treated simultaneously as an ends.

An alternative principle-based approach focuses on autonomy and the right to self-determination. The argument is that some children do have decisional capacity and their decisions should be respected [29, 84–86]. However, even some who believe that some children have the cognitive ability to understand the risks and benefits of donation may not want to give them independent authority to consent to organ donation on the grounds that decisional capacity is necessary but not sufficient for affording adolescents' decisional authority in the context of organ transplantation [26, 87]. Although there are some empirical studies that suggest that adolescents have similar cognitive abilities to adults on research surveys performed in a safe environment [88, 89], we believe that adequate cognitive ability is necessary but not sufficient to empower minors to decide about serving as an organ donor [26]. There are at least two arguments to support our position against giving teenagers decisional authority. First, it places too much emphasis on the adolescent's short-term goals rather than long-term goals. For example, it ignores the possibility that this minor may become a parent and may have a child who needs an organ in the future especially in the case of a familial cause of kidney failure. The adolescent's limited life experience means that his/her decisions are not part of a well-conceived life plan [26]. Second, there is no validated test to distinguish between those adolescents who are mature from those who are not [26]. This means that our judgment of maturity is quite subjective and may be more reflective of agreement with the decision than with the maturity of the reasons and reason processes. This is not to deny the need to show respect for the adolescent's developing autonomy, which should be determinative in cases of an adolescent dissent, but to acknowledge the limitations of the adolescent's judgment about what is in his/her best interest, and to be respectful of the rights and responsibilities of his/her parents [26]. Likewise, Capitaine and colleagues argue against empowering minors on the grounds that children may have cognitive maturity, but often lack its psychosocial counterpart [87]. Citing work by Steinberg and colleagues [90, 91], Capitaine et al. point to four relevant psychosocial factors: (1) susceptibility to social coercion; (2) risk perception; (3) future orientation; and (4) impulsivity [87]. They conclude that minors should not have the autonomy to consent by themselves to serve as an organ donor. Again, this does not mean that the adolescent's opinion should be ignored, only that it is not sufficient. To participate as an organ donor to a sibling, parental permission is necessary.

1.5 Exceptional Cases

1.5.1 The Case of Identical Twins

One must remember that when the courts were hearing the earliest cases of minor donations, living donation was the only option for individuals in end stage renal disease. Chronic dialysis did not exist until the early 1960s and was pretty sparse until the United States Congress passed legislation in 1972 authorizing Medicare coverage to Americans if they had chronic kidney failure [92]. The program was launched on July 1, 1973 [92], and dialysis programs and transplant programs quickly expanded. Because immunosuppression was in its infancy, successful transplants were often between identical twins.

Today, although chronic dialysis exists and can be used successfully even in small infants [93], kidney transplantation is the treatment of choice [94], particularly for children who fail to grow and have cognitive learning problems on dialysis [95, 96]. This is part of the basis for the special priority that children get on the deceased donor waitlist.

An identical twin donor offers an additional special benefit to the recipient because he/she will not need immunosuppression. Does this justify ignoring the third criterion/safeguard enumerated by both the Live Consensus Panel and the AAP—the criterion that minors should only participate as a last resort? [5, 22]. In 1996, Spital surveyed transplant centers regarding their attitude toward minor donation [97]. One-third of responding centers would allow a monozygotic twin minor to donate a kidney to his or her twin, but only 21 % would allow a non-twin minor to donate to a sibling [97]. In 2007, Joseph and colleagues surveyed members of the American Society of Transplantation (AST) and the AAP about their attitude toward organ donation by minors [98]. Respondents were asked whether they would permit kidney donation between 15 year-old fraternal and identical twins. Overall, a minority of respondents were willing with 32 % allowing for a donation by a minor fraternal twin and 39 % permitting the donation by a minor identical twin (p = NS) [98]. Willingness increased with increasing wait time and decreased when the age of the children decreased, although the changes were not statistically significant. The data at best suggest that a minority of health care professionals support allowing minors to donate, even in the setting of minor identical twins. Interestingly, the data from Delmonico and Harmon found that only seven (12 %) of the 60 live donors between 1987 and 2000 were identical twins and only 24 (40 %) donated to another minor [13].

What do the guidelines say? In a review of guidelines from all over the world, Thys and colleagues found that most guidelines prohibit all minors from donating. However, "in South Korea and some Canadian provinces, minors above a certain age may be considered competent to consent to living kidney donation" [18 at p. 957]. In addition, guidelines from the World Health Organization and the Netherlands only allow kidney donation by minors to their identical twin [99, 100]. In the US, the Live Donor Consensus Group cited identical twins as a case in which donor and recipient are both highly likely to benefit (criterion 1 above), suggesting that this would be one scenario in which donation by minors could meet their criteria [22]. The AAP did not include this example in its adoption of the four criteria because (1) being an identical twin did not ensure the donor twin benefited; and (2) having an identical twin did not mean that one should ignore condition 3 which only allows a minor to serve as a living donor as a last resort. As the AAP holds, "If it is ethically impermissible for a minor to serve as living donor to a sibling because of the risks or because the child cannot make a voluntary and informed decision, the same standards should hold if the potential donor is an identical twin" [5 at p. 457]. This is our position.

1.5.2 Domino Donors

The living domino donor offers an exception to the restrictive policy of minors serving as living donors. By "domino donor" we refer to the donation of an organ by an individual who needs to have an organ (usually a liver or heart) removed in order for the individual to receive his/her own transplant. Since the alternative is to discard the organ, the domino donation involves no additional risk to the donor. For example, in the 1990s, children with cystic fibrosis with respiratory failure were sometimes transplanted with a heart-double lung graft. Their native hearts were sometimes healthy enough that they could be transplanted into a third party who needed a solitary heart transplant [101]. Today, there is interest in treating metabolic conditions such as maple syrup urine disease (MSUD) with liver transplantation. Children with MSUD have decreased levels of branched chain keto acid dehydrogenase (BCKA) enzyme activity which mainly but not exclusively is made in the liver. The MSUD can be cured by a liver transplant. Excluding their inability to produce BCKA, these livers are otherwise functional (and have comparable graft survival with grafts from expanded criteria donors) and can be transplanted into patients in liver failure who do not suffer from MSUD because other cells in their body will make the BCKA enzyme. To-date, a handful of domino liver transplants have been performed using the livers of children with MSUD [102, 103]. Domino donation by children is ethically permissible because it entails no risk to the child living donor.

1.5.3 Minor Parents

Another exception to the restrictions on allowing a minor to act as a living donor is the minor who is a parent and seeks to donate to his or her child. In most but not all states, the minor who is a parent is an emancipated minor and legally has the right to consent for him or herself [104].

1.6 Concluding Remarks

Although the US-based guidelines by both the Consensus group and the AAP permit children to serve as living donors, this practice is meant to be rare and strict safeguards are in place. Although the Consensus group sought to make an exception about identical twins, the AAP did not, nor did most of the health care providers surveyed. Despite the growing gap between supply and demand, the solution should not come on the backs of children. More needs to be done to reduce the health conditions responsible for organ failure (obesity, diabetes, hypertension and alcoholism), to increase the number of deceased donors, and to develop alternative therapies.

Acknowledgements This chapter was supported by a Robert Wood Johnson Investigator Award in Health Policy, (Principal investigators: Lainie Friedman Ross and J Richard Thistlethwaite, Jr), A Case-Based Approach to the Ethical and Policy Issues Raised by Living Donor Transplantation.

References

1. Merrill JP, Murray JE, Harrison JH, Guild WR. Successful homotransplantations of the human kidney between identical twins. J Am Med Assoc. 1956;160:227–82.
2. Foster v Harrison, 68674 Eq Mass Sup Jud Ct (1957).
3. Huskey v Harrison, 68666 Eq Mass Sup Jud Ct (1957).
4. Masden v Harrison, 68651 Eq Mass Sup Jud Ct (1957).
5. Ross LF, Thistlethwaite Jr, The American Academy of Pediatrics Committee on Bioethics. Minors as Living Solid-Organ Donors. Pediatrics. 2008;122:454–61.
6. Hart v Brown, 29 Conn Sup Ct 368, 289 A2d 386 (1972).
7. Living Donor 101. On the web at: http://www.livingdonor101.com/kidney-failure-risk.shtml.
8. Mjøen G, Hallan S, Hartmann A, Foss A, Midtvedt K, Oven O, et al. Long-term risks for kidney donors. Kidney Int. 2014;86:162–7.
9. Muzaale AD, Massie AB, Wang MC, Montgomery RA, McBride MA, Wainright JL, et al. Risk of end-stage renal disease following live kidney donation. JAMA. 2014;311:579–86.
10. MacDonald D, Kukla AK, Ake S, Berglund D, Jackson S, Issa N, et al. Medical outcomes of adolescent live kidney donors. Pediatr Transplant. 2014;18:336–41.
11. Gibney EM, Parikh CR, Garg AX. Age, gender, race, and associations with kidney failure following living kidney donation. Transplant Proc. 2008;40:1337–40.
12. Steiner RW. 'Normal for now' or 'at future risk': a double standard for selecting young and older living kidney donors. Am J Transplant. 2010;10:737–41.
13. Delmonico FL, Harmon WE. The use of a minor as a live kidney donor. Am J Transplant. 2002;2(4):333–6.
14. Morel P, Kadry Z, Charbonnet P, Bednarkiewicz M, Faidutti B. Paediatric living related intestinal transplantation between two monozygotic twins: a 1-year follow-up. Lancet. 2000;355:723–4.
15. Berney T, Genton L, Buhler LH, Raguso CA, Charbonnet P, Pichard C, et al. Identical 13 year old twins have donated small bowel; five year follow-up after pediatric living related small bowel transplantation between two monozygotic twins. Transpl Proc. 2004;36:316–8.
16. Aulakh R. Too young to save his mother's life? Hospital changes policy for teen. Thestar.com. July 12, 2011. On the web at: http://www.thestar.com/life/health_wellness/news_research/2011/07/12/too_young_to_save_his_mothers_life_hospital_changes_policy_for_teen.html.
17. US Department of Health and Human Services (DHHS), Health Resources and Services Administration (HRSA), Organ Procurement and Transplantation Network. National Data. On the web at: http://optn.transplant.hrsa.gov/latestData/step2.asp?. Data updated through August 15, 2014.
18. Thys K, Van Assche K, Nobile H, Siebelink M, Aujoulat I, Schotsmans P, et al. Could minors be living kidney donors? A systematic review of guidelines, position papers and reports. Transpl Int. 2013;26:949–60.
19. Ethics Committee of the Transplantation Society. The consensus statement of the Amsterdam forum on the care of the live kidney donor. Transplantation. 2004;78:491–2.
20. Delmonico F, on behalf of the Council of the Transplantation Society. A report of the Amsterdam forum on the care of the live kidney donor: data and medical guidelines. Transplantation. 2005;79(6 Suppl):S53–66.
21. Barr ML, Belghiti J, Villamil FG, Pomfret EA, Sutherland DS, Gruessner RW, et al. A report of the Vancouver forum on the care of the live organ donor: lung, liver, pancreas, and intestine data and medical guidelines. Transplantation. 2006;81:1373–85.
22. Abecassis M, Adams M, Adams P, Arnold RM, Atkins CR, Barr ML et al. for the Live Organ Donor Consensus Group. Consensus statement on the live organ donor. JAMA. 2000;284:2919–26.
23. Raia S, Nery JR, Mies S. Liver transplantation from live donors. Lancet. 1989;2(8661):497.

24. Strong RW, Lynch SV, Ong TH, Matsunami H, Koido Y, Balderson GA. Successful liver transplantation from a living donor to her son. N Engl J Med. 1990;322:1505–7.
25. Broelsch CE, Whitington PF, Emond JC, Heffron TG, Thistlethwaite JR, Stevens L et al. Liver transplantation in children from living related donors. Surgical techniques and results. Ann Surg. 1991;214:428–37; discussion 437–429.
26. Ross LF. Children, families and healthcare decision making. Oxford UK: Oxford University Press; 1996.
27. Crouch R, Elliott C. Moral agency and the family: the case of living related organ transplantation. Camb Q Healthc Ethics. 1999;8:275–87.
28. Sells RA. Consent for organ donation: what are the ethical principles. Transplant Proc. 1993;25(1 Pt 1):39–41.
29. Zinner S. Cognitive development and pediatric consent to organ donation. Camb Q Healthc Ethics. 2004;13:125–32.
30. Holm S. The child as organ and tissue donor: discussions in the Danish Council of Ethics. Camb Q Healthc Ethics. 2004;13:156–60.
31. Parikh ND, Ladner D, Abecassis M, Butt Z. Quality of life for donors after living donor liver transplantation: a review of the literature. Liver Transpl. 2010;16:1352–8.
32. Trotter JF, Hill-Callahan MM, Gillespie BW, Nielsen CA, Saab S, Shrestha R et al. for the A2ALL Study Group. Severe psychiatric problems in right hepatic lobe donors for living donor liver transplantation. Transplantation. 2007;83:1506–8.
33. Johnson EM, Anderson JK, Jacobs C, Suh G, Humar A, Suhr BD, et al. Long-term follow-up of living kidney donors: quality of life after donation. Transplantation. 1999;67:717–21.
34. Rice SG, The American Academy of Pediatrics Council on Sports Medicine and Fitness. Medical conditions affecting sports participation. Pediatrics. 2008;121:841–8.
35. Army Regulation 40–501. Medical services standards of medical fitness. Rapid action revision (RAR) issue date: 4 August 2011. Washington DC: Headquarters of the US Army. On the web at: http://armypubs.army.mil/epubs/pdf/r40_501.pdf.
36. Andersen MH, Bruserud F, Mathisen L, Wahl AK, Hanestad BR, Fosse E. Follow-up interviews of 12 living kidney donors one year after open donor nephrectomy. Clin Transplant. 2007;21:702–9.
37. Watson JM, Behnke MK, Fabrizio MD, McCune TR. Recipient graft failure or death impact on living kidney donor quality of life based on the living organ donor network database. J Endourol. 2013;27:1525–9.
38. Kroencke S, Nashan B, Fischer L, Erim Y, Schulz KH. Donor quality of life up to two years after living donor liver transplantation: a prospective study. Transplantation. 2014;97:582–9.
39. Weizer N, Weizman A, Sapira Z, Yussim A, Munitz H. Suicide by related kidney donors following the recipients' death. Psychother Psychosom. 1989;51:216–9.
40. Schoeman F. Parental discretion and children's rights: background and implications for medical decision-making. J Med Philos. 1985;10:47–8.
41. Rhodes R, Burrows L, Reisman L. Mt. St. Anonymous: the adolescent living related donor. HEC Forum. 1992;4(5):314–23.
42. Klepper H. Incompetent organ donors. J Soc Philos. 1994;25:241–55.
43. American Academy of Pediatrics Committee on Bioethics. Informed consent, parental permission, and assent in pediatric practice. Pediatrics. 1995;95:314–7.
44. Harrison C, The Canadian Paediatric Society (CPS), Bioethics Committee. Treatment decisions regarding infants, children and adolescents. Paediatr Child Health. 2004;9:99–103.
45. Office of the United Nations High Com-mission for Human Rights, Convention on the Rights of the Child Adopted and Opened for Signature, Ratification and Accession by General Assembly resolution 44/25 of 20 November 1989 Entry into Force 2 September 1990, in Accor-dance with Article 49, at Article 3.1.
46. Buchanan AE, Brock DW. Decid-ing for others: the ethics of surrogate decision making. New York: Cambridge University Press; 1990.
47. Kopelman LM. The best-interests standard as threshold, ideal, and standard of reasonableness. J Med Philos. 1997;22:271–89.

48. Salter EK. Deciding for a child: a comprehensive analysis of the best interest standard. Theor Med Bioeth. 2012;33:179–98.
49. Little v Little. 576 S.W.2d 493; (Tex App 1979).
50. In re Richardson. 284 So. 2d 185 (La Ct App. 1973).
51. Lausier v Pescinski. 8226 N.W.2d 180 (Wisc. 1975).
52. Fleck LM. Children and organ donation: some cautionary remarks. Camb Q Healthc Ethics. 2004;13:161–6.
53. Ladd RE. The child as living donor: parental consent and child assent. Camb Q Healthc Ethics. 2004;13:143–8.
54. Schoeman F. Rights of children, rights of parents, and the moral basis of the family. Ethics. 1980;91(1):6–19.
55. Blustein J. Doing the best for one's child: satisficing versus optimizing parentalism. Theor Med Bioeth. 2012;33:199–205.
56. Goldstein J, Freud A, Sol-nit AJ. Before the best interests of the child. New York: The Free Press; 1979.
57. Downie RS, Randall F. Parenting and the best interests of minors. J Med Philos. 1997;22:219–31.
58. Nathan v. Farinelli. Eq. No. 74–87 (Mass July 3, 1974).
59. Dwyer J, Vig E. Rethinking transplantation between siblings. Hastings Cent Rep. 1995;25 (5):7–12.
60. Strunk v Strunk. 445 S.W.2d 145 (Ky 1969).
61. Department of Health and Human Services (DHHS). 45 CFR 46—Additional protections for children involved as subjects in research. Federal Register 48(46);9814–20, March 8, 1983. Last updated January 15, 2009. On the web at: http://www.hhs.gov/ohrp/humansubjects/guidance/45cfr46.html.
62. Dindo D, Demartines N, Clavien PA. Classification of surgical complications: a new proposal with evaluation in a cohort of 6336 patients and results of a survey. Ann Surg. 2004;240:205–13.
63. Mjøen G, Øyen O, Holdaas H, Midtvedt K, Line PD. Morbidity and mortality in 1022 consecutive living donor nephrectomies: benefits of a living donor registry. Transplantation. 2009;88:1273–9.
64. Lentine KL, Segev DL. Better understanding live donor risk through big data. Clin J Am Soc Nephrol (CJASN). 2013;8:1645–7.
65. Lentine KL, Patel A. Risks and outcomes of living donation. Adv Chronic Kidney Dis. 2012;19(4):220–8.
66. Muzaale AD, Dagher NN, Montgomery RA, Taranto SE, McBride MA, Segev DL. Estimates of early death, acute liver failure, and long-term mortality among live liver donors. Gastroenterol. 2012;142:273–80.
67. Yamaoka Y, Morimoto T, Inamoto T, Tanaka A, Honda K, Ikai I et al. Safety of the donor in living-related liver transplantation—an analysis of 100 parental donors. Transplantation 1995;59:224–6.
68. Grewal HP, Thistlethwaite JR Jr., Loss GE, Fisher JS, Cronin DC, Siegel CT et al. Complications in 100 living-liver donors. Ann Surg. 1998;228:214–9.
69. Sterneck M, Nischwitz U, Fischer L, Malago M, Rogiers X, Raedler A et al. Evaluation and morbidity of the living liver donor in pediatric liver transplantation. Transplant Proc. 1995;27:1164–5.
70. Kitai T, Higashiyama H, Takada Y, Yamamoto Y, Nakamura Y, Ikai I et al. Pulmonary embolism in a donor of living-related liver transplantation: estimation of donor's operative risk. Surgery. 1996;120:570–3.
71. Renz JF, Roberts JP. Long-term complications of living donor liver transplantation. Liver Transpl. 2000;6(6 Suppl 2):S73–6.

72. Olthoff KM, Abecassis MM, Emond JC, Kam I, Merion RM, Gillespie BW et al, for the Adult-to-Adult Living Donor Liver Transplantation Cohort Study Group. Outcomes of adult living donor liver transplantation: comparison of the adult-to-adult (A2ALL) living donor liver transplantation cohort study and the national experience. Liver Transpl. 2011;17:789–97.
73. Abecassis MM, Fisher RA, Olthoff KM, Freise CE, Rodrigo DR, Samstein B et al, for the Adult-to-Adult Living (A2ALL) Donor Liver Transplantation Cohort Study Group. Complications of living donor hepatic lobectomy—a comprehensive report. Am J Transplant. 2012;12:1208–17.
74. Marubashi S, Nagano H, Wada H, Kobayashi S, Eguchi H, Takeda Y et al. Donor hepatectomy for living donor liver transplantation: learning steps and surgical outcome. Dig Dis Sci. 2011;56:2482–90.
75. Uchiyama H, Shirabe K, Nakagawara H, Ikegami T, Toshima T, Soejima Y, et al. Revisiting the safety of living liver donors by reassessing 441 donor hepatectomies: is a larger hepatectomy complication-prone? Am J Transplant. 2014;14:367–74.
76. Marcos A, Fisher RA, Ham JM, Shiffman ML. Sanyal AJ. Luketic VA et al. Liver regeneration and function in donor and recipient after right lobe adult to adult living donor liver transplantation. Transplantation. 2000;69:1375–9.
77. Abraham EC, Wilson AC, Goebel J. Current kidney allocation rules and their impact on a pediatric transplant center. Am J Transplant. 2009;9:404–8.
78. Kidney Transplant Committee of the Organ Procurement Transplantation Network (OPTN)/ United Network for Organ Sharing (UNOS). Proposal to substantially revise the National Kidney Allocation System. On the web at: http://optn.transplant.hrsa.gov/PublicComment/pubcommentPropSub_311.pdf.
79. Reese PP, Feldman HI, McBride MA, Anderson K, Asch DA, Bloom RD. Substantial variation in the acceptance of medically complex live kidney donors across us renal transplant centers. Am J Transplant. 2008;8:2062–70.
80. Becker LE, Susal C, Morath C. Kidney transplantation across HLA and ABO antibody barriers. Curr Opin Organ Transplant. 2013;18:445–54.
81. Ross LF, Rubin D, Siegler M, Josephson MA, Thistlethwaite JR Jr, Woodle ES. Ethics of a paired-kidney-exchange program. N Engl J Med. 1997;336:1752–5.
82. Veale J, Hil G. The national kidney registry: transplant chains—beyond paired kidney donation. Clin Transpl. 2009:253–64.
83. Marks SD. Should children ever be living kidney donors? Pediatr Transplant. 2006;10:757–9.
84. Brierley J, Larcher V. Organ donation from children: time for legal, ethical and cultural change. Acta Paediatr. 2011;100:1175–9.
85. Webb NJA, Fortune P-M. Should children ever be living kidney donors. Pediatr Transplant. 2006;10:851–5.
86. Campbell M, Wright L, Greenberg RA, Grant D. How young is too young to be a living donor? Am J Transplant. 2013;13:1643–9.
87. Capitaine L, Thys K, Van Assche K, Sterckx S, Pennings G. Should minors be considered as potential living liver donors. Liver Transpl. 2013;19:649–55.
88. Grisso T, Vierling L. Minor's consent to treatment: a developmental perspective. Prof Psychol. 1978;9:412–27.
89. Weithorn L, Campbell S. The competency of children and adolescents to make informed treatment decisions. Child Dev. 1982;53:1589–98.
90. Steinberg L, Scott ES. Less guilty by reason of adolescence: developmental immaturity, diminished responsibility, and the juvenile death penalty. Am Psychol. 2003;58:1009–18.
91. Steinberg L, Cauffman E, Woolard J, Graham S, Banich M. Are adolescents less mature than adults?: minors' access to abortion, the juvenile death penalty, and the alleged APA 'flip-flop'. Am Psychol. 2009;64:583–94.
92. Social Security Amendments of 1972 (P.L. 92–603).
93. Carey WA, Talley LI, Sehring SA, Jaskula JM, Mathias RS. Outcomes of dialysis initiated during the neonatal period for treatment of end-stage renal disease: a North American pediatric renal trials and collaborative studies special analysis. Pediatrics. 2007;119:e468–73.

94. Abecassis M, Bartlett ST, Collins AJ, Davis CL, Delmonico FL, Friedewald JJ. Kidney transplantation as primary therapy for end-stage renal disease: a national kidney foundation/kidney disease outcomes quality initiative (NKF/KDOQI™) conference. Clin J Am Soc Nephrol (CJASN). 2008;3:471–80.
95. Gulati A, Sarwal MM. Pediatric renal transplantation: an overview and update. Curr Opin Pediatr. 2010;22:189–96.
96. McDonald SP, Craig JC, The Australian and New Zealand Paediatric Nephrology Association. Long-term survival of children with end-stage renal disease. New Engl J Med. 2004;350:2654–62.
97. Spital A. Should children ever donate kidneys? Views of U.S. transplant centers. Transplantation. 1997;64:232–6.
98. Joseph JW, Thistlethwaite JR Jr, Ross LF. An empirical investigation of physicians' attitudes towards intrasibling kidney donation by minor twins. Transplantation. 2008;85:1235–9.
99. Health Council of the Netherlands. New Options for Organ Donation. The hague: health council of the Netherlands, 2003: publication no. 2003/01, ISBN 90-5549-462-3. On the web at: http://gr.nl/sites/default/files/summary_orgaandonatie.pdf.
100. World Health Organization. WHO Guiding Principles on Human Cell, Tissue and Organ Transplantation. 2010. On the web at: http://www.who.int/transplantation/Guiding_PrinciplesTransplantation_WHA63.22en.pdf.
101. Klepetko W, Wolner E. Domino transplantation of heart-lung and heart: an approach to overcome the scarcity of donor organs. J Heart Lung Transplant. 1991;10:129–31.
102. Khanna A, Hart M, Nyhan WL, Hassanein T, Panyard-Davis J, Barshop BA. Domino liver transplantation in maple syrup urine disease. Liver Transpl. 2006;12:876–82.
103. Mazariegos GV, Morton DH, Sindhi R, Soltys K, Nayyar N, Bond G, et al. Liver transplantation for classical maple syrup urine disease: long-term follow-up in 37 patients and comparative united network for organ sharing experience. J Pediatr. 2012;160:116–21.
104. Schlam L, Wood JP. Informed consent to the medical treatment of minors: law and practice. Health Matrix. 2000;10(2):141–74.

Chapter 2
Living Donors for Fulminant Hepatic Failure in Children

Maja Segedi, Genieve Dhani, Vicky L. Ng and David Grant

Keywords Pediatric acute liver failure (PALF) · Living donors · Ethics · Liver transplantation

Abbreviations

ALF	Acute liver failure
DDLT	Deceased-donor liver transplant
HRQOL	Health-related quality of life
KCHC	Kings College Hospital Criteria
LDLT	Living donor liver transplant
LKD	Living kidney donors
NPV	Negative predictive value
NRD	Non-related donors
PALF	Pediatric acute liver failure
PELD	Pediatric end-stage liver disease score
PPV	Positive predictive value
PRISM	Pediatric risk score of mortality
SPLIT	Studies of pediatric liver transplant
UNOS	United network for organ sharing

M. Segedi (✉) · G. Dhani · D. Grant
Multi-Organ Transplant Program, University Health Network, Toronto, ON, Canada
e-mail: Maja.Segedi@vch.ca

V.L. Ng · D. Grant
Liver Transplant Program, Hospital for Sick Children, Toronto, ON, Canada

M. Segedi
Liver Transplant Program, Vancouver General Hospital, Vancouver, BC, Canada

R.A. Greenberg et al. (eds.), *Ethical Issues in Pediatric Organ Transplantation*,
International Library of Ethics, Law, and the New Medicine 66,
DOI 10.1007/978-3-319-29185-7_2

2.1 Introduction

This chapter reviews and addresses the ethical aspects of living donor liver transplantation (LDLT) for pediatric acute liver failure (PALF) patients. PALF is a rare, rapidly progressing and life-threatening diagnosis for which liver transplantation is often the only life-saving treatment. Once the diagnosis of PALF and its aetiology are established, there are very limited directed medical treatments available. Thus, the risk of rapid progression and poor natural history in PALF drives decision making about early liver transplantation. There is no guarantee that a deceased-donor liver transplant (DDLT) will become available in time, so LDLT is considered as an alternative when available. However, the use of LDLT requires careful consideration of the ethical issues. The probability of a good outcome with liver transplantation (LT) may not be high enough to justify the risks to the living donor—but who should make this judgement? Is it fair to ask potential donors to make such a big decision in a time-pressured situation? The high mortality from PALF can also exert internal and external pressure on the candidate donor leading to potentially coercive decisions. As will be discussed, these issues are best managed by a collaborative interdisciplinary team with extensive experience and success in LDLT.

This chapter will discuss the following issues:

1. Overview of prevalence and etiologies of PALF;
2. The role of transplantation and indications for transplantation in PALF;
3. Types of transplants used for treatment of PALF including deceased whole liver, deceased split liver, and living donor livers, and their relative advantages and disadvantages;
4. Discussion of wait list issues for children with PALF;
5. Explanation of the evaluation process for living donors for PALF patients;
6. Ethical arguments for and against LDLT in PALF.

2.2 Overview of Pediatric Acute Liver Failure (PALF)

PALF presents as a multisystem disorder with severe liver injury, with or without encephalopathy, and no previous history of liver disease. Without treatment, PALF often progresses to death from multisystem organ failure and/or cerebral anoxia. Supportive care includes the provision of intravenously administered vitamin K to correct abnormal INR, occasional and cautious administration of fresh frozen plasma, cryoprecipitate or platelet infusions to decrease risk of bleeding at time of interventional procedures, and treatment with prophylactic antibiotics to prevent sepsis [1]. It is vital to try to determine the cause of liver failure in order to provide as early as possible the cause-directed treatments that are discussed below. Children with PALF should also be referred to an experienced pediatric liver transplant centre for coordinated evaluation and management can be conducted. Considering the above issues, the transplant team's role is to:

1. Expeditiously ensure a thorough work-up has been completed
2. Determine the cause of PALF
3. Ensure cause-directed medical therapy is provided whenever possible
4. Rule out PALF etiologies that have a poor outcome with LT
5. List for transplantation if this is deemed the best treatment option, and
6. Explore the option of live donation, when appropriate.

2.3 Role of Liver Transplantation in the Treatment of PALF

2.3.1 Decision to Transplant and Recipient Selection Criteria

The LT decision is complex and should proceed deliberately and thoughtfully despite the inherent uncertainties associated with the fact that PALF is a clinically diverse, multi-faceted and dynamic condition. Treatment generally supportive unless a specific diagnosis responsive to targeted therapy is identified. The dynamic nature of PALF challenges our ability to predict outcome. All scoring systems used to predict mortality in non-transplanted patients with acute liver failure, including the Kings College Hospital Criteria (KCHC), the Clichy score, Pediatric Risk Score of Mortality (PRISM), and Pediatric End-Stage Liver Disease (PELD) score have limitations. Thus, when applied to PALF transplant candidacy decisions, these measures provide helpful clinical guidance, but the final decision to offer transplantation requires experienced clinical judgment [2].

LT holds the greatest lifesaving potential, but it carries with it potential complications related to surgery and immunosuppression [3, 4]. Albeit rarely, some children listed for LT who would have received an organ had one become available, are removed from the list due to clinical improvement. It is difficult to predict which children may improve clinically, and thus need not be exposed to the risk of transplantation. Vigilance must be maintained to ensure that children requiring LT survive, and ongoing efforts to identify clinical models that reliably predict patient death and survival. The challenges of when to apply living donor liver transplantation in PALF are herein discussed [5–7].

2.3.2 Organ Allocation and Different Surgical Options for Liver Transplantation

Estimates of annual incidence rate of PALF in the United States are at 2800 cases per annum with an average mortality rate of up to 50 % [8, 9]. Outside of countries that solely use living donation for all transplantation, the main source of liver transplants are deceased-donor organs. Unfortunately, the deceased-donor pool is always insufficient to meet all demands and the lack of control over the timing of obtaining a good quality organ for a child with PALF can result in death while on the wait list.

Deceased donor liver allocation for children with PALF is evolving. In North America these decisions are based on a recipient disease severity score, or on PELD score for children with chronic liver disease. Children with acute liver failure are listed with Status 1A by United Network for Organ Sharing (UNOS). The criteria for 1A listing are as follows:

1. Onset of hepatic encephalopathy within eight weeks of the first symptoms of liver disease,
2. Absence of pre-existing liver disease,
3. AND one of:
   (a) Ventilator dependence
   (b) Need for dialysis, continuous venous hemofiltration, or continuous venovenous hemodialysis
   (c) INR > 2.0

Based on a weekly reassessment, the child can be either kept on the deceased-donor waiting list, removed from the list if he/she becomes too ill to undergo transplantation or improves, or can be demoted from Status 1A and listed according to the disease severity score.

Several different types of donor grafts can be used to transplant pediatric patients with ALF. Sized-matched pediatric grafts from deceased organs provide excellent outcomes. However, advancements in surgical techniques have also made it possible to transplant different parts of the liver from an adult. Options include segments 2 and 3—also known as the left lateral segment liver graft, the entire left lobe or the right lobe transplants. The use of these techniques increases graft availability for all recipients, thereby improving survival rates in PALF patients [10]. Following live donation, the donor's remaining liver regenerates to a near normal volume and the recipient liver also undergoes hyperplasia (growth) or apoptosis (shrinkage) in volume to match the recipient's size. Partial deceased-donor grafts and live donor grafts have higher or comparable patient survival rates to that of whole deceased-donor grafts [11].

2.3.3 *Shortage of Donors and Magnitude of Death on the Wait List*

The deceased-donor pool remains insufficient to meet all demands. The lack of control over the availability of a good quality organ for a child with PALF can result in death while on the wait list. This can be ameliorated by using living donor grafts. A large series of pediatric liver transplant cases in the United States revealed that of 442 pediatric transplants performed between 1993 and 2006, a total of 49 or 11 %, were from living donors [12].

More than 6000 liver transplants are performed annually in the United States. According to the 2012 US Organ Procurement and Transplantation Network report, each year over 600 children were listed for liver transplant, and on average 540 received a transplant. However, annually, more than 40 children die waiting for a donor liver and approximately 14 became too sick to transplant [13]. This resulted in pre-transplant mortality of 5.8 deaths per 100 wait-list years [13].

Patients with acute liver failure have the highest pre-transplant mortality rates of all patients with end stage liver disease [13]. Of 1621 pediatric patients listed for transplantation, 11.2 % (n = 182) had PALF and about 18 % of these children died waiting for a donor organ [13]. Despite these statistics, LDLT only account for a small proportion of all liver transplants performed [13]. Success rates, concerns, and barriers to LDLT are discussed below.

2.4 Overall Outcomes of Liver Transplantation in Children

Thomas Starzl performed the first liver transplant from a deceased-donor in 1963. Steady advances in surgical technique and improvements in outcomes ensued in the decades to follow. The first reduced-size deceased-donor liver graft was transplanted into a child patient in Belgium in the 1980s. The first pediatric living donor liver transplants were carried out in Brazil in 1989 [14], followed by transplants in the United States and Japan [15, 16].

During the 2000s, the PELD allocation system was introduced, along with increased regulation of living donor (LD) organ allocation, with specific United Network for Organ Sharing (UNOS—a private, non-profit organization that manages the US's organ transplant system under contract with the federal government) centre requirements for LDLT and government oversight of living donor transplantation [17].

Liver transplantation with a cadaveric whole and split or living donor organ is a highly successful treatment for PALF. Survival rates of children with PALF who do not undergo a transplant range from 10 to 30 %, whereas children with PALF who undergo a liver transplant have survival rates in excess of 80 % [18]. The outcomes of transplantation for PALF have been steadily improving. The Studies of Pediatric

Liver Transplant (SPLIT) consortium reported 4-year post-transplant patient survival rates of 69.4 % for children with PALF [18, 19]. In a 2010 report of 57 more recent transplants the survival rate exceeded 80 % at 10 years [5]. Several retrospective series have identified adverse factors that affect outcome of transplantation for PALF [3, 7, 20, 21]. These factors include, but are not limited to: age less than one year, need for pre-transplant dialysis, and idiopathic PALF etiology [7, 22]. Morbidity-related complications of pediatric liver transplantation include infection, hepatic arterial thrombosis, and biliary strictures.

Children with PALF who undergo a liver transplant have worse outcomes than those who are transplanted for chronic liver diseases. There are several reasons for this, including their poorer condition at the time of transplantation, and the fact that postoperative infections and the risk of aplastic anemia are more common than in other transplant recipients, perhaps as a result of intrinsic immune deficits [23, 24].

2.5 Outcomes of Living Donor Liver Transplantation

2.5.1 Recipient Outcomes

The principle advantage of live liver donation is the opportunity to avoid or minimize the risks of waiting for a deceased-donor organ [3]. Other advantages of LDLT for PALF include thorough donor screening leading to a high quality liver graft and brief preservation times [25–27].

LDLT provides excellent outcomes in children [7, 28, 29, 30]. Most of the surgical techniques for these procedures were developed and refined in Japan and Korea where deceased donation is uncommon or unavailable. A study by Baliga and the SPLIT group from centers in the US and Canada, suggests that LDLT accounts for about 14 % of all pediatric liver transplants in the United States and Canada [19]. Liu et al. in Hong Kong reported that right lobe LDLT offers superior survival rates to DDLT in adults [28, 29]. A study by Bourdeaux et al. [28] compared 1010 children who received LDLT to 135 children treated with a DDLT. Although only 16 of the patients were treated for PALF, this study showed a superior survival rate at 1 and 5 years of 94 and 92 % in LDLT group, versus 93 and 91 % after whole DDLT group, 83 and 79 % after reduced-size DDLT, and 90 and 83 % after split DDLT, respectively ($p = 0.169$) (see Fig. 2.1). The chance to obtain a high-quality graft with LDLT is counterbalanced by a slightly higher rate of vascular and biliary complications. Because of these offsetting benefits and risks, transplantation graft survival rates are similar for DDLT and LDLT after ten years [12].

A recent decision analysis compared the costs of DDLT and LDLT. The authors concluded that the addition of LDLT to a standard waiting list for DDLT prevents waiting list deaths and improves recipient survival and, but at greater cost [31]. The cost of being listed for DDLT only was quantified with an Incremental Cost-Effectiveness Ratio (ICER) of $35,000. In comparison, being listed for DDLT

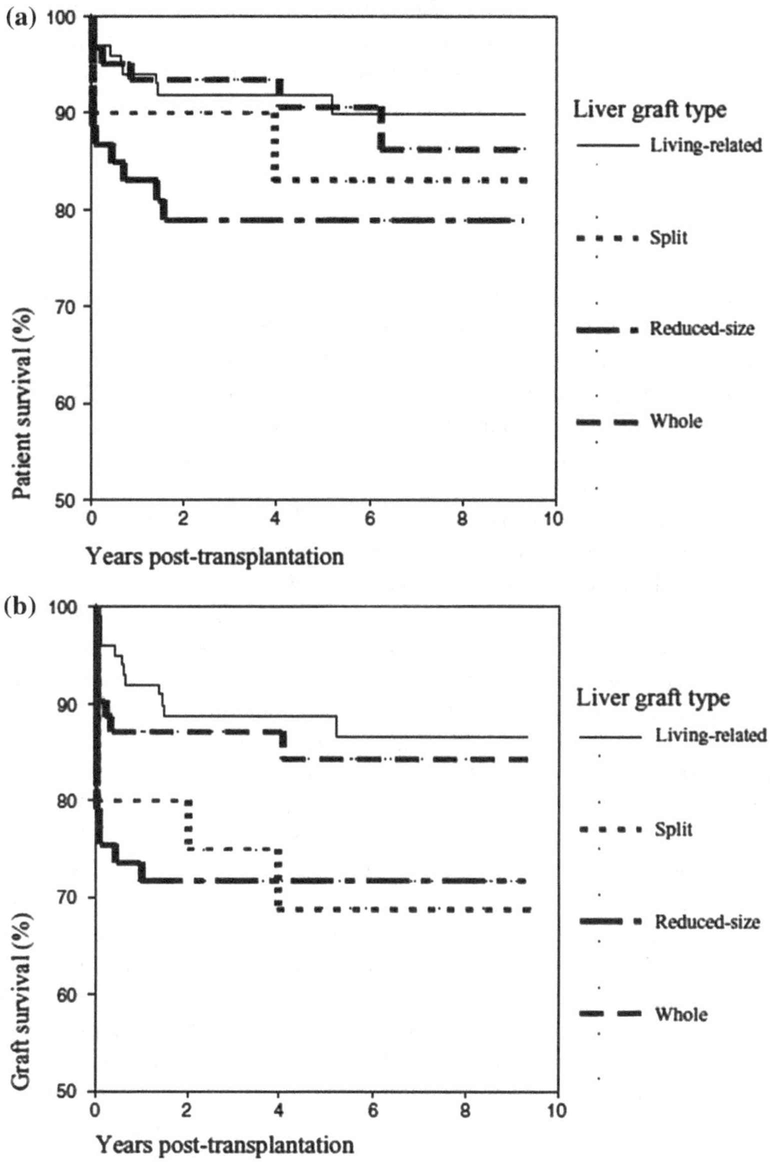

Fig. 2.1 Similar post transplant overall survival but superior graft survival from living donor liver transplantation in 235 children who received either living donor (n = 100) or deceased-donor (n = 135) liver transplant from 1993 to 2002 in a Single Centre. Adapted from Bourdeaux et al. [28]. *Note* Five-year patient survival rates were 92 and 85 % for groups LD and DD, respectively ($p = 0.181$), the corresponding graft survival rates being 89 and 77 % ($p = 0.033$)

with the LDLT option carried an ICER of $106,000, both measured over a 10-year time frame [31]. These costs examined in this study included costs to the donor, the recipient, the health care system, as well as the costs of complications and loss of life.

2.5.2 Donor Outcomes

Living donation (LD) is an imperfect solution to the problem of organ shortage, because of the risks for the donor. Even in the best of hands, this surgery is associated with a small but significant risk of permanent morbidity or mortality [32–36]. By offering "unnecessary" surgery to a healthy person, LD violates the classic Hippocratic oath of doctors to "abstain from doing harm" [37]. In the modern era, the focus of medical ethics is on the efforts to prevent or minimize harm while trying to help the patient. A key concept is the principle of non-maleficence, defined by Beauchamp and Childress, as not inflicting evil or harm [38]. With live donation, the risk of serious complications or death is directly proportional to the amount of liver that is removed. The estimated mortality risk for a donor of a left lateral liver segment (about 25 % of the liver volume) is about 0.15 %, whereas the estimated mortality risk for a donor of a right lobe liver donation (about 68 % of the liver volume) is 0.30 %. Thus, serious complications or death are rare after LD but when they do occur, the outcomes are devastating for the donor, recipient, the family and the clinicians involved.

The largest review of living donor hepatectomy outcomes examined a total of 214 studies on all adult and child LDLT performed up to year 2004 [36]. The authors estimated that approximately 6000 living donor liver transplantations had been performed worldwide, with a mortality rate of 12–13 in 6000, or 0.2 %. Of ten early donor deaths, 1 was in a donor of left lateral segment, 1 of left lobe, and 5 of right lobe donations. Specifically, donor mortality in adult to child LDLT ranged from 3 to 7 per 3500, or 0.09 to 0.2 % [36]. This study reported a median donor morbidity rate of 16.1 % (ranging from 0 to 100 %), with biliary complications and infections being the most common, at 6.2 % (range 0–38.6 %). Other complications included infections at a median rate of 5.8 %, and need for transfusion at a median rate of 1.9 %. The most common morbidities were biliary leaks and strictures (median 6 %; range 0–39 %) and infections (median 6 %; range 0–29 %). Other reported complications included incisional hernias, pleural effusion requiring intervention, and neuropraxia [39]. Nearly all donors had returned to normal function by three to six months.

A more recent large study examined late complications among 4111 LDs in the United States from 1994 to 2012 and compared their mortality rates to the participants in the National Health and Nutrition Examination Survey III (NHANES III). The risk of long-term mortality did not differ from that of healthy, matched individuals over a mean of 7.6 years [40].

A key ethical issue is whether the risks to the donor described above are outweighed by the benefits of helping a loved one and preventing death. We will analyze this ethical issue in more detail in the following sections.

2.5.3 Urgent Live Donor Work-Ups for PALF Recipients: The Challenges

LDLT for acute liver failure presents many serious ethical, medical, logistic, and economic issues. In cases where the recipient has subacute liver disease, a live donor work-up takes 4–6 weeks to complete. This provides the donor physicians with ample time to complete all of the required tests and provides the potential donor with a "cooling-off period" to decide if he/she truly wishes to proceed with this operation [41]. In contrast, the donor and recipient work-up for a LDLT for PALF must be completed within hours or a few days. The rapidity of this work-up makes it more difficult to obtain voluntary and fully informed consent [42]. Notwithstanding this concern, we believe that an ethical and timely work-up of a potential donor is possible if several considerations are met. The first step is to ensure the recipient is a good candidate for liver transplantation. The second step is to confirm that a transplant is the best treatment option. The third step is to undertake a well-defined protocol for the donor work-up that includes multiple safety checklists to ensure that nothing is overlooked or missed, including both clinical and ethical issues, including informed consent and prevention of coercion. Several such protocols have been developed and used worldwide and a sample protocol is shown in Fig. 2.2 [43–47]. Undertaking this work-up requires adequate hospital resources and personnel to complete a thorough medical and psychosocial evaluation; and to provide the potential donor with detailed information about all possible surgical complications including mortality, morbidity, and long-term quality-of-life impact, as well as the risk of recipient mortality despite LDLT. Because of the physician's *prima facie* duty to his or her patient—in this case a child with PALF, it is important that there are separate clinical teams working up to the donor and the recipient in order to avoid perceived or real conflicts of interest [48] (Table 2.1).

The assessment of a living donor for a PALF recipient requires confirmation of a compatible blood group, excellent general global health (physical, mental, and social well-being), the absence of any medical co-morbidities that might increase the risks of surgery, exclusion of transmissible malignancy or infections, and confirmation of suitable liver anatomy. Determining the psychosocial status of the donor and social support is a particularly important part of this process. The presence of inherited metabolic disease in the recipient child may not preclude parental donation when the donor is an asymptomatic heterozygote carrier. An estimation of steatosis ("fattiness") of the liver is essential in the selection process for donors with a higher than normal body mass index (BMI); with modern imaging and liver function tests a liver biopsy is only required when screening tests are abnormal.

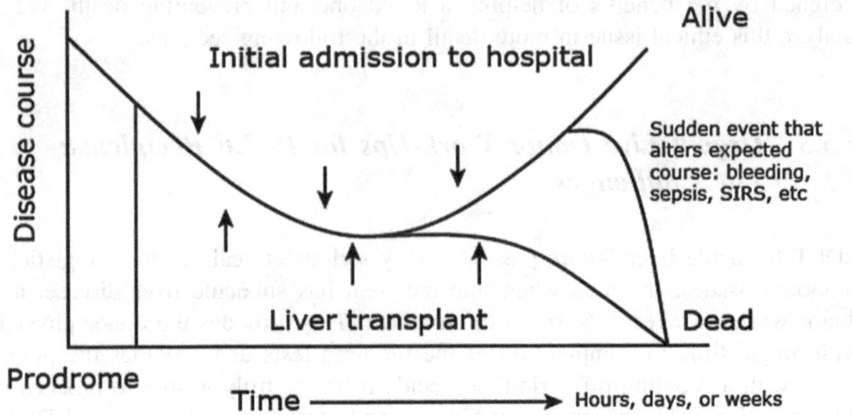

Fig. 2.2 Schematic model of the natural course and outcome of acute liver failure in children. *SIRS* Systemic inflammatory response syndrome. Reproduced from Squires and Alonso [49]. Copyright © 2012 Cambridge University Press

Table 2.1 Pros and cons of living donor liver donation to children with PALF

PRO	CON
Autonomy of the donors is supported by donation if they have strong emotional links to the sick child with PALF (i.e. are parents, adult siblings or other family members and care-givers)	Post-operative complications and potential mortality to donors
Quality of life of the child and the family is improved with LDLT if no deceased organ is available	Increased costs and difficulty coping with child's illness for parent donors who are also healing from donor surgery
Utilitarian argument—LDLT decreases wait time in a rapidly deteriorating child with PALF and wait times of other children on the deceased organ wait list are not affected	Additional costs and affects to other family members in the home, such as siblings, if parents are unable to care for them
Not donating might impart long-term negative effects on the parents and family members if the child with PALF was a transplant candidate but did not receive an organ and had serious sequellae including mortality	The challenge of accurate and objective quantification of risk-benefit ratio for the donor, including their psychosocial and physical postoperative risks, the effects that not donating might impart long term, and the effects that postoperative complications of the recipient child can have

When considering a donor for a child with PALF, the goal is to complete the donor work-up within 24–48 h [47]. A suggested evaluation protocol for a potential living donor for a child with PALF is shown in Table 2.2. During this process, the physician responsible for the donor work-up must be readily available to provide education and psychological support, and to ensure that the assessment process is free of pressure or coercion to donate. The donor assessment team is completely independent from the recipient assessment team. Challenges in communication between donor and recipient parties should follow the pre-established program guidelines and consider the guiding principles discussed in Sect. 2.7. With these strict criteria and well-defined processes, donor rejection rates are significant, ranging from 29 to 86 % [47, 48, 50, 51].

Table 2.2 Suggested evaluation protocol for potential living liver donors for children with PALF[a]

Step 1	
	– History and physical examination – Laboratory investigations: blood group, hematological tests, chemistry, coagulation profile, C-reactive protein, pregnancy test in females – Serology for Hepatitis A, B, C, HIV, CMV, HSV, EBV – Cross sectional imaging with CT scan and MRI to delineate vascular and biliary anatomy – Liver biopsy (in select cases based on suspicion for steatosis or other liver pathology) – Psychosocial evaluation – If no contraindications proceed to Step 2
Step 2	
	– Surgical risk assessment for comorbidities: ECG, chest X-ray – Laboratory: thyroid function tests (TSH, T3, T4), immunoglobulins IgA, IgG, IgM, iron, transferrin, ferritin, alpha-1-antitrypsin, ceruloplasmin, tumor markers (CEA, AFP, Ca19-9), factors V, VII and VIII, protein C and S, APCR, and urine sediment – Pulmonary function tests, echocardiography, stress testing if suspicion for cardiopumonary disease – Select consultations if needed – Hepatitis B vaccination – If no contraindications proceed to Step 3
Step 3	
	– Surgeon assessment – Hepatologist assessment – Second psychosocial assessment – If no contraindications proceed to Step 4
Step 4	
	– Second Hepatitis B vaccination – Anaesthesia consultation – Ethics board evaluation – Final informed consent
Optional steps	Autologous blood donation, liver function tests: galactose, indocyanine clearance test, lidocaine

HIV human immunodeficiency virus; *CMV* cytomegalovirus; *HSV* herpes simplex virus; *EBV* Epstein-Barr virus; *CT* computed tomography; *ECG* electrocardiogram; *TSH* thyroid-stimulating hormone; *T3* triiodothyronin; *T4* thyroxin; *CEA* carcinoembryonic antigen; *AFP* alpha-fetoprotein; *APCR* activated protein C resistance

[a]Adapted from Valentin-Gamazo et al. [47]

2.6 Ethical Issues

2.6.1 Donor Health-Related Quality of Life and Psychosocial Health After Live Donor Surgery PALF

Physical complications developing following donor hepatectomy to donors are well documented in many publications, but only a few studies have assessed the health related quality of life (HRQOL) and the psychosocial outcomes after adult-to-child donation. The goals of the donor psychosocial evaluation are to identify issues that might contribute to ill health or decision regret after donation and to ensure that the decision to donate is well informed. The limited studies on psychosocial aspects of adult-to-child LDLT show that the HRQOL for living donors is as good as, or better than that of, the general population. Studies looking at outcomes for donors who are parents versus those who are relatives versus anonymous donors are much needed to explore this further. Interview assessments with family and parent donors for children with non-acute liver failure show nearly uniform absence of regret [52, 53]. In cases where the transplant failed, a re-transplant was needed, or the child recipient died, many donors were comforted in knowing that they did all that could possibly be done to save the child [52]. In these studies the majority of donors were parents, and a much smaller proportion was represented by relatives and grandparents [53]. In contrast, one study assessed the frequency and risk factors for donor ambivalence in adult-to-adult donors, and found that 34 % of donors had some regrets [53]. In this study, donor ambivalence was defined as "simultaneous and contradictory attitudes and/or feelings toward a person (the recipient) or an action [of right hepatic lobe donation]" [54]. Of a total of 45 donors studied, one was a parent, three were friends or relatives, and the remaining 41 were adult children donating to parents [54]. Those donors who reported ambivalence were more likely to be male, over 35 years old, with a higher level of education, and donating to a recipient with hepatitis C or alcohol-related liver disease [54]. Similar findings have not been documented in parent-to-child LDLT.

Living donor liver donation for babies, infants, and young children with PALF most commonly requires resection of a small segment of liver, such as a left lateral lobe, from parents or relatives. Only larger, older children require a larger right-sided graft. The available evidence on psychosocial aspects and consequences of donation, post-donation quality of life, and decisional regret suggest that when the donors are parents or close relatives and are emotionally invested in the well-being of the child, LDLT for PALF has minimal medical/surgical risks, no negative effects on the donor HRQOL, and negligible decisional regret. In situations where right or left lobe donation is needed, a larger segment of liver is resected for an older child with PALF. Notably, most of the positive studies on absence of decisional regret in parent-to-children donors and high levels of post-donation HRQOL are of donors of small segments of liver. Therefore, it is particularly important that the slightly higher morbidity and mortality risks and potentially greater effect on HRQOL in donors to older children who require right lobe transplants are fully disclosed to potential donors.

2.6.2 *Unique Pediatric Issues Relevant to Living Donation for Children with Acute Liver Failure*

There are several unique social and ethical aspects related to living liver donation for a child with PALF. Unlike adult medicine and surgery, which generally focuses on individual outcomes, caring for a child with PALF requires an understanding of how this condition impacts on the entire family. In this setting caregivers must provide strong psychosocial support to all family members, be aware of the extreme stresses imposed by the need to make vital but expedient healthcare decisions; and recognize the challenges associated with making decisions for a patient who may be unable to provide any independent input, either because of young age, general ill health, or hepatic encephalopathy.

2.7 Ethical Arguments for and Against Urgent Living Donor Liver Transplantation for PALF in Children

In the LDLT assessment process, physicians assessing the donor and recipients must be mindful of their moral and legal fiduciary responsibilities. Fiduciary obligations flow from the relationship of trust between physician and patient and the implicit or explicit understanding that physicians will focus, almost exclusively, on the best interests of the patient using their knowledge and expertise. Thus, caregivers must not only do their best to minimize harm to the donor and recipient. As moral agents in this process, they must also ensure patients are well informed and determine for themselves if the net benefits of LDLT outweigh the net risks. A more detailed analysis of the ethical issues associated with LDLT is provided below.

2.7.1 *Autonomy, Time Pressure and Coercion of Donors*

The principle of autonomy provides a strong ethical foundation supporting urgent LDLT for PALF. We live in a society that allows people to participate in high-risk activities like hang-gliding and enlists young men and women in high-risk military activities. In this context, it seems reasonable to also offer individuals the option of live donation, with relatively low associated risks, in order to save the life of child.

Surveys of the general population and of past donors support that offering LDLT in PALF appeals to the autonomy of the parent donors [53, 55, 56]. In a survey of

150 people recruited at the time of their presentation for a routine medical visit at an internal medicine clinic, thresholds for donation to a loved one were quantified. This was based on hypothetical and not real-life scenarios. This study showed that most laypersons have extremely high thresholds for donation risk and mortality. Sixty percent of the respondents suggested they would prefer to donate and die and have the transplant recipient live rather than forgo donation and have the potential transplant recipient die of liver failure. A personal survival threshold after LDLT was only 79 % and a threshold survival for the loved one after transplantation was only 55 % for them to agree to donate. Only a small proportion of correspondents had a survival threshold that exceeded the estimated risk of donation, as shown in Fig. 2.3. The majority of respondents (82 %) believed that the potential donor, not a physician, should have the final say regarding candidacy for living donation [55]. Survey studies and populations surveyed may have unrecognized biases, so these data should be interpreted cautiously. In addition, the survey was given to people who were not in fact dealing with the realities of these decisions, and who were answering hypothetical questions. Nevertheless, these data suggest that the potential donors' threshold and tolerance of risk is much higher than the actual risks of donation for PALF (and much greater than risks that would be deemed acceptable by most health care professionals). Thus, when evaluating the candidacy of a potential donor, clinicians must recognize the very high value that most people place on taking all possible steps to ensure the survival of a patient with organ failure.

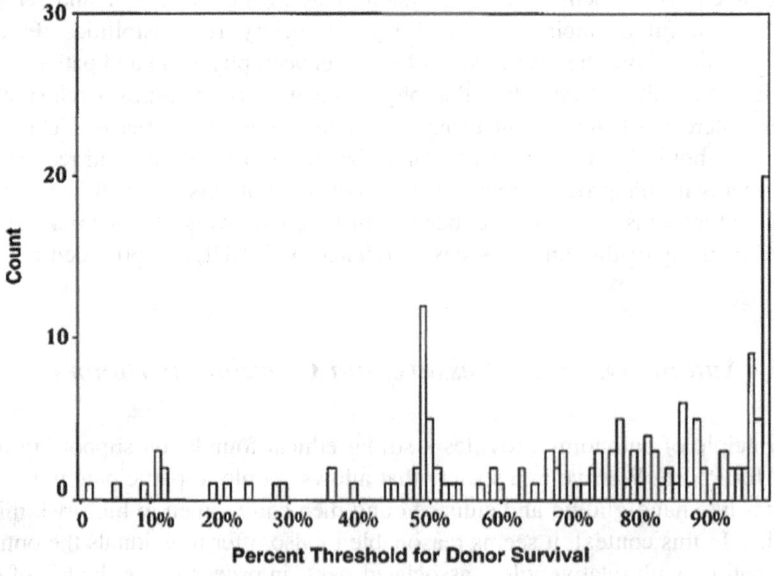

Fig. 2.3 Survey of 150 non-donor persons recruited from an internal medicine clinic suggests that only a small proportion reported their threshold for mortality for donation to a loved one to exceed the actual risk of donation. Adapted from Cotler et al. [55]

2.7.2 Principles of Non-maleficence and Beneficence and as They Apply to LDLT for PALF

The principle of non-maleficence, or doing no harm to the donor, is potentially violated if the measures of harm are solely the physical consequences, such as pain, scarring, temporary lack of mobility and potential development of post-operative complications, including death. On the other hand, it can also be argued that these are minor risks compared with the harm that might be posed to parents, close family members, or friends if they were forced to stand by watching a child die knowing that his or her life might be saved through live donation.

The concept of beneficence also generally supports live liver donation. The principle of beneficence refers to the moral obligation to act for the benefit of others, and not merely refrain from harming them [38]. In a society that does not support deceased donation, the principle of beneficence with LDLT is easily met because the acutely ill child with PALF will otherwise die. At first glance, the principle of beneficence is more difficult to invoke when offering LDLT for a sick child with PALF if the option of deceased donation exists. However, LDLT may still be justified if the probability of getting a suitable donor organ in time is small.

LDLT has several other attributes supporting the principle of beneficence. First, live donation offers an opportunity for families to forego the sense of helplessness that is associated with waiting for a deceased-donor organ by taking active control of this process. Most often the donor is a parent or very close relative and so the potential benefit psychologically is high. Second, when donating a part of one's liver to a child with PALF, the donor benefits from the continued survival of the recipient and from the increased self-esteem derived from actively contributing to a child's survival. One study documenting these benefits examined the stress for donors in urgent situations by comparing living donors for hepatocellular carcinoma (HCC) versus those for recipients with ALF. Before surgery, the donors for recipients with ALF had much lower health related quality of life (HRQOL) scores than the donors for recipients with HCC, indicating that they were under severe stress [56]. After surgery, the ALF donors showed the largest improvement in HRQOL scores, supporting the notion that LDLT for PALF helps relieve stress and brings fulfillment to the donor. In addition, LDLT for a child with PALF provides benefit to the family unit, other parent, and the siblings, which further increases the benefit to the donor.

2.7.3 Utilitarian Perspective

Utilitarian arguments also favour LDLT for PALF. LDLT not only directly benefits the sick child with PALF, but it also allows children on the waiting list and their families who do not have a live donor to benefit from the addition of an extra donor organ that would not otherwise be available. In addition, the risk of donation for living donors is much lower than the risk of waiting for a deceased organ for a child with PALF. Also, the benefit of the donation is higher when the donors are parents or anyone emotionally invested in the well-being of the sick child.

Indeed, using statistical models, Durand et al. [57] found that LDLT is most effective for overcoming the consequences of organ shortage when performed in patients at high risk of death on the waiting list, including those who are acutely ill but can still withstand the stress of transplantation. On one hand, it can be argued that putting living donors at risk should only be justified if the expected benefit for the recipient is maximal. In that view, low-risk patients, those whose condition is less severe, should be preferentially oriented to LDLT while the sicker individual should be kept on the waiting list for DDLT with a reduced waiting time (strategy 1). On the other hand, it can also be argued that the potential risk for the donor would be better justified if the recipient's risk of death on the waiting list for DDLT were high. In this view, the sickest patients should be preferentially oriented to LDLT (strategy 2). They embarked on a study to determine which of the two strategies was superior in reducing death on wait lists and maximizing survival after transplantation. After comparing these two strategies using the existing mortality statistics on both high-risk and low-risk patients, they found that LDLT was more effective for overcoming the consequences of organ shortage when used in high-risk patients [57]. Findings included a superior 3-year survival for strategy 2 of 17 % versus 8 % for strategy 1 and strategy 2. Even though the one-year survival after LDLT was lower for high-risk patients their one-year survival benefit with LDLT was 75 % versus zero survival benefit in low-risk patients who received LDLT. Notably, this study was performed on adult patients so its applicability to PALF patients might be limited. Nonetheless, it does challenge the notion that patients at high-risk of death on the waiting list who are acutely ill should receive a timely transplant, in order to provide the maximum utility and survival benefit for all patients on the wait list.

In the absence of deceased organs for a child with PALF, LDLT is generally accepted as the best way to overcome death on the waiting list. The rationale for this is its benefit of overcoming death for the child with PALF, which is much higher when waiting for a deceased organ. Due to the high case fatality rate in a shorter period of time for children with PALF compared to those with more chronic conditions needing transplant, providing a living organ when a deceased organ is not available can be life-saving, and delays can result in rapid deterioration and death. While waiting for a deceased organ, children can die from liver failure, or develop conditions that will make them no longer suitable for transplantation, such as advanced encephalopathy, cerebral oedema or sepsis.

2.7.4 Minimizing the Risk of Coercion and Exploitation

Opponents of LDLT for PALF raise concerns about the potential for coercion of vulnerable donors, especially in light of likely time pressures because of the acuity in the context of PALF diagnosis. Fully informed consent requires that a competent person is given full disclosure of the procedure and all options and their consequences; that he or she understands what is disclosed; and that consent is voluntary [58]. We believe that these requirements can be satisfied with LDLT to PALF provided the donor team uses a robust assessment process that includes intensive psychosocial assessment and support and provides confidential opportunities to opt-out.

We reject the view of those who suggest that coercion is inevitable with live donation for PALF because the situation itself presents families and friends with bleak choices, which are in turn misconstrued as coercion. Situations in which people may feel as though they have no real freedom to choose may present bleak choices, but these choices do not inherently result in coercion. Indeed, studies show that most donors who come forward have made up their mind long before the formal consent process begins [59]. In addition, it is uncommon for potential donors to change their mind during the consent process. Hawkins and Emanuel [60] point out that dire situations not of anyone's doing, such as in the case of a child with acute liver failure, are not a moral problem and are best solved by active informed decisions.

In order for consent to be fully informed, full disclosure of all options and short-term and long-term risks and benefits to the donor of each option must be discussed in detail with the donor. This is a difficult task and the best ways to do this are unclear because there are few studies about the impact of disclosure practices on the outcomes of live liver donation [61]. Survey studies on living kidney donors (LKD) document significant variation in the information disclosed, and wide centre differences in the weight put on certain elements of this discussion. For instance, an international study of 221 transplant professionals from 177 US centers found greater emphasis on disclosure of financial burdens, expenses and loss of income, than non-US centers [62]. According to a study of consent processes for LKD in 132 US centers showed that programs varied in the extent to which all elements of disclosure were discussed with donor candidates, with only 42 % disclosing all elements [63].

Recommended disclosure elements include: (1) description of evaluation, surgical procedure and recovery process; (2) Potential donor mortality and morbidity; (3) medical uncertainties, including potential for long-term complications; (4) short-and long-term follow-up; (5) recipient-specific risks and benefits; (6) expected outcome of transplantation for recipient; (7) expenses for the donor; (8) potential impact on donor's candidacy for health and life insurance; (9) potential impact on donor's lifestyle and prospects for future employment; (10) potential benefits to the donor; (11) alternatives to donation and different donation procedures; (12) center-specific statistics on donor and recipient outcomes; (13) alternative treatments available to the recipient. Beavers et al. [64] found that 64 % of right lobe donors reported complications but stated that they would still choose to undergo the procedure. Of all partial liver donors in their series, 40 % reported adverse events that did not appear in the medical record, meaning that they were not collected or recorded in the medical record but were experienced by the donors.

The findings in this study highlight the need to agree upon the specific adverse events or risks that must be disclosed during the consent process [64].

Based on these findings, we support the view that all elements of informed consent for a potential LD for a child with PALF should be standardized, and practices consistently reevaluated on an institutional or government level. These elements include determination of competence and voluntariness; standardized and explicit disclosure, recommendation and documentation of understanding; and documentation of decision and authorization of the chosen plan [38]. One such example is the US Charlie W. Norwood Living Organ Donation act of 2007, that calls for the Secretary of the Department of Health and Human Services to submit a yearly report detailing the long-term health effects of living organ donation and the disclosure process of informed consent [65].

2.7.5 Optimal "Cooling off" Period

In an effort to safeguard against undue inducement, a mandatory "cooling off' period of one to two weeks from the time of coming forward to donate has been proposed and used in LDLT for chronic liver disease. However, this lengthy period is not feasible for LDLT to most children with PALF [42]. Furthermore, there is little evidence that this period is necessary or makes the donor work-up process more robust. Evidence from qualitative studies of adult-to-child donors for non-acute indications provides helpful insights into donor perceptions and experiences around urgency prior to donation. In a study of adult-to-child living donors by Crowley-Matoka et al. [52] in-depth interviews conducted 3–10 years after donation provided retrospective comments on their decisions to donate. The study participants were mostly parents of the recipients. They perceived their child's need for transplant as an acute crisis. They reported that this perception of urgency prompted an overwhelming desire to save the life of the child, and made the decision to donate an easy choice that did not require a prolonged evaluation of all considerations [52]. In situations where the child did not survive despite LDLT, the donors derived great comfort from the knowledge that they did their best by directly participating in the attempt to heal the child. At the same time, however, many donors acknowledged that they had not fully appreciated the recipient's future need for chronic medical care and life-long follow up. Even so, none of the donors in this study regretted their decision to donate [52]. The findings of this study further substantiate the argument that a lengthy cooling-off period is needed.

2.7.6 Significance of the Donor's Relationship to the Child

In a study of 20 adults who came forward to be potential liver donors for friends, children, parents or siblings, in-depth interviews revealed important insights into the donation decision-making process and need for psycho-emotional support surrounding the decision [59]. The study authors classified the subjects into either the "openly ambivalent" or the "openly motivated" type. The motivated donors idealized their relationship with the recipient: they linked their desire to donate with their wishes and expectations for the relationship after the surgery. On the other hand, the ambivalent donors were more likely to view their personal relationships critically, seeing little potential for improvement of the relationships in the future [59]. Moreover, the authors found that "openly motivated" donors idealized both the recipient and the relationship as a subconscious means of protecting the self. Potential donors face a complex medical, ethical and social situation, marked by high levels of familial, institutional and normative pressures. As a result, they might deny the significance of the harmful effects of donation to their own physical integrity. Based on these findings, and findings of other studies on donor ambivalence, we conclude that the emotional bond, the perception of the relationship that the donor has toward the recipient, and the high level of perceived expectations are important factors clinicians should not ignore [54].

The importance of repetitive interviews and conversations with donors prior to consent is that these opportunities to exchange ideas allow for shared decision-making, in an environment that is non-judgemental, and supports either of the two alternative decisions (to donate or not to donate). These interviews should allow for the potential donors to voice their concerns about treatment and complications. Additionally, the clinicians, social workers, nurses and psychologists involved with these interviews should have special training in communication skills in the preoperative interview. It is important that they provide empathy when needed, but also note when help with decision-making is needed, as the level of anxiety and psychological pressure is very high [59].

2.7.7 LDLT from Anonymous or Non-related Donors

Other sources of living donor organs for LDLT are non-related donors (NRDs), who are defined as those donors with neither a genetic nor emotional relationship with the recipient [66]. NRDs can donate in either a non-directed or directed manner. In non-directed donation, a person donates his or her organ unconditionally to the general pool of recipients on the waiting list. In non-related directed donation, a person designates the particular recipient or group of recipients to whom his or her organ is to be given [67]. The published experience with NRD for LDLT is limited [67, 68]. The few centres that do perform NRD transplants have yet to reporting using this option for PALF.

An extensive discussion of NRD in the context of pediatric liver transplantation is beyond the scope of this chapter. However, as it pertains to PALF, there are some unique ethical issues that arise, mainly related to the fairness of subjecting these donors to the increased stress of a rapid work-up and the reduced opportunity for beneficence given the absence of a direct emotional bond with the recipient.

Thus, we conclude that the use of NRD organs for children with PALF and other acute situations should be used with caution, if at all, until more studies and evidence is obtained on the long-term quality of life and decisional regret in this unique and small donor population.

2.8 Key Concepts

- It is important to determine the cause of PALF as quickly as possible to determine the best treatment and avoid the need for a transplant if possible.
- LDLT is driven by respect for donor autonomy and the opportunity for beneficence. The donor program has a fiduciary responsibility to ensure that conflicts of interest are avoided, informed consent is obtained, and potential harm to the donor is minimized (non-maleficence).
- The risk of coercion can be minimized by experienced multidisciplinary assessment, the use of well-developed protocols and safety checklist, and an extensive donor education process.
- Most live liver donors recover fully and have no regrets about donation. LDLT reduces the risks of death on the waiting list and provides long-term graft outcomes that are comparable to DDLT.

References

1. D'Agostino D, Diaz S, Sanchez MC, Boldrini G. Management and prognosis of acute liver failure in children. Curr Gastroenterol Rep. 2012;14(3):262–9.
2. Squires RH Jr. Acute liver failure in children. Semin Liver Dis. 2008;28(2):153–66.
3. Squires RH Jr, Shneider BL, Bucuvalas J, Alonso E, Sokol RJ, Narkewicz MR, et al. Acute liver failure in children: the first 348 patients in the pediatric acute liver failure study group. J Pediatr. 2006;148(5):652–8.
4. Devictor D, Desplanques L, Debray D, Ozier Y, Dubousset AM, Valayer J, et al. Emergency liver transplantation for fulminant liver failure in infants and children. Hepatology. 1992; 16(5):1156–62.
5. Mohamed El Moghazy W, Ogura Y, Mutsuko M, Harada K, Koizumi A, Uemoto S. Pediatric living-donor liver transplantation for acute liver failure: analysis of 57 cases. Transpl Int. 2010;23(8):823–30.
6. Bucuvalas JC, Campbell KM, Cole CR, Guthery SL. Outcomes after liver transplantation: keep the end in mind. J Pediatr Gastroenterol Nutr. 2006;43(Suppl 1):S41–8.
7. Yamashiki N, Sugawara Y, Tamura S, Nakayama N, Oketani M, Umeshita K, et al. Outcomes after living donor liver transplantation for acute liver failure in Japan: results of a nationwide survey. Liver Transpl. 2012;18(9):1069–77.
8. Ostapowicz G, Fontana RJ, Schiodt FV, Larson A, Davern TJ, Han SH, et al. Results of a prospective study of acute liver failure at 17 tertiary care centers in the United States. Ann Intern Med. 2002;137(12):947–54.
9. Devictor D, Tissieres P, Durand P, Chevret L, Debray D. Acute liver failure in neonates, infants and children. Expert Rev Gastroenterol Hepatol. 2011;5(6):717–29.

10. Uribe M, Gonzalez G, Alba A, Godoy J, Ferrario M, Hunter B, et al. Living donor liver transplantation in pediatric patients with acute liver failure: safe and effective alternative. Transpl Proc. 2008;40(9):3253–5.
11. Shouval DS, Mor E, Avitzur Y, Shamir R, Bar-Nathan N, Steinberg R, et al. Living-related donor liver transplantation for children with fulminant hepatic failure in Israel. J Pediatr Gastroenterol Nutr. 2009;48(4):451–5.
12. Hong JC, Yersiz H, Farmer DG, Duffy JP, Ghobrial RM, Nonthasoot B, et al. Longterm outcomes for whole and segmental liver grafts in adult and pediatric liver transplant recipients: a 10-year comparative analysis of 2988 cases. J Am Coll Surg. 2009;208(5):682–9 (discusion 9–91).
13. Kim WR, Smith JM, Skeans MA, Schladt DP, Schnitzler MA, Edwards EB, et al. OPTN/SRTR 2012 Annual Data Report: Liver.
14. Raia S, Nery JR, Mies S. Liver transplantation from live donors. Lancet. 1989;2(8661):497.
15. Bourdeaux C, Tri TT, Gras J, Sokal E, Otte JB, de Ville de Goyet J, et al. PELD score and posttransplant outcome in pediatric liver transplantation: a retrospective study of 100 recipients. Transplantation. 2005;79(9):1273–6.
16. Tanaka K, Uemoto S, Tokunaga Y, Fujita S, Sano K, Nishizawa T, et al. Surgical techniques and innovations in living related liver transplantation. Ann Surg. 1993;217(1):82–91.
17. Reding R, de Goyet Jde V, Delbeke I, Sokal E, Jamart J, Janssen M, et al. Pediatric liver transplantation with cadaveric or living related donors: comparative results in 90 elective recipients of primary grafts. J Pediatr. 1999;134(3):280–6.
18. Devlin J, Wendon J, Heaton N, Tan KC, Williams R. Pretransplantation clinical status and outcome of emergency transplantation for acute liver failure. Hepatology. 1995;21(4):1018–24.
19. Baliga P, Alvarez S, Lindblad A, Zeng L. Posttransplant survival in pediatric fulminant hepatic failure: the SPLIT experience. Liver Transpl. 2004;10(11):1364–71.
20. Uribe M, Buckel E, Ferrario M, Godoy J, Blanco A, Hunter B, et al. Epidemiology and results of liver transplantation for acute liver failure in Chile. Transplant Proc. 2003;35(7):2511–2.
21. Baliga P, Alvarez S, Lindblad A, Zeng L. Studies of pediatric liver transplantation research G. Posttransplant survival in pediatric fulminant hepatic failure: the SPLIT experience. Liver Transpl. 2004;10(11):1364–71.
22. Bucuvalas JC, Zeng L, Anand R. Studies of pediatric liver transplantation research G. Predictors of length of stay for pediatric liver transplant recipients. Liver Transpl. 2004;10 (8):1011–7.
23. Narkewicz MR, Dell Olio D, Karpen SJ, Murray KF, Schwarz K, Yazigi N, et al. Pattern of diagnostic evaluation for the causes of pediatric acute liver failure: an opportunity for quality improvement. J Pediatr. 2009;155(6):801–6 e1.
24. Liu CL, Fan ST, Lo CM, Tam PK, Saing H, Wei WI, et al. Live donor liver transplantation for fulminant hepatic failure in children. Liver Transpl. 2003;9(11):1185–90.
25. Squires RH, Ng V, Romero R, Ekong U, Hardikar W, Emre S, et al. Evaluation of the pediatric patient for liver transplantation: 2014 practice guideline by the American Association for the study of liver diseases, American Society of transplantation and the North American Society for pediatric gastroenterology, hepatology and nutrition. Hepatology. 2014;60(1):362–98.
26. Sundaram SS, Alonso EM, Narkewicz MR, Zhang S, Squires RH. Pediatric acute liver failure study G. Characterization and outcomes of young infants with acute liver failure. J Pediatr. 2011;159(5):813–8 e1.
27. Saidi RF, Jabbour N, Li Y, Shah SA, Bozorgzadeh A. Outcomes in partial liver transplantation: deceased donor split-liver vs. live donor liver transplantation. HPB. 2011;13(11):797–801.
28. Bourdeaux C, Darwish A, Jamart J, Tri TT, Janssen M, Lerut J, et al. Living-related versus deceased donor pediatric liver transplantation: a multivariate analysis of technical and immunological complications in 235 recipients. Am J Transpl. 2007;7(2):440–7.
29. Liu CL, Fan ST, Lo CM, Yong BH, Fung ASM, Wong J. Right-lobe live donor liver transplantation improves survival of patients with acute liver failure. Br J Surg. 2002;89 (3):317–22.

30. Ueda M, Oike F, Ogura Y, Uryuhara K, Fujimoto Y, Kasahara M, et al. Long-term outcomes of 600 living donor liver transplants for pediatric patients at a single center. Liver Transpl. 2006;12(9):1326–36.
31. Northup PG, Abecassis MM, Englesbe MJ, Emond JC, Lee VD, Stukenborg GJ, et al. Addition of adult-to-adult living donation to liver transplant programs improves survival but at an increased cost. Liver Transpl. 2009;15(2):148–62.
32. Hippocrates. The hippocratic oath (updated 2002; cited 2014 Oct 3 2014). Available from: http://www.nlm.nih.gov/hmd/greek/greek_oath.html.
33. Kousoulas L, Emmanouilidis N, Klempnauer J, Lehner F. Living-donor liver transplantation: impact on donor's health-related quality of life. Transplant Proc. 2011;43(10):3584–7.
34. Gruttadauria S, Marsh JW, Vizzini GB, di Francesco F, Luca A, Volpes R, et al. Analysis of surgical and perioperative complications in seventy-five right hepatectomies for living donor liver transplantation. World J Gastroenterol. 2008;14(20):3159–64.
35. Kousoulas L, Becker T, Richter N, Emmanouilidis N, Schrem H, Barg-Hock H, et al. Living donor liver transplantation: effect of the type of liver graft donation on donor mortality and morbidity. Transplant Int. 24(3):251–8.
36. Middleton PF, Duffield M, Lynch SV, Padbury RT, House T, Stanton P, et al. Living donor liver transplantation-adult donor outcomes: a systematic review. Liver Transpl. 2006;12 (1):24–30.
37. Hippocrates. The hippocratic oath. Jones WHS, editor. Cambridge: Cambridge University Press; 1868. 258 p (1 Mar 1825).
38. Beauchamp TL, Childress JF. Principles of biomedical ethics. 6th ed. New York: Oxford University Press; 2009 417 p.
39. Ghobrial RM, Freise CE, Trotter JF, Tong L, Ojo AO, Fair JH, *et al.* Donor morbidity after living donation for liver transplantation. Gastroenterology. 2008;135(2):468–76.
40. Muzaale AD, Dagher NN, Montgomery RA, Taranto SE, McBride MA, Segev DL. Estimates of early death, acute liver failure, and long-term mortality among live liver donors. Gastroenterology. 2012;142(2):273–80.
41. Cronin DC 2nd, Millis JM, Siegler M. Transplantation of liver grafts from living donors into adults–too much, too soon. N Engl J Med. 2001;344(21):1633–7.
42. Abecassis M, Adams M, Adams P, Arnold RM, Atkins CR, Barr ML, *et al.* Consensus statement on the live organ donor. JAMA. 2000;284(22):2919–26.
43. Araujo CC, Balbi E, Pacheco-Moreira LF, Enne M, Alves J, Fernandes R, et al. Evaluation of living donor liver transplantation: causes for exclusion. Transpl Proc. 2010;42(2):424–5.
44. Baker A, Dhawan A, Devlin J, Mieli-Vergani G, O'Grady J, Williams R, et al. Assessment of potential donors for living related liver transplantation. Br J Surg. 1999;86(2):200–5.
45. Pascher A, Sauer IM, Walter M, Lopez-Haeninnen E, Theruvath T, Spinelli A, et al. Donor evaluation, donor risks, donor outcome, and donor quality of life in adult-to-adult living donor liver transplantation. Liver Transpl. 2002;8(9):829–37.
46. Sauer P, Schemmer P, Uhl W, Encke J. Living-donor liver transplantation: evaluation of donor and recipient. Nephrol Dial Transpl (official publication of the European Dialysis and Transplant Association—European Renal Association). 2004;19(Suppl 4):iv11–5.
47. Valentin-Gamazo C, Malago M, Karliova M, Lutz JT, Frilling A, Nadalin S, et al. Experience after the evaluation of 700 potential donors for living donor liver transplantation in a single center. Liver Transpl. 2004;10(9):1087–96.
48. Spital A. More on parental living liver donation for children with fulminant hepatic failure: addressing concerns about competing interests, coercion, consent and balancing acts. Am J Transplant. 2005;5(11):2619–22.
49. Squires RH, Alonso EM. Acute liver failure in children. In: Suchy FJ, Balistreri WF, editors. Liver disease in children. 4th ed. New York: Cambridge university Press; 2012.
50. Marcos A. Right lobe living donor liver transplantation: a review. Liver Transpl. 2000;6(1): 3–20.

51. Pomfret EA, Pomposelli JJ, Gordon FD, Erbay N, Lyn Price L, Lewis WD, et al. Liver regeneration and surgical outcome in donors of right-lobe liver grafts. Transplantation. 2003;76(1):5–10.
52. Crowley-Matoka M, Siegler M, Cronin DC 2nd. Long-term quality of life issues among adult-to-pediatric living liver donors: a qualitative exploration. Am J Transplant. 2004;4 (5):744–50.
53. Fukuda A, Sakamoto S, Shigeta T, Uchida H, Hamano I, Sasaki K, et al. Clinical outcomes and evaluation of the quality of life of living donors for pediatric liver transplantation: a single-center analysis of 100 donors. Transpl Proc. 2014;46(5):1371–6.
54. Simpson MA, Kendrick J, Verbesey JE, Morin DS, Dew MA, Trabucco A, et al. Ambivalence in living liver donors. Liver Transpl. 2011;17(10):1226–33.
55. Cotler SJ, McNutt R, Patil R, Banaad-Omiotek G, Morrissey M, Abrams R, et al. Adult living donor liver transplantation: Preferences about donation outside the medical community. Liver Transpl. 2001;7(4):335–40.
56. Erim Y, Beckmann M, Kroencke S, Valentin-Gamazo C, Malago M, Broering D, et al. Psychological strain in urgent indications for living donor liver transplantation. Liver Transpl. 2007;13(6):886–95.
57. Durand F, Belghiti J, Troisi R, Boillot O, Gadano A, Francoz C, et al. Living donor liver transplantation in high-risk vs. low-risk patients: optimization using statistical models. Liver Transpl. 2006;12(2):231–9.
58. Levine RJ. Informed consent in research and practice: similarities and differences. Arch Intern Med. 1983;143(6):1229–31.
59. Walter M, Papachristou C, Danzer G, Klapp B, Frommer J. Willingness to donate: an interview study before liver transplantation. J Med Ethics. 2004;30(6):544–50.
60. Hawkins J, Emanuel E. Clarifying confusions about coercion. Hastings Cent Rep. 2005;35 (5):16–9.
61. Gordon EJ. Informed consent for living donation: a review of key empirical studies, ethical challenges and future research. Am J Transpl. 2012;12(9):2273–80.
62. Parekh AM, Gordon EJ, Garg AX, Waterman AD, Kulkarni S, Parikh CR. Living kidney donor informed consent practices vary between US and non-US centers. Nephrol Dial Transpl (official publication of the European Dialysis and Transplant Association—European Renal Association). 2008;23(10):3316–24.
63. Rodrigue JR, Pavlakis M, Danovitch GM, Johnson SR, Karp SJ, Khwaja K, et al. Evaluating living kidney donors: relationship types, psychosocial criteria, and consent processes at US transplant programs. Am J Transplant. 2007;7(10):2326–32.
64. Beavers KL, Sandler RS, Fair JH, Johnson MW, Shrestha R. The living donor experience: donor health assessment and outcomes after living donor liver transplantation. Liver Transpl. 2001;7(11):943–7.
65. Charlie W. Norwood living organ donation act, 121 STAT., 1813 Public Law no: 110-144 (2007).
66. Dew MA, Jacobs CL, Jowsey SG, Hanto R, Miller C, Delmonico FL, et al. Guidelines for the psychosocial evaluation of living unrelated kidney donors in the United States. Am J Transpl. 2007;7(5):1047–54.
67. Reichman TW, Fox A, Adcock L, Wright L, Abbey SE, Levy G,et al. Anonymous living liver donation: donor profiles and outcomes. Am J Transpl. 2010;10(9):2099–104.
68. Wright L, Ross K, Abbey S, Levy G, Grant D. Living anonymous liver donation: case report and ethical justification. Am J Transpl. 2007;7(4):1032–5.

Chapter 3
Answers to Questions that Weren't Asked: The Ethical Complexities of Identifying Misattributed Paternity During the Transplant Evaluation Process

Michael A. Freeman and Lisa S. Parker

Abstract Blood typing and Human Leukocyte Antigens (HLA) matching, while playing a vital role in improving allograft survival following organ transplantation, also have the potential to identify circumstances in which paternity or other degrees of genetic relatedness are misattributed. The decision of whether to disclose a finding of misattributed paternity is ethically complex and has been an issue of concern within the transplant community. In this chapter, we will consider the incidental identification of misattributed paternity within the broader context of incidental findings (IFs) in general. We will then review the prevailing arguments for and against disclosure of a finding of misattributed paternity during the transplant evaluation process. We will consider specific ethical complexities of such disclosure that arise within the context of pediatric organ transplantation, complexities that we feel make the disclosure of misattributed paternity ethically inappropriate within that setting. Finally, we will conclude by recommending a model policy to prevent some of these ethical challenges from arising in the clinical setting.

Keywords Transplant evaluation · Incidental findings · Misattributed paternity

3.1 Introduction

Blood typing and human leukocyte antigen (HLA) matching has served a vital role in the development of modern organ transplantation. The ability to assess the likelihood that an organ will be initially compatible has allowed medicine to move

M.A. Freeman (✉)
Penn State College of Medicine, Penn State Hershey Children's Hospital,
500 University Drive, Hershey, PA 17033, USA
e-mail: MFreeman3@hmc.psu.edu

L.S. Parker
Graduate School of Public Health, Center for Bioethics & Health Law, University of Pittsburgh, 4200 Fifth Avenue, Pittsburgh, PA 15260, USA

R.A. Greenberg et al. (eds.), *Ethical Issues in Pediatric Organ Transplantation*,
International Library of Ethics, Law, and the New Medicine 66,
DOI 10.1007/978-3-319-29185-7_3

beyond the early successes in organ transplantation between monozygotic twins and opened the doors to both living and deceased donor transplantation within the broader populace. However, in addition to providing this crucial information regarding organ compatibility, such testing also has the potential to reveal other, frequently unanticipated, information regarding the degree of biological relatedness (or lack thereof) between a potential recipient and a potential donor.

The HLA genes are a set a genes found on chromosome 6 which help the body distinguish between native and foreign cells [1]. Although there are more than 2000 identified alleles of the six primary HLA genes, these genes are grouped closely together on chromosome 6 and are generally transmitted as a complete set to one's offspring. Therefore, in the vast majority of cases an individual will inherit two sets of HLA genes, one set from his or her mother and one from his or her father. If these sets of HLA genes do not match, it suggests that the presumed parent-child relationship is misattributed—i.e. the presumed father (or very infrequently, the mother) is not the child's actual biological parent. Although the degree of expected HLA matching between siblings and other more distant relatives is more variable, if multiple family members are tested, HLA matching has the ability to identify other instances of misattributed familial relationships as well.

This is not a speculative concern. In 2009, Young et al. [2] published the results of an analysis of information contained within the United Network for Organ Sharing (UNOS) and Canadian Organ Replacement Registry (CORR) detailing the HLA testing for all living kidney donors from 1992–2006. After examining all cases in which the participants reported a father-child relationship for which HLA data were recorded and adjusting for the degree of presumed error in HLA typing, Young et al. identified a prevalence of misattributed paternity of between 1 and 3 %. These cases accounted for between 0.25 and 0.5 % of all living kidney donations performed during this time, making misattributed paternity within the kidney donation context an uncommon, but not exceedingly rare occurrence.

Despite the fact that misattributed paternity is identified relatively frequently during the transplant evaluation process, there is a lack of consensus within the transplant community about how to best address this finding. As part of their study, Young et al. [2] also surveyed transplant professionals, potential donors and potential recipients regarding whether information of misattributed paternity should be disclosed. As a whole, the group was almost evenly split, with 24 % of respondents strongly disagreeing with disclosure and 23 % strongly agreeing that the information should be disclosed. Despite the lack of a unanimity within the transplant community regarding whether a finding of misattributed paternity should be disclosed to potential donors or recipients, there have been multiple calls to produce policy statements to help guide individual transplant programs through the complicated ethical considerations entailed in electing or declining to disclose findings of misattributed paternity [3–6].

Unfortunately, the literature examining the ethical implications of disclosure or non-disclosure within the pediatric transplant setting is relatively sparse. The majority of the pertinent ethics literature arose out of the consideration of identification of misattributed paternity within the context of genetic testing for pre-natal genetic counseling, genetic testing for an adult onset disease discovered in childhood, or genetic testing for research. While this literature can be informative, we would argue that there are several key differences between the context of genetic testing and HLA compatibility testing which makes the direct application of these arguments problematic. Indeed, we feel that there are several concerns arising in the context of pediatric transplantation which have a profound impact on the ethical analysis.

We begin this chapter with a brief consideration of the meaning of the concept of an incidental finding and how that applies within the context of transplantation and HLA matching. We then consider arguments both for and against disclosure of a finding of misattributed paternity within the transplant process. Next, we consider specific ethical complexities which arise within the context of a pediatric transplantation, complexities that we feel make the disclosure of misattributed paternity ethically inappropriate in the pediatric transplant setting. Finally, we will conclude by recommending a policy to prevent some of these ethical challenges from arising in the clinical setting.

3.2 Incidental Findings: Definitions, Ethics, and Implications for Our Analysis

3.2.1 What Are Incidental Findings?

Discovery of information that was not specifically sought in the course of medical testing is neither new nor peculiar to transplantation or to any one type of testing or healthcare context. Two contexts have prompted substantial literature regarding the definition and management of such findings: genetic testing and diagnostic imaging. Initially described as unanticipated or unexpected findings, these are now known to occur frequently enough that they should neither be unanticipated nor unexpected. Therefore, these findings are now generally termed "incidental findings" (IFs) because their prevalence is sufficiently high that practitioners, at least in these domains, should anticipate or expect them and plan for their management vis-à-vis their disclosure to patients or clients. At the same time, for the individual patients or clients, IFs are typically unexpected/unanticipated, though their significance is not incidental in the sense of being unimportant. Though such information is substantially incidental to the reason for testing, it may have important meaning for individuals' understandings of themselves and their relationships, as well as for their health or health risks. Nevertheless, IFs have come to be conceptualized in relation to the reason or aim of testing, for example as "a finding concerning an individual research participant that has potential health or reproductive importance and is discovered in the course of conducting research but is beyond the aims of the study" [7, p. 219] or as "results that are not related to the indication for ordering the sequencing but that may nonetheless be of medical value or utility to the ordering physician and the patient" [8, p. 565].

3.2.2 *Ethical Justifications for Reporting Incidental Findings*

Contemporary clinical bioethics is strongly focused on questions of patient autonomy as manifest through the process of granting informed consent for medical testing and treatment. IFs challenge the process of informed consent as they are, by their very nature, at least somewhat unexpected. If the IF is closely related to reason that testing is being performed, such as identifying a second illness when investigating the symptoms of the first, a relatively strong presupposition that the person would like to receive that information may exist. In other circumstances, such as when an incidental finding is identified in a "healthy" control participant in a research study, that presupposition may be weaker. However, patients can and do elect to forgo testing or treatment for a variety of reasons however, so it cannot simply be assumed that all patients desire that all information be provided. Therefore, unless the possibility of a given IF was discussed before-hand, the decision to disclose the IF cannot be justified based on support for a patient's autonomy alone.

As a result, polices (or proposed policies) that support disclosure of IFs frequently find further justification in the principle of beneficence as well [7, 8]. Disclosure of these IFs to the affected parties may be supported by the principle of beneficence on the assumption that by conveying information that may be important to the current or future health of those to whom the information applies.

It is important to note, however, that not all IFs are equivalent and that utilizing a single framework for their management may not be appropriate. Some IFs, such as identifying a tumor on imaging for an acute issue, represent information that is vitally important to that patient's personal health. As such, there is a strong argument for conveying such information. While it is possible that the tumor may be benign or that health effects arising from the tumor may be delayed many years hence, the potential for a more immediate effect on the patient's health reinforces the need for disclosure. Only if the patient knows of the finding can he or she make an informed choice about further investigation of it or future care.

However, some findings result in a more remote or probabilistic impact on a patient's health. An ethical argument grounded in beneficence for the necessity of conveying such information is much less clear. One example of this second class of IFs can be illustrated by the scenario of the identification of a genetic susceptibility to breast cancer due to the BRCA1 gene in the context of a child undergoing whole exome sequencing for another genetic condition. While surveys demonstrate that parents of children undergoing genetic testing are generally supportive of receiving IFs, there is ambivalence regarding the disclosure of information regarding the risk of genetic conditions that for the child may arise in the distant future, particularly because the child may, as an adult, choose whether to seek such risk information [9, 10]. This ambivalence makes the argument that disclosure of such information supports, rather than impedes, patient or parental autonomy more tenuous.

A finding of misattributed paternity may constitute a third, distinct class of incidental finding. Knowledge of a presumed father's genetic relationship with the potential recipient has no immediate impact on the patient's health. While such information can impact the recipient's understanding of his or her risk for genetic disease, it can only do so by reducing the risk for genetic conditions which "run on my father's side". This does not mean that there is no value to either the child or the presumptive father being notified of misattributed paternity, simply that the general arguments for disclosure within the context of optimizing a patient's health do not apply as immediately or strongly as they do in circumstances when testing reveals a propensity (or lack thereof) towards developing a genetic disease. The relevance of other potential benefits or harms, including the changed understanding of familial relatedness itself, that can result from the disclosure of a finding of misattributed paternity will be explored in greater detail as we proceed.

3.2.3 Implications for Our Analysis

Modeled on the tenets of preventive medicine, a "preventive ethics model" for addressing IFs has emerged [11–13]. A preventive ethics approach requires recognizing recurrent events which raise ethical problems and taking steps to prevent those problems from arising or to mitigate their harms [11]. With regard to IFs this approach has two implications. First, having recognized that IFs are a recurrent problem, professionals should develop a policy to address it; thus there is a need for the sort of policy regarding an IF of misattributed paternity we will recommend at the conclusion of this chapter. Second, respect for patient autonomy and well-being suggests that in most cases patients should be informed of both the potential problem—the prevalence of discovery of IFs—and the policy for addressing it. Such policies may, in fact, be specific to a medical domain (e.g. transplantation or genomic sequencing); an institution, clinic or group practice; or an individual or group of individuals (e.g. physicians or nurses). In the case of transplantation, which involves teams of medical personnel, it would be inappropriate for different personnel to be subject to different disclosure policies, so that disclosure depended on which professional learned of the IF. Therefore, we feel that a medical domain or institutional policy model is most appropriate in regards to management of an IF of misattributed paternity in the transplant evaluation process. The conclusion of the chapter will outline such a policy for the pediatric transplantation context.

It has also been noted that any obligation to disclose IFs must be considered within the context of the particular physician-patient (or professional-patient, investigator-participant) relationship [14–16]. In the prenatal genetic testing context, a counselor may have two clients—e.g., a mother and presumed father who are seeking information for reproductive planning following birth of their first child with a genetic condition. They may have substantially different interests in a finding of misattributed paternity with regard to their first offspring; though relevant to calculating the risk that a future child will be affected by the genetic disease in question, the finding has substantially different implications for each parent individually. Though here both husband and wife may be considered equally the clients of the counselor, in other cases, such as when testing several family members to calculate disease risk through linkage analysis, perhaps only one family member may properly be said to be in a professional-patient relationship in virtue of which professional obligations are owed to that one person. This point may be relevant in the transplantation context, where the potential recipient is the patient; family members undergoing HLA matching as potential donors do not stand in a professional-patient relationship in the same way either qualitatively or temporally (i.e., they likely have no long-standing relationship to the team). The nature and duration of a professional-patient relationship has been considered relevant to developing policy regarding disclosure of IFs in other contexts [7] and may be relevant in transplantation. Generally, those with whom one has a professional relationship are owed greater duties than those with whom one does not. Whether HLA typing a prospective organ donor alone constitutes establishment of such a relationship may be salient.

3.3 An Overview of Prevailing Arguments

When considering the specific question of disclosure of the finding of misattributed paternity in the pediatric transplant setting, it is useful to begin with a survey of the arguments both for and against such disclosure within the broader transplant setting. As noted previously, the body of literature addressing the specific question of disclosing misattributed paternity within the context of organ transplantation is relatively limited. While authors such as Schroder [4], Soderdahl et al. [5], and Wright et al. [6] have advanced their arguments within the context of individual cases of a finding of misattributed paternity, it must also be noted that they present many of the same overarching ethical themes. To provide background for the course of action and policy that we recommend, we present some of the primary arguments and considerations and cite the authors responsible for specific formulations of the arguments when applicable.

3.3.1 Arguments for the Disclosure of Misattributed Paternity

The primary arguments for the disclosure of a finding of misattributed paternity within the transplant evaluation process are grounded in: (1) respect for the autonomy of the donor and informed consent, (2) the importance of truth-telling, (3) an individual's right to his or her genetic information, and (4) avoidance of paternalism. We will discuss each of these in turn.

The notion of respect for the autonomy of the donor is relatively straightforward. Although the principle of respect for autonomy should not be limited to the singular issue of informed consent, it is clear that the two concepts are closely related and that appropriate respect for autonomy requires adherence to appropriate procedures for obtaining informed consent [17]. Such informed consent is reliant on the ability of the individual to understand the risks and benefits of various interventions or courses of treatment and to decide between them (i.e. competence), which in turn is dependent on the disclosure of pertinent information by the medical team [17]. In the scenario of a presumed father coming forward to donate an organ to his child, if misattributed paternity is identified but not revealed pertinent information is withheld and therefore the requirements of appropriate informed consent are not met. The donor may have come forward due to a sense of obligation based, in part or in total, on the familial relationship between the donor and the recipient. The prospective donor's understanding of 'family relationship' and indeed 'paternity' or 'fatherhood' may rest to varying degrees on a belief in genetic relationship to the patient. For some the importance of social relatedness may far outweigh the genetic tie, though in cases where misattributed relatedness is an unexpected IF, it is likely that some combination of genetic and social connection informed the notion of "family" that grounded the prospective donor's sense of obligation. Alternatively, the donor may feel that he will be a "better match" than other potential living or deceased donors due to the presumed genetic relationship and feel the desire or obligation to maximize the likelihood of transplantation success. If the donor elects to proceed with organ donation, he does so with the belief that said genetic relationship exists. If it does not, the consent to proceed with the procedure is not fully informed. This is not meant to say that the lack of a genetic relationship would preclude all social or psychological sense of obligation to serve as a donor, but simply that it is important that the potential donor be informed of a lack of genetic relationship prior to proceeding.

The second argument is that truth-telling has an inherent moral value and therefore the action of telling the truth with regard to IFs is, on its face, justified. This is not an argument which is unique to medical ethics, as a similar assertion has been made within numerous moral and ethical systems throughout history. Additionally, within the context of the physician-patient relationship, truth-telling takes on an even deeper meaning and value. Much of the therapeutic value of the physician-patient relationship is predicated upon a foundation of trust between the two parties, trust that the physician has the patient's best interests at heart and the medical knowledge to advise the patient in regards to those interests as well as the physician's trust that the patient is an honest and willing participant in the care process. If medical information is unilaterally withheld by the physician, the trust that serves as the foundation of the physician-patient relationship is weakened.

The third argument is that an individual has a right to his or her genetic information. Within the modern Western medical tradition, there is a presumption that a patient has the right to receive all information regarding his or her health obtained during the course of care unless he or she expressly waives that right. Moreover, some may also believe that genetic information occupies a privileged space within the broader category of an individual's health information based on the idea that each individual's genetic code is unique and constitutes part of what serves to make that individual himself or herself. As such, this knowledge is considered intensely personal. While the interpretations of the meaning of one's genetic code may shift over time, the content of one's genetic code does not [18]. An individual's genetic code is relatively unchanging and therefore information about one's genetic code has a certain permanence, in contrast to other types of health information, which frequently reflect transient or shifting conditions. Finally, in contrast to the physical symptoms which manifest as a result of health issues on a more macro level within one's body, in most cases an individual cannot easily perceive his or her genetic characteristics. As a result, this personal genetic information is often unknowable by an individual without the intervention of a third (scientifically trained) party. Although this is changing with the advent of direct-to-consumer genetic testing, it raises the question of whether physicians or other health care practitioners have a special obligation to make the findings of genetic studies available to their patients and clients.

The fourth argument frequently made in support of disclosure of a finding of misattributed paternity (or other incorrectly presumed genetic relationship) is that arguments against such disclosures suffer from a common flaw, namely that they presume a paternalistic understanding of what is "best" for the patient [19]. This paternalistic understanding is predicated upon both the belief that the state of affairs as it exists (e.g., a belief that paternity is correctly attributed) is the preferred state and a belief that by revealing the information of misattributed paternity this preferred state will be disrupted in a negative manner. Proponents of this fourth argument contend that because it is generally impossible for the veracity of these beliefs to be established, withholding information out of paternalistic motives is misguided, as one cannot know what would, in fact, promote the person's good [19].

3.3.2 Arguments for the Non-disclosure of Misattributed Paternity

The primary arguments for the non-disclosure of the finding of misattributed paternity on the part of a potential donor are grounded in four considerations: (1) the technical limitations of HLA matching during a transplant evaluation make it unsuitable for addressing the question of paternity, (2) the view that information regarding paternity is not pertinent to the purpose of HLA testing to assess donor-recipient compatibility, (3) the goal of family preservation, and (4) protection of the confidentiality of the recipient's mother. We will examine each of these arguments in greater detail.

Although HLA testing is generally considered a viable method for determining paternity [20], there are several, rare circumstances which can result in a false determination of misattributed paternity [21]. Conditions such as genetic mosacism or chimerism can result in the finding of additional genetic information due to the expression of multiple genetic lineages within a single individual. The results of testing for genetic relatedness to individuals affected by these non-typical expression patterns can be complex and may be easily misinterpreted. Conditions of uniparental disomy, in which both chromosomes possessed by an individual are obtained from a single parent, have been described in the literature as well [22]. Given the potential for such scenarios, the argument has been made that it would be inappropriate for a transplant center to disclose a finding of misattributed paternity based on the results of pre-transplant HLA typing alone [21]. Instead, when the evaluation process determines that the donor is a poor match, due to either blood type or HLA allele matching, the transplant center should simply report the results of blood type and HLA matching and that the donors are not compatible. If asked whether these findings disprove genetic parentage, the transplant center should reply 'no' due to technical complexities noted above. If the donor-recipient pair request more information, they should then be referred for more comprehensive genetic testing by a genetic professional.

The second argument for non-disclosure is not dependent on the quality of the information obtained, but on the nature of the information itself. Proponents contend that given the lack of immediate health benefit to those tested and the harms which could result, the principle of beneficence does not support a duty to convey this information [19, 23]. Unlike the finding of misattributed paternity within the context of assessing the genetic cause of a child's illness or testing for the purpose of family planning, information regarding the familial relationship between the potential donor and recipient has no bearing on the question at hand. The question of compatibility between a potential donor and recipient pair is addressed directly and a poor HLA match is a poor match regardless of the underlying causes. Similarly, a good HLA match between the donor and the recipient predicts that the outcome of an organ transplantation between these two individuals will be acceptable, regardless of paternity. Thus there is no compelling reason to convey the finding of misattributed paternity.

The third common argument for the non-disclosure of misattributed paternity is straightforward: conveying the finding of misattributed paternity may disrupt the family and therefore is in opposition to the principle of non-maleficence. Although a family may remain intact despite the revelation that there is no biologic kinship between a father and a child, in almost all circumstance the information will be, at the very least, highly disruptive. In more severe circumstances it may result in the severing of parental/filial bonds, either temporarily or permanently, separation or divorce between spouses, or domestic violence. While it may be faulty to presume that the status quo represents the optimal state, nevertheless there is both a visceral and an intellectual understanding that the range of familial disruption is harmful at least in the short term [24–26]. The finding of misattributed paternity during the pediatric transplant evaluation process may be especially harmful because it is experienced within a present context of significant personal, social and familial stress brought about by the child's severe illness. Given the potential for harm that can occur within this context, proponents of this argument claim that revealing the finding of misattributed paternity is in direct opposition to the broader goals of optimizing the child's physical and mental health and should therefore not be done [23].

The fourth argument against the disclosure of a finding of misattributed paternity highlights that misattributed paternity is not a simple medical fact for which there is a primary interested party, but instead exists at a nexus of complex familial relationships. Within this argument, disclosure of misattributed paternity should be deferred, not because of the primary effect it has on the child and the non-biologic father, but because it violates the privacy—and may be contrary to the interests—of the child's mother. This argument has clear roots within the context of genetic testing and family planning, in which both parents are clear and distinct interested parties. Within that context, the obligations of the genetic counsellor are clearly divided. It is important to note that the importance of preserving the mother's privacy within this context has been expressly challenged by some critics [19], but nevertheless, it remains a frequently cited concern [6]. Within the context of testing for HLA compatibility for a potential organ transplant between a patient and a family member, the obligation to protect the mother's privacy is less clear. Nevertheless, although the care team has no direct fiduciary obligation to the mother, those concerns cannot be completely jettisoned. The impact of the finding of misattributed paternity is experienced by many parties. The magnitude of that impact in particular families is difficult to predict accurately; however, because there are so many parties with near-equal interests in this case, disregarding the concerns of any of the participants is difficult to justify.

3.4 The Ethical Complexities of the Finding of Misattributed Paternity Within the Pediatric Transplant Setting

There are two constellations of specific concerns which arise when the affected offspring is a severely ill minor in need of an organ transplant. First are questions of autonomy and appropriate consent within this population. Second are those concerns related to the special care that the health care team must take to adhere to the requirements of beneficence and non-maleficence while caring for pediatric transplant patients. We feel that appropriately balancing the weight of these considerations makes it difficult to justify the disclosure of a finding of misattributed paternity within the specific setting of organ compatibility testing for a pediatric transplant patient.

3.4.1 The Question of Autonomous Consent

In attempting to meet the demands of respecting patient autonomy, the notion of appropriate informed consent is often treated as a panacea. When the patient is a competent adult who can be provided with the pertinent information and who is capable of understanding it, the patient both is best able to determine the optimal course of action in accordance with his or her goals and values and has the right to do so. In much of the published literature addressing the finding of misattributed paternity within the transplant process, one potential solution offered is that individuals participating in HLA matching testing be informed prior to proceeding of the potential that this testing may identify misattributed paternity [5, 6]. The participants are then asked whether they would like that information if available. This approach, while laudable in some ways, becomes much more difficult within the context of a pediatric recipient, where obtaining meaningful consent regarding disclosure from the participants is highly problematic. The notion of autonomy as it applies to minor children has always been one of subtle shadings and complexity. The ability to make autonomous decisions in a philosophically meaningful way exists on a spectrum, with younger children being less able to exert meaningful autonomy and with children gradually achieving greater degrees of autonomy as they mature. Nevertheless, many of the most controversial issues in pediatric ethics surround questions of the apportionment of decision-making between minors and their parents. As applied to the question of the identification of misattributed paternity, it becomes very difficult to envision a circumstance in which a pediatric patient can give meaningful autonomous consent. For young children, identifying and understanding the complexities of the implications of a discovery of misattributed paternity will likely be impossible. Even for older children, the issue remains problematic. As part of the formation of their adult identity, adolescents often go through a phase in which they seek to establish themselves in contrast to their parents and their parents' values [27]. This process of identity formation may preclude older adolescents from evaluating the complex ramifications of the disclosure of a finding of misattributed paternity. Given the fact that a frequently cited prerequisite for autonomous decision-making is possession of a stable set of values *over time*, a decision which so closely relates to what is for most people an episodic period of personality formation is therefore somewhat ethically troubling.

Moreover, beyond the interests any child has in the preservation of a well-functioning family, a child in need of an organ transplant may correctly perceive that it is in his or her interests that both parents feel obligations grounded in biological relatedness to offer themselves as potential donors; therefore, the potential for revelation of any finding that would weaken those ties or obligations would not be in the child's interest. (Mothers who may recognize the risk of a finding of misattributed paternity may seek to avoid having their husbands undergo HLA matching. Children have an interest in not having such a finding undermine their fathers' agreeing to follow through with donation if they are found to be suitable donors.) Children too immature to perceive these interests can be thought incapable of making a decision regarding disclosure that reflects their own interests.

Critics of this line of argument may note that concerns about the changing values and incomplete understanding could be applied to almost all adolescent (and to much adult) decision-making within the medical field. If one presumes that an adolescent's values are rapidly changing, why don't we disregard all assertions of adolescent autonomy? When we elect to heed the decisions made by adolescents, it is because we feel that the net benefit of respecting the adolescents' developing autonomy exceeds the potential harms. As such, two criteria need to be examined. First, what is the degree of capacity exhibited by the child? Second, how serious are the potential harms resulting from the decision to engage in or forgo the intervention (e.g. treatment, disclosure) in question? It is only by comparing these two factors that we determine if a given adolescent has capacity to a make a given decision; relatively trivial decisions require less evident capacity whereas decisions of more gravity require a greater degree of capacity. Although different jursidictions may have different thresholds for determining which children demonstrate appropriate capacity to grant or deny consent for a given medical decision, the overall tenet is the same. Decisions about whether or not receive information regarding a finding of misattributed paternity are serious and should be deemed to require a corresponding degree of capacity, as such a disclosure presents non-trivial burdens and unclear net benefit. However, in contrast to many other difficult ethical decisions regarding an adolescent's capacity (such as the prototypical case of an adolescent forgoing treatment for malignancy), there is no urgent need for the adolescent or child to make a decision regarding the disclosure of misattributed paternity. While both the transplant and the HLA matching of prospective donors are medically necessary, the disclosure regarding paternity is not. As such, prudence suggests that such a determination be deferred until the child reaches the age of majority and so the child will be granted the opportunity to allow their understanding of the risks and benefits of a disclosure of misattributed paternity to deepen as he or she matures.

The next logical step in this line of argument is to ask whether the prospective donor should be allowed to decide if findings of misattributed paternity can be disclosed, independent of the child's consent. For the vast majority of healthcare interventions involving pediatric patients, parental consent on their behalf is sufficient. What makes the question of the disclosure of a finding of misattributed paternity different? Parental exercise of surrogate decision-making for their minor children is predicated on the understanding that the decisions made by the parents are in the child's best interests. Although parents are often given wide latitude to reach such determinations, their right to make such decisions is often suspended when the decision made is clearly contrary to the child's interests. The difficulty with questions regarding the finding of misattributed paternity, however, is that each parent has a personal and self-directed interest in whether a finding of misattributed paternity will be revealed. No matter how well-meaning that parent may be, it is unlikely that he or she will be able to make a separate determination of whether disclosure is in the best interests of the child.

Critics of the 'best interests' standard for pediatric decision-making may argue that a singular focus on the child's interests as an individual may miss the broader question of the child's interests as a member of a social framework with his or her parents, siblings and families [28]. These broader interests are expressed in the prevailing model of "Family Centered Care" which seeks to consider the needs and interests of the family as a whole, with the child's interest serving as the core [29].

While this may be true in general, in the scenario of a disclosure of misattributed it is complicated because the ancillary interests of various family members, such as the child's siblings and parents, may not align; moreover, the interests of the family as a whole are far from clear. Given this fact—as well as the presumption that the disclosure of misattributed paternity will likely result in some, at least temporary, harm to the patient—the ability of parents to appropriately consent to the disclosure of this information is in question. If a determination of paternity is absolutely necessary to proceed with the transplant evaluation (for example, if a prospective donor's decision to donate is expressly predicated on the establishment of a hereditary link to the recipient), then that (nonmedical) necessity in conjunction with the consent of parties involved may provide sufficient grounds to warrant disclosure. Nevertheless, considerations of the child's privacy and other interests may argue against transplant programs acceding to such a demand on the part of the child's parent. In sum, when disclosure is neither medically necessary nor practically necessary (as a condition of donation), reliance on either the parent's surrogate consent or a child's suboptimal consent as a justification to disclose is ethically suspect.

There are additional ethical complexities to be considered. Given the sensitive nature of the decision to receive or not receive information regarding potential misattributed paternity and the lack of necessity that such information be obtained, each party should ideally have the opportunity to consider his or her decision at leisure. However, given the sense of relative urgency that generally accompanies transplant in the pediatric setting, it may be difficult to engage in such measured consideration.

Not only does the receipt of information regarding misattributed paternity have the potential to disrupt social relationships, the opportunity to decide whether to receive or decline such information or the need to consent to a policy of routine disclosure in order to be evaluated for HLA matching could also be disruptive. A father expressing interest in information regarding misattributed paternity may be interpreted as his having suspicions of infidelity. A child who expresses interest in such information may engender an increased emotional distance from his or her parents, at a time when familial support is crucial. If a policy of routine disclosure is implemented, a mother who discourages her husband from being evaluated as a donor—perhaps out of concern that he may become debilitated and unable to provide for other children, or because she prefers that another relative be evaluated first—may be mistakenly thought to be worried about the possibility of misattributed paternity. If participants are asked to give their consent to receive such information in the presence of other interested parties, they may be pressured to give the answer that "should" be given, rather than a true reflection of their preference. If participants are allowed to give or decline consent confidentially, it may prove difficult for the medical team to maintain that confidentiality in situations in which the lack of unity of the opinions is revealed by the subsequent disclosure or lack thereof.

3.4.2 The Question of Non-maleficence

Although all health professionals are charged with protecting their patients from harm, those professionals charged with caring for pediatric patients bear special responsibilities. In contrast to adults, who are presumed to be able to weigh the relative risks and benefits of many interventions and who are therefore given wide latitude in determining whether a given intervention is "harmful," pediatric patients are subject to no such presumption. Thus, providers of pediatric care are charged with a responsibility to identify actions and decisions that may result in harm to their patients and to serve as advocates for their patients' best interests. The question of determining "best interests" is complex, and many critics of the standard have noted that a duty to provide for the optimal outcome for children is both infeasible and ethically suspect [30]. In contrast, the importance of avoiding significant harm is much more thoroughly grounded. Therefore, it is essential that we consider whether disclosing misattributed paternity presents special harms within the context of an evaluation for living donor transplantation for pediatric patients.

First, as discussed previously, numerous harms are risked by such disclosure. There is potential disruption of interparental relationships, potential disruption of parent-child relationships, and further fractures of the child's support network. The child may face loss of financial and medical insurance support as well. While healthcare providers may not be able to discern the extent of these harms prior to disclosure, there is certainly a reasonable expectation that some of these harms are likely to occur. Children are uniquely vulnerable to such harms. Relationships between children and their parents are inherently asymmetrical due to children's emotional, social and economic dependence on their parents. Children are therefore at great risk if these familial bonds are disrupted. These potential harms are magnified in the setting of a child with a critical illness, who would be expected to have increased need for emotional and financial support. While it is possible that benefits may accrue as the result of the disclosure of misattributed paternity—such as providing relief to other sources of tension within family relationships or by allowing the child to form new bonds if the child's genetic father is identified—these benefits are unlikely to be realized immediately.

Second, the child may face the immediate harm of a parent deciding not to donate an organ after the finding of misattributed paternity is disclosed, even in those circumstances in which the organ would otherwise be considered an acceptable match. A relative of the child's father might similarly withdraw upon learning paternity was misattributed. If information regarding the misattributed paternity of a potential donor is disclosed, the intimacy of the relationship between the potential donor and the potential recipient of the organ may be disrupted, regardless of the fact that neither party is culpable for said disruption.

This changing relationship may concurrently affect both the desire of the potential donor to proceed with donation and the strength of any obligation that the donor may feel to proceed in such circumstances. While perfect altruism would impel any individual who is an appropriate match for organ donation to serve as a donor, such perfect altruism is not expected by any but the most stringent of ethical systems and is not a tenet of ethics or policy surrounding organ donation. We generally view the decision to provide an organ or tissue for transplantation as a supererogatory rather than obligatory one. Although many authors [31–34] have discussed the special ethical obligations (or perceived obligations [35]) between friends, family members and other intimate acquaintances, most concede certain limitations on those obligations. The first is that any such special obligation is not absolute. Although a heightened degree of ethical responsibility may exist within the context of intimate relationships, this responsibility must be weighed against the other ethical responsibilities held by the potential donor, including those to self and other intimates. Second, it is asserted that this obligation is not derived solely, or even primarily, from the degree of genetic relatedness between to the parties, but is instead based on the intimacy of the relationship. This understanding of the importance of such an intimate relationship has served as the basis for the American Academy of Pediatrics position on sibling donation of hematopoetic stem cells (HSC) and has been used to argue against allowing HSC donation within the context of donation between siblings living in different families [36, 37]. The revelation of a finding of misattributed paternity is not simply a statement regarding genetic relatedness however, it is a statement that has the potential to have a profound effect on those intimate relationships that bind a family, including the relationship between the potential donor and potential recipient. As such, it may be quite reasonable for a potential donor with a misattributed genetic relationship to the recipient to elect not to proceed with organ donation in this circumstance. If that would occur, the child would experience a real and measureable harm, a harm which was the proximate result of the disclosure of misattributed paternity.

3.5 The Next Step: Institutional Policies

Ultimately, the pediatric patient has very uncertain expectations of benefit from disclosure of a finding of misattributed paternity during the transplant evaluation process and faces a very real risk of harm. Further, no participating agent has clear standing by which to offer consent to such disclosure on behalf of the minor. For these reasons, we feel that the disclosure of a finding of misattributed paternity in the setting of testing for organ donation to a pediatric transplant recipient is not appropriate. Moreover, the question being asked by HLA matching is a straightforward one: "Am I suitable organ donor for this child?" This can be answered clearly and comprehensively without discussion of issues of misattributed paternity, and should be answered as clearly and straightforwardly as possible.

As a result of the considerations discussed above, we feel that healthcare institutions or transplantation programs should establish—and make explicit in advance—a policy of not reporting findings indicating misattributed paternity. Instead the policy should be to report results regarding the quality of the match alone, not the broader genetic implications of those findings. While we generally recommend that policies for the disclosure of IFs be discussed with the parties undergoing testing beforehand, in this particular circumstance discussion of the policy would likely be counterproductive and runs the risk of participants considering a poor match tantamount to a disclosure of misattributed paternity. If participants independently confront the transplant team about HLA or blood type matching results and any implications these results may have for questions of misattributed paternity, we feel that Dr. Ross' suggestion [21] that they be referred to a genetic counsellor for more thorough testing is both helpful and appropriate. Such referral would afford participants the ability to consider the advantages and disadvantages of pursuing such testing at greater length, as well as access to providers who have greater experience counselling individuals at the time of such disclosures.

References

1. Howell WM, Carter V, Clark B. The HLA system: immunobiology, HLA typing, antibody screening and crossmatching techniques. J Clin Pathol. 2010;63(5):387–90.
2. Young A, Kim SJ, Gibney EM, Parikh CR, Cuerden MS, Horvat LD, et al. Discovering misattributed paternity in living kidney donation: prevalence, preference, and practice. Transplantation. 2009;87(10):1429–35.
3. Tong A, Chapman JR, Wong G, Craig JC. Living kidney donor assessment: challenges, uncertainties and controversies among transplant nephrologists and surgeons. Am J Transplant. 2013;13(11):2912–23.
4. Schroder NM. The dilemma of unintentional discovery of misattributed paternity in living kidney donors and recipients. Curr Opin Organ Transplant. 2009;14(2):196–200.
5. Soderdahl DW, Rabah D, McCune T, Colonna J, French R, Robey E, et al. Misattributed paternity in a living related donor: to disclose or not to disclose? Urology. 2004;64(3):590.
6. Wright L, MacRae S, Gordon D, Elliot E, Dixon D, Abbey S, et al. Disclosure of misattributed paternity: issues involved in the discovery of unsought information. Semin Dial. 2002;15 (3):202–6.
7. Wolf SM, Lawrenz FP, Nelson CA, Kahn JP, Cho MK, Clayton EW, et al. Managing incidental findings in human subjects research: analysis and recommendations. J Law Med Ethics. 2008;36(2):219–48.
8. Green RC, Berg JS, Grody WW, Kalia SS, Korf BR, Martin CL, et al. ACMG recommendations for reporting of incidental findings in clinical exome and genome sequencing. Genet Med. 2013;15(7):565–74.
9. Sapp J, Dong D, Stark C, Ivey L, Hooker G, Biesecker L, et al. Parental attitudes, values, and beliefs toward the return of results from exome sequencing in children. Clin Genet. 2014;85 (2):120–6.
10. Kleiderman E, Knoppers BM, Fernandez CV, Boycott KM, Ouellette G, Wong-Rieger D, et al. Returning incidental findings from genetic research to children: views of parents of children affected by rare diseases. J Med Ethics. 2014;40(10):691–6.
11. Forrow L, Arnold RM, Parker LS. Preventive ethics: expanding the horizons of clinical ethics. J Clin Ethics. 1993;4(4):287.
12. Parker LS. Ethical concerns in the research and treatment of complex disease. Trends Genet. 1995;11(12):520–3.
13. Parker LS. The future of incidental findings: should they be viewed as benefits? J Law Med Ethics. 2008;36(2):341–51.

14. Miller FG, Mello MM, Joffe S. Incidental findings in human subjects research: what do investigators owe research participants? J Law Med Ethics. 2008;36(2):271–9.
15. Berland LL, Silverman SG, Gore RM, Mayo-Smith WW, Megibow AJ, Yee J, et al. Managing incidental findings on abdominal CT: white paper of the ACR incidental findings committee. J Am Coll Radiol. 2010;7(10):754–73.
16. Lucast EK. Informed consent and the misattributed paternity problem in genetic counseling. Bioethics. 2007;21(1):41–50.
17. Beauchamp TL, Childress JF. Principles of biomedical ethics. 7th ed. New York: Oxford University Press; 2013. xvi, 459 pp.
18. Shirts BH, Parker LS. Changing interpretations, stable genes: responsibilities of patients, professionals, and policy makers in the clinical interpretation of complex genetic information. Genet Med. 2008;10(11):778–83.
19. Ross LF. Disclosing Misattributed Paternity. Bioethics. 1996;10(2):114–30.
20. Zachary AA, Leffell MS. Response: good ethics require good science: why transplant programs should not disclose misattributed parentage. Am J Transplant. 2011;11(1):179.
21. Ross LF. Good ethics requires good science: why transplant programs should not disclose misattributed parentage. Am J Transplant. 2010;10(4):742–6.
22. Engel E. Uniparental disomies in unselected populations. Am J Hum Genet. 1998;63(4): 962–6.
23. Palmor M, Fiester A. Incidental findings of nonparentage: a case for universal nondisclosure. Pediatrics. 2014;134(1):163–8.
24. Kleinsorge C, Covitz LM. Impact of divorce on children: developmental considerations. Pediatr review/Am Acad Pediatr. 2012;33(4):147–54 (quiz 54–5).
25. Troxel WM, Matthews KA. What are the costs of marital conflict and dissolution to children's physical health? Clin Child Fam Psychol Rev. 2004;7(1):29–57.
26. Nakonezny PA, Rodgers JL, Nussbaum JF. The effect of later life parental divorce on adult-child/older-parent solidarity: a test of the buffering hypothesis. J Appl Soc Psychol. 2003;33(6):1153–78.
27. Lewis M. Child and adolescent psychiatry : a comprehensive textbook. 3rd ed. Philadelphia: Lippincott Williams & Wilkins; 2002. xxiv, 1495 pp.
28. Walter JK, Ross LF. Relational autonomy: moving beyond the limits of isolated individualism. Pediatrics. 2014;133(Suppl 1):S16–23.
29. Committee on hospital care and institute for patient and family-centered care. Patient- and family-centered care and the pediatrician's role. Pediatrics. 2012;129(2):394–404.
30. Diekema DS. Parental refusals of medical treatment: the harm principle as threshold for state intervention. Theor Med Bioeth. 2004;25(4):243–64.
31. Scheffler S. Relationships and responsibilities. Philos Public Aff. 1997;26(3):189–209.
32. Glannon W, Ross LF. Do genetic relationships create moral obligations in organ transplantation? Camb Q Healthc Ethics. 2002;11(02):153–9.
33. Jecker NS. Conceiving a child to save a child: reproductive and filial ethics. J Clin Ethics. 1990;1(2):99–103.
34. Zeiler K, Guntram L, Lennerling A. Moral tales of parental living kidney donation: a parenthood moral imperative and its relevance for decision making. Med Health Care Philos. 2010;13(3):225–36.
35. Zeiler K. Just love in live organ donation. Med Health Care Philos. 2009;12(3):323–31.
36. Committee on Bioethics. Children as hematopoietic stem cell donors. Pediatrics. 2010;125 (2):392–404.
37. Diekema DS, Joffe S, Vandeven AM, Lantos JD. Bone marrow donation between siblings living in different families. Pediatrics. 2011;127(1):158–62.

Part II
Deceased Donation

Chapter 4
The Ethics of Persuasion: Evaluating the Ethical Limits on Attempting to Persuade Families to Donate the Organs of Deceased Family Members

Jennifer A. Chandler and Vanessa Gruben

Abstract This chapter examines the ethics of using persuasive techniques to encourage families to consent to organ donation by their deceased loved ones. We review the literature on the psychological effects of donation on families, and consider when and how the treatment of families in the requesting and donation process might put their future psychological well-being at increased risk. We also consider lawsuits brought against healthcare personnel for the negligent infliction of psychiatric injury on family members in several end-of-life contexts including organ donation. Although successful lawsuits arising from the manner in which families are asked to consent are unlikely for several reasons, there are lawsuits in partly analogous contexts that do shed light on the types of treatment that may raise ethical, if not legal, concerns. We do not attempt to propose boundaries for the use of persuasive techniques, as the question of how much risk to families is justified by the potential benefits sought for others (e.g. transplant recipients) is a difficult problem that exceeds the scope of this chapter. Nonetheless, we identify some persuasive practices as ethically justifiable and others that require further ethical consideration.

Keywords Organ donation · Family consent · Ethics of persuasion · Nudging

4.1 Introduction

This chapter examines the ethics of asking families to donate the organs of their deceased loved ones. We focus on the common situation where there is no prior registered consent by the deceased patient and families are asked to consent on

---

J.A. Chandler (✉) · V. Gruben
Faculty of Law, University of Ottawa, Ottawa, Canada
e-mail: jennifer.chandler@uottawa.ca

V. Gruben
e-mail: vanessa.gruben@uottawa.ca

R.A. Greenberg et al. (eds.), *Ethical Issues in Pediatric Organ Transplantation*, International Library of Ethics, Law, and the New Medicine 66,
DOI 10.1007/978-3-319-29185-7_4

behalf of their loved one. This conversation, between families on the one hand and healthcare personnel (HCP) or organ donation organization representatives on the other, is an ethically complicated one. The interests and well-being of multiple groups are affected by how this discussion is carried out, as well as by the decision that is reached. While the nature of any ongoing interests of the deceased is a matter of debate, society recognizes a continuing legal obligation to give effect to the pre-mortem wishes of a deceased person in several contexts such as the disposition of property as well as the pre-mortem wishes in relation to organ and tissue donation. Family well-being is also at stake, and family members have an interest in being treated in a way that does not put their psychological well-being at increased risk. The potential recipients of tissues and organs evidently have an interest in the increased availability of transplants. All members of society have a collective interest in the healthcare cost savings associated with transplantation rather than dialysis to address kidney failure [1]. Finally, the personnel asking families to consent should not be put in a position that makes it difficult to navigate these competing interests. This suggests a need for guidance on the relevant ethico-legal considerations, as well as adequate time and resources to treat families with care.

Within the domain of health communications generally, and to some extent within discussions of how to ask families to consent to donation, there is interest in the efficacy and ethics of persuasive techniques and "nudging." We examine the ethics of using persuasive techniques to encourage families to consent to organ donation by their deceased loved ones. While we acknowledge the importance of all of the interested groups mentioned above, our central focus here is on the psychological well-being of families due to our perception that the literature on the ethics of organ donation emphasizes the rights and interests of individual donors and the eventual recipients, with less attention to the well-being of families. The ethics of using persuasive techniques in this context is applicable to interactions with families in general, but it can be expected to be important in pediatric cases. In the following sections, we highlight studies of the psychological reactions of families in pediatric donation cases to illustrate some of the sensitivities that may be particular to such cases.

We also focus on donation decisions that are made after a declaration of death on neurological criteria ("brain death"). There is less available research on the psychological experiences of family members in the context donation after circulatory death ("DCD"). We speculate that the ethical issues raised by persuasion or "nudging" may be more acute in the DCD context. This is because families must make a decision to withdraw life-sustaining treatment from a living person close in time to a decision about organ donation, and so may be more susceptible to feelings of guilt or uncertainty about their decisions. In addition, there is a risk that the excessive use of persuasive techniques may cause families to be concerned that the desire to obtain organs for transplant may have affected the prior medical advice about the withdrawal of life-sustaining treatment. As a result, we believe that the questions and issues we raise here will exist, perhaps with even more importance, in the context of DCD.

Our approach is to review the literature on the psychological effects of donation on families, and in particular whether the treatment of the family in the requesting and donation process might put their future psychological well-being at increased risk. We also consider existing legal cases as a source of normative information. These cases involve claims by families that they have suffered psychiatric harm as a result of various kinds of actions by HCP in the context of organ donation, autopsies and end-of-life decision-making. Only a couple of these cases included complaints about "persuasive" techniques. Such legal claims are unlikely to succeed in cases where reasonable "persuasive" techniques are used to encourage families to consent. Nevertheless, a review of the existing cases does shed light on the kind of harms some families may suffer, as well as on what those families regard as wrongful. Even if the legal claims may be unlikely to succeed, ethical obligations may still exist. As a result, even legally unsuccessful cases are still of interest in assessing the moral boundaries on persuasion in the context of organ donation requests. We conclude with suggestions about the kinds of cases and persuasive techniques that we feel are ethically acceptable, and others that are riskier and should be subject to further ethical consideration.

The ethical approach that we use is utilitarian consequentialism, although this is evidently not the only ethical framework that could be applied. In particular, we assess persuasive efforts according to their effects on the psychological well-being of family members. Our ethical assessment is limited in that we look at family well-being primarily, and do not take a position on the difficult question of the extent to which harm to family well-being may be justified by the potential benefit to transplant recipients or to society. This is a challenging question that exceeds the scope of this chapter. Accordingly we also do not consider whether other moral principles or legal human rights protections may set absolute limits on the amount of harm to family members that may be justified by the objective of securing consent to donation.

4.2 Psychological Impact on Families of the Organ Donation Decision

The ethical analysis of how families are asked to donate the organs and tissues of a deceased loved one must include the question of whether the request harms families, benefits them, or both.[1] This section analyzes how families might be harmed or benefited by the request and the opportunity of organ donation.

[1]As noted above, the interests of the deceased (which are recognized as having a continuing value after death in various contexts), of transplant recipients, and of the broader society must also be considered, and we do not take a position here on which should be weighted more heavily or how any conflicting interests should be balanced.

The evidence of the psychological impact on families of the organ donation decision is generally encouraging, as the studies summarized below suggest. Many families are satisfied with their decisions (for or against donation), and those that donate often report psychological benefits of deciding to do so. However, some families also suffer psychological harms flowing from the organ donation decision, and it is possible that some methods of asking families may increase the risk of subsequent psychological suffering.

If the families at risk of serious psychological complications are identifiable, or aspects of the way families are treated by medical and organ donation organization personnel that increase the risk of complications are known, a corresponding ethical obligation to minimize these risks arises. This does not necessarily mean that risk should be minimized by refraining from asking families to donate, although the balancing of family well-being against other competing interests may suggest this is the appropriate course in extreme cases. More likely, it will set limits on the ways that families should be approached and asked to donate.

Several lines of recent research are helpful in understanding the psychological impact of the donation request and decision on families. These include studies of the prevalence of decision regret (i.e. how many families regret their decision to donate or their refusal to donate), and qualitative interview-based studies of the psychological impact of donation.

The studies of decision regret indicate that most families are satisfied with their decisions, but that families that refuse to donate have a somewhat greater tendency to regret that decision [2–10]. One caveat in interpreting these results is that the results may be affected by the respondents' desire to avoid the discomfort of regret, perhaps leading them to overvalue their prior decision and to discount the possible benefits of the alternative choice. This psychological phenomenon may protect many families against later distress regardless of the decision reached on whether to donate, and may partly explain the high rates of satisfaction with the decision.

Several of the studies of decision regret or instability were conducted in the pediatric context. In a sample of 74 American parents, Rodrigue et al. [2] found that most parents were satisfied with their prior decisions, but 10 % of consenting families regretted their decision to donate and 16 % of refusing families regretted their refusal [2]. Bellali and Papadatou [11] found that in their sample of 22 Greek parents only 18 % (2 of 11) of the consenting parents and the same proportion of the 11 refusing parents regretted their decisions [11]. In this study, the consenting couple that regretted agreeing to the donation of the organs of their brain-dead child believed in the value of organ donation, "but held traumatic memories of the insensitive approach of the [pediatric intensive care unit] personnel and still wondered whether they had caused the child's death by giving their consent" [11, p. 893].

Ultimately, this type of study does not indicate whether the organ donation decision increased or decreased family distress at the time of death or subsequently, as there is no comparison with the grief responses of families who were not asked to make this decision. However, one can perhaps conclude that organ donation either helped (or did not significantly harm) most families who consented, given that most would do the same thing again. As for families who refused, it is more difficult to draw conclusions. Most would refuse again, but a sizeable number later regretted their decisions. It is not possible to determine if this group would have drawn benefits from donating, but what does seem clear is that they have suffered psychological harm in the form of regretting their decision.

Studies that ask more specific questions of families, such as whether they felt that the donation was comforting to them, offer another useful source of information on the psychological effects of donation on grieving families. There is some evidence that consenting families perceive donation to have been comforting [12], albeit perhaps only modestly so [9]. For example, in Merchant et al.'s [9] study, the mean response to the question of whether donation was comforting was 1.9 on a scale between 1 (total agreement) and 5 (total disagreement) in a sample of 73 Canadian families who consented to donation. Although these studies suggest some families may think donation assists with grief, several studies of adverse psychological symptoms following bereavement failed to find significant differences between families who donate and those who refuse to donate [13, 14].

Other studies document specific positive and negative feelings flowing from donation [15–17]. Positive consequences associated with consenting to donate included feelings of peace and calm, spiritual achievement, social approval, happiness in helping the recipient and allowing their loved one to assist others, and the sense that their loved one was living on in some way. Negative consequences of donation included anxiety and uncertainty over brain death and difficulty in accepting the death, feelings of guilt or shame for having given up too early on their loved one or having allowed harm to come to their loved one, anxiety about the deceased's experience in the afterlife due to disfigurement of the body, and tensions due to social and family disapproval of their decision. Cultural and religious variation may limit the extent to which these studies, which were conducted Iran, Hong Kong and Taiwan, may be translated elsewhere. On the other hand, multicultural societies may well encounter diverse views and family reactions.

In the pediatric context, Bellali and Papadatou [11] found that none of the 22 Greek parents they interviewed spontaneously mentioned donation as a factor that positively or negatively affected their grief process. When the question was put to them, some said it had an impact whereas others felt it neither helped nor hindered their grieving, saying their suffering was the same. The authors suggest that donation may assist families (e.g. by positive emotions derived from helping another person to live), but may also lead to complicated grief reactions (e.g. due to the perception that the child continues to live in the recipient, impeding acceptance of the death or encouraging attachment to the recipient).

Given that donation can produce benefits and harms to families, is there any way to identify more precisely those who stand to benefit and those more likely to be harmed?

Several studies suggest that families with a poor understanding of brain death may be at particular risk of psychological suffering after they consent to donation. This would suggest care should be taken when trying to convince uncertain or confused families to consent. One such study is Smudla et al. [18], who studied 29

Hungarian families who had consented to donation. They found that 41.4 % of the families remained uncertain that the brain death diagnosis was reliable. Families with less confidence in the brain death diagnosis had more intense grief reactions, and more serious depressive symptoms. They concluded that it is essential for medical staff to provide an adequate explanation of brain death to families to decrease the risk of negative psychological consequences for families. Similarly, Manzari et al. [15] noted that some of the consenting families who did not understand brain death suffered uncertainty, guilt and shame for possibly having given up too early or allowed harm to come to their loved ones.

Harm may come not just from confusion over brain death, but from other aspects of the donation request and donation process. Exley et al. [19] quoted one consenting family member who said she would now refuse if she had to make the decision again because she had been "treated so bad[ly] by the coordinator—I still suffer from the way she hurried me." Kesselring et al. [20] noted that some families had traumatic memories of the organ donation decision. This was more likely to be the case with families who had had a poor interaction with the medical personnel (i.e. discussions that were perceived as rushed and focused on organ procurement) and/or were ambivalent about the decision. Merchant et al. [9] investigated the levels of depression, post-traumatic stress, and bereavement symptoms in a set of 73 Canadian families who had donated. Upsetting or negative experiences during donation[2] were associated with increased post-traumatic stress, but depression was reduced where there were positive experiences related to the donation.[3]

As noted above, some refusing families also suffer regret. Sque et al.'s [21] study of 26 British family members who declined donation found that some felt guilty and selfish for having declined. Manzari et al. [15] recount the misgivings of one Iranian family who thought they were right in refusing but then later worried about having foregone possible spiritual rewards for donating. It would thus seem important and humane to avoid as far as possible encouraging doubts that may cause later suffering to refusing families, but instead to help them to reach a carefully considered decision that is less likely to produce longer-term suffering.

In summary, the evidence for positive or negative psychological consequences of the donation decision is mixed. For many families, significant and continuing comfort is drawn from the opportunity for some good to come to others from their loss. For others, the reaction may be more complicated, perhaps raising subsequent feelings of guilt for being unable at the time to bring themselves to donate (for those who do not donate), or later feelings of having betrayed a loved one by "giving up too early" or later difficulties in letting go (for those who do donate). This suggests that it is not in itself unethical to put the burden of this decision on emotionally

[2]These included uncertainty about the recipient and whether the transplant was successful, skepticism or confusion about brain death, feeling rushed once the donation decision had been made, and keeping a loved one "alive" until donation could occur.

[3]These included the receipt of letters from recipients, a thank you ceremony from the transplantation organization. Many also cited the donation as positive in itself as it allowed something good to come from a tragic event or that it gave effect to their loved one's wishes.

distressed families, particularly since they may thereby be denied an opportunity that seems to provide many with psychological benefit. Instead, it suggests we should focus on identifying those who are at risk of negative consequences, as well as how to best reduce the risks of those negative consequences and to increase the likelihood of benefits for all families.

4.3 The Use of Persuasive Techniques in Requesting Consent to Organ Donation

The growing shortage in the availability of organs for transplantation has led to increasing concern and interest in how to raise donation rates. Among the approaches considered have been attempts to encourage more families to consent by modifying the manner in which families are approached and asked to donate. For example, among other modifications to the organ donation system, the "Organ Donation Breakthrough Collaborative" methodology in the United States encouraged a more aggressive pursuit of donation, including "reapproaching refusing families" [22]. Another suggestion is the deliberate use of a pro-donation rather than a value-neutral stance in asking families to consent, as well as the suggested use of techniques of social persuasion [23, 24]. This shift raises two issues, the first being whether it is advisable or ethical to declare that the dominant objective is to maximize organ donation in all cases. Second, what techniques or methods of encouraging consent are acceptable, and which should be avoided and why.

It seems unwise to declare that the single dominant objective of those interacting with families is to maximize donation. Cases vary both in the circumstances and deaths of the patient as well as in family personalities and beliefs, and the risks of harm to families will also vary. Williams et al. [25] argue that it is "ethically problematic if consent is the primary outcome measure of the donation process," as this is likely to deflect attention from the needs of individual families as well as to increase the risk of actual or perceived conflicts of interest (e.g. between obtaining organs and providing optimal end-of-life care for patients). Along similar lines, Streat [26] argues that those requesting that families donate should take a neutral position (i.e. offering the option of donation and focusing on family needs rather than seeking family consent to donation as the primary objective), as this frees them to ensure clinical and interpersonal aspects of the process are performed to the highest standards, and it is more likely to properly reflect the social acceptability of organ donation even if it does not produce more donors.

There seems to be an association between long-term psychological harm and the family's perception that HCP are focused on procuring organs. Kesselring et al. [20] interviewed the relatives of 33 brain-dead patients in Switzerland. They identified two risk factors for traumatic memories in families: poor interaction with professionals (100 % of families who described professional behavior as focused on organ procurement had signs of traumatic memories versus 8 % of those who described professional behaviour as focused on the needs of the patient and family) as well as ambivalence about the donation (29 % of families who made clear and unambiguous decisions had traumatic memories, while 67 % of those who were ambivalent had traumatic memories).

Others may also be affected by their perceptions of how bereaved families are treated. The demonstration of continuing empathetic care for refusing families may also reassure others about the humanity of the organ procurement system. Shih et al. [27] report an interesting story of a father who donated in part because of witnessing the handling of another bereaved family who happened to be present at the same time. The father referred to the caring and respectful treatment of the deceased boy and the humane handling by the medical staff of his father who initially consented but then changed his mind and refused donation.

Thus, the idea that success is to be measured by consent to donation alone is a risky approach, given variation amongst families. It must be balanced with proper attention to the needs and interests of those families.

We turn now to the question of when and how persuasive approaches may nonetheless be justified. Here again, blanket condemnation of persuasive approaches also strikes us as unwise, as there may be situations where this is beneficial to families, or where the risk of harm to families is sufficiently low to justify preferring the interests of recipients and society in increased donation rates.

There may be cases where families are unable, due to the emotional distress at the time of the death of their loved one, to make the decision that would ultimately provide them with psychological benefits. In such cases, a paternalistic argument could be advanced that it is ethical to "nudge" them toward the decision that they would have taken but for the difficult circumstances [28]. However, not everyone is confident that families should be nudged toward consent—particularly in situations where we are not sure that this is the decision that families would have reached but for the immediate stress of the situation. Truog [29] suggests that it is disingenuous and puts public confidence at risk if we are not "open and honest with families about our aims and intentions" [29, p. 44]. In his view, if the organ donation system's dominant objective is maximizing donation rather than helping families to make the best decisions for their loved one and themselves, then we should be clear about that. Nonetheless, there are likely situations where the best decision for highly stressed families is donation (e.g. where they are likely to later draw psychological comfort from the donation), and an attempt to guide them toward that decision may be understood as helping those families while also promoting the interests of potential transplant recipients and the public generally.

There is a broad range of "persuasive techniques," with a large gray area between some that seem acceptable or unacceptable. Sensitive and supportive treatment of families is generally associated with consent [3, 6, 27, 30–32] and so it is recommended as a means to improve donation rates. It is clearly ethical to treat families kindly notwithstanding that there may be mixed motives in play (i.e. concern for family well-being, as well as hope that families will consent to donation). At the other end of the spectrum, it strikes us as unethical to make attempts to shame families or to induce feelings of guilt in order to extract consent to donation, as the risks of increased psychological suffering at a time of great distress and in the future are too high. What of techniques that fall between these extremes?

One possible approach is that those requesting that families consent to donation adopt a confident, pro-donation attitude, and avoid nervous, tentative or apologetic attitudes. When requestors display anxiety or reluctance (e.g. "I'm legally required to ask…") or take a neutral or ambivalent approach to donation families are less likely to consent [31, 33, 34]. This is perhaps unsurprising as these approaches suggest to families that donation is something that others find objectionable. One might argue that an unduly apologetic or reluctant approach is itself unethical as it might dissuade families who might otherwise have later drawn some comfort from donating. As a result, a positive and confident approach strikes us as ethical and advisable, but this is not the same thing as a "presumptive approach" that assumes all families wish to donate. The concerns with the presumptive approach have been outlined above, and have to do with the reduced emphasis on the sensitive handling of family objections and misgivings and the risk that family well-being as a goal of the request process will be obscured.

A range of other persuasive techniques are discussed in the literature, such as deliberately making offers of assistance before asking for donation—a strategy described as appealing to the "norm of reciprocity" [35]. Zink and Wertlieb [24] recommend that requestors avoid questions that may elicit negative responses and derail or halt the donation discussion, and that requestors should be advocates for donors and recipients. Other techniques include a "credibility strategy" (opening the discussion by establishing credentials and suggesting the family wanted or expected to be contacted about donation), the "foot-in-the-door tactic" (asking for a specific type of tissue or tissue for a specific use followed by a larger request), or attempts to evoke feelings of guilt by emphasizing the needs of recipients or the need for donation from specific racial or ethnic groups [36–38]. Anker and Feeley [1, 37] provide detail of the specific persuasive strategies employed by 102 American organ procurement organization (OPO) coordinators interviewed between 2008 and 2009 [37, 38]. These often have to do with educating families about donation and the benefits of donation to recipients and families. However, a range of other tactics that seem more manipulative because they are less transparent are also employed such as "source homophily" (i.e. attempting to make the requestor seem similar to the family by modifying language and clothing to match the family, and explicitly identifying shared interests between requestor and family etc.).

Arguments in favour of "nudging" families to consent to donation depend largely on the benefits of donation to others (recipients and society generally), although it may also benefit families who may later draw psychological benefits from the donation even if they were unable at the time to make that decision without being nudged. Some persuasive techniques may be always justified because they are highly beneficial to all families in addition to their tendency to increase consent (e.g. compassionate and supportive treatment of families), or because they present organ donation positively while still recognizing and handling family reluctance sensitively. However, there are good reasons to be cautious with persuasive techniques that are solely or predominantly aimed at securing consent. In addition to a general ethical concern about respecting the autonomy of the family decision-makers, there are the risks of (1) ignoring the characteristics of the family that may suggest they fall within the group who are at risk of later regretting the donation and suffering serious psychological complications, (2) eroding the trust of families in medical practitioners and "turning off" families who might otherwise have consented if they felt that the interests of the patient and family were not being ignored in the pursuit of the donation.

4.4 Legal Obligations Toward Families in the End-of-Life Context

In this section we consider legal claims that have been made by families in various end-of-life contexts, as illustrations of the kinds of behaviour that some families perceive as harmful and wrongful, drawing on reported cases in Canada, the United States and the United Kingdom. These cases involve claims that the conduct of HCP has resulted in the negligent infliction of psychiatric harm[4] to families in the context of organ donation, autopsies and end-of-life care. We regard legal liability in the context of persuading families to consent to organ donation to be unlikely given the immunity clauses in organ donation legislation in many jurisdictions and, even in the absence of immunity clauses, (e.g. *Trillium Gift of Life Act*, Sect. 9 [39]) because of the fairly high threshold required to establish a viable claim. However, while most of the reported cases that we describe have been unsuccessful, they are relevant to our exploration of the ethics of requesting. They illustrate the expectations families have with respect to their treatment in partly analogous contexts, and so offer insight into the ethical obligations toward bereaved families in the requesting process. In our view, the law sets a minimum standard of behaviour that HCP should satisfy. However ethical obligations may exceed this basic standard so that even if legal liability does not exist, the conduct complained of may nevertheless be unethical.

4.4.1 Examples of Conduct Resulting in a Complaint of Negligent Infliction of Psychiatric Harm in the End-of-Life Context

Families have brought claims for psychiatric harm resulting from the conduct of HCP in various end-of-life care contexts. Although these legal claims do not all arise out of the manner in which the organ request has been made, they illustrate the

[4]The courts use multiple names for this tort, including variants of the negligent infliction of nervous shock, mental or emotional distress, or psychological or psychiatric harm.

kind of behavior that has been perceived as harmful and provide important insight into whether such a claim might be successful.

In the broader context of end-of-life care, families have alleged that psychiatric harm has resulted where an HCP allegedly ripped out a patient's feeding tubes in order to coerce a family member into unwanted non-emergency surgery [40], where the process for terminating life support of a loved one was overly complicated [41], where a hospital buried premature twin babies in a communal grave without authorization [42], and where HCPs and hospitals have not adequately prepared family members prior to viewing their loved one's bodies after death [43, 44].

Claims of psychiatric harm have also been made in the context of organ and tissue donation, where the most frequent type of complaint brought by families is that the HCP acted without consent or contrary to the consent given for the removal of their loved one's tissues and organs. In several cases, families alleged that the unauthorized removal of tissues, organs and bones from their loved one without consent resulted in psychological injury [45–51]. In these cases, families have argued that the HCP were aware of the decedent or family's wishes regarding donation, that they disregarded these wishes and that the resulting mishandling of the loved one's body by the HCP resulted in emotional harm or psychiatric injury to family members.

Further, there are a number of cases where families have brought claims for negligent infliction of psychiatric harm on the basis that the HCPs did not provide sufficient or accurate information regarding the removal or donation of their loved one's organs. For example, in *A B and Others v Leeds NHS Hospital,* several families claimed they suffered psychological harm when they learned of a long-standing practice of removing tissue from their deceased children at or after post-mortem examinations [52]. The families claimed that although they had consented to the post-mortem examinations, they had not been advised that tissue would be removed as part of these examinations.[5] Similarly, in *Carey*, the parents consented to the donation of their organs and were erroneously advised that their son's organs and tissues had been successfully transplanted [54]. The parents claimed that the misinformation provided to them, coupled with the disposal of the tissue without authorization, resulted in emotional harm.

Finally, there are cases in which families have brought claims for psychiatric harm relating to the request for a loved one's organs. For example, in *Bushlow*, the claimants' daughter had been fatally injured in a motor vehicle accident [55]. While she was in hospital, the parents were repeatedly asked by different HCP whether they had considered organ donation and they declined each time. Eventually, the father agreed, testifying that "once he was informed that his daughter's heart could be used to save another child's life, he felt "compelled" to do what he could to save this child's life." After their daughter's death, they brought an unsuccessful claim

[5]Similarly, in *Joyner*, a mother claimed that health officials had led her to believe that no organs would be removed from her child's body during autopsy, and that their subsequent removal resulted in emotional distress [53].

alleging psychiatric harm arising from the numerous and persistent requests made by various HCP, arguing that the actions were unreasonable, overbearing and outrageous, and as such not entitled to the protection of the immunity clause.

Similarly, in *Siegel*, the parents of a fatally injured teenager claimed they suffered psychiatric injury from the use of their daughter's tissues without consent [51]. Further, they argued that the manner in which their consent was sought was not made in good faith (and thus that the legislated immunity from liability was inapplicable given that it was limited to good faith actions). In seeking consent to organ donation, the HCP stated that "there is a very good chance that Jessica could be a hero through the gift of donation" [51, para 13]. The HCP also explained that the daughter's organs, bones and tissue could be used to "really change the lives of several families that are in need right now," and asked the father, "is that the kind of gift that you would like to honor your daughter Jessica with?" [51, para 13]. The plaintiff unsuccessfully argued that the HCP's use of the terms "hero" and "honor" in relation to their daughter, the use of language which implied that they would consent to organ donation, and her failure to read portions rather than the entire consent form indicated that the HCP's actions were not in good faith and thus should not be shielded from liability.

4.4.2 *The Legal Analysis*

4.4.2.1 Immunity Clauses in Organ Donation Legislation

Many jurisdictions provide immunity from legal liability for HCP working in the context of organ donation. For example, s.9 of Ontario's *Trillium Gift of Life Act* states that no legal action shall be brought against HCP for any acts performed or omissions committed in the performance of duties or functions under the *Act* as long as the HCP were acting in good faith [39]. The rationale for providing immunity against legal claims to the HCP involved in the organ donation process is to encourage organ donation and to eliminate uncertainty regarding their legal liability [52, 54]. However, the immunity is not unlimited. By requiring that the individual claiming immunity have acted in good faith, we believe that the law seeks to strike a balance between encouraging donation and ensuring the respectful treatment of human remains.

American courts have applied the legal immunity clauses to block claims brought by family members [51, 55]. In these cases, the courts have defined good faith as an "honest belief, the absence of malice and the absence of design to defraud or to seek an unconscionable advantage" [51, para 10, 54, 55]. However, bad faith is possible in the organ donation context. An example is provided in *Perry*, which involved a motion to dismiss a claim prior to trial. In such a proceeding, a court decides whether the plaintiff's allegations would constitute a valid legal claim assuming the allegations to be true. In *Perry,* the court concluded that the nurse who had acted without the consent of the family had not acted in good faith because, if the allegations were proven, she had intentionally misled the family into signing the consent [45]. As such, the court concluded that her alleged conduct, if proven, would have been "an abuse of the organ donation process and a betrayal of a grieving family's trust" and thus could not be protected by a statutory immunity clause [45, p. 29].

As we have noted above, although this type of immunity clause has not been applied by a Canadian court to our knowledge, we feel it is likely that immunity clauses (where they exist) and the difficulties of establishing the claim for damages for the negligent infliction of psychiatric harm (as we discuss in the next section) make successful suits unlikely in the context of attempts to persuade families to donate. Despite these immunity clauses, however, the record demonstrates that claims are still brought and the aforementioned decisions indicate that they could conceivably succeed.

4.4.2.2 The Tort of Negligent Infliction of Psychiatric Injury

Historically, the courts in common law jurisdictions were reluctant to recognize claims for damages arising from mental suffering standing alone (rather than flowing from a physical injury) but they have increasingly recognized tort liability for this type of harm in several jurisdictions [56, 57]. To succeed in Canada, the party bringing the claim (the plaintiff or claimant) must establish four things: first, that the defendant owed him or her a duty of care; second, that the defendant's behavior breached the standard of care; third, that the plaintiff has sustained damage; and, fourth, that the damage was caused by the defendant's breach of that standard [58].

The first element that the family members, as plaintiffs, must establish is that the HCP owed a duty of care to family members of the deceased. Canadian law makes it clear that liability may be imposed for psychiatric injury to close family members that results from some injury or mistreatment to their loved ones [59]. The duty of care owed to family members covers behaviour that poses a foreseeable risk of producing psychiatric injury [58]. The risk will be considered to be foreseeable if the behaviour is of a type that can be expected to produce psychiatric injury in a person of ordinary resilience [58, 60]. However, where it is actually known that a particular person is of less than ordinary fortitude, courts may regard the risk of injury in that case as reasonably foreseeable [58]. As is discussed below, the courts also require that the level of harm be fairly serious, amounting to "serious trauma or illness" rather than lesser degrees of emotional upset.

Several courts have found that a duty of care exists between physicians and family members in the context of the unauthorized removal of tissue. For example, in *A B and Others*, the courts found that a duty of care exists between physicians and family when obtaining consent for a post-mortem examination of their deceased loved one [52]. In *Perry*, the court concluded that a duty of care might well arise between HCP and families who are asked to consent to the donation of the deceased's organs [45]. In that case, the court described a nurse who had asked the family for consent to donate their deceased loved one's organs as occupying "a position of trust" with an "emotionally vulnerable family" [45, p. 30].

The second element that the plaintiffs must prove is that the HCP's behavior breached the standard of care [58]. This is a very case-specific inquiry, but it essentially measures the HCP's behaviour against what the hypothetical reasonable HCP would have done in the circumstances to avoid the reasonably foreseeable risks of serious psychiatric harm to families.

The cases supply useful illustrations of the kinds of actions that might be taken to breach the standard of care in the context of organ donation. This standard will likely be breached where HCP have failed to comply with the wishes of the family member, for example where a family has either refused to consent to donation or where HCP exceed the limited consent given [45, 52]. This standard will likely also be breached where HCP act in an aggressively coercive manner. For example, in *Washington*, the court concluded that, if proven to have occurred, the standard of care would be breached where an HCP tried to coerce a family member into agreeing to unwanted non-emergency surgery by ripping out the patient's feeding tubes [40]. The standard of care will likely also be breached where HCP do not disclose all relevant information when seeking consent to donation in order to avoid objections from the family. For example, in *A B and Others*, the court found that the standard of care was breached where HCP failed to fully explain the nature of the post mortem examination, in particular that such an examination included the removal and retention of organs [52].

On the other hand, the standard will not be breached simply on the basis that HCP have solicited or sought the family members' consent to organ donation [55]. Nor will this standard be breached where a HCP is simply "insensitive" when communicating with a patient's family [61, para 93] or where the HCP is not "tactful" [50, p. 477]. The HCP behaviour in *Bushlow* and *Siegel* (mentioned above) is closer to our topic of interest in this chapter—namely the manner in which families are asked to consent [51, 55]. Unfortunately for our purposes, the courts did not rule on whether the repetitive requesting or the persuasive attempts complained of in those cases fell below the required standard of care. This was because the cases were dismissed due to the application of immunity clauses.

Third, the claimant must establish that he or she has suffered legally recognized damage as a result of the HCP's breach of the standard of care. Courts have historically had "substantial difficulties" accepting claims for damages for psychiatric harm [56, p. 497]. These "difficulties" stem from the courts' skepticism about claims for mental harm including concerns about the making of fraudulent claims; pragmatic concerns about proving such harm; and, fears that, once recognized, there would be a proliferation of such claims [62]. Although it is now well-settled that the law recognizes these claims for damages, the courts, in part to address these concerns, have applied a demanding standard to the type of psychological harm that is compensable. The Supreme Court of Canada recently described the nature of this harm as follows:

> [P]sychological disturbance that rises to the level of personal injury must be distinguished from psychological upset. Personal injury at law connotes serious trauma or illness….The law does not recognize upset, disgust, anxiety, agitation or other mental states that fall short of injury. I would not purport to define compensable injury exhaustively except to say *that it must be serious and prolonged and rise above the ordinary annoyances, anxieties and fears* that people living in society routinely, if sometimes reluctantly, accept. [58, para 9] [emphasis added].

Prior to this decision, the courts required the claimant to establish a "recognizable psychiatric illness" [63, para 1, 64]. Some commentators suggested that the Supreme Court's ruling was actually a relaxation of the prior standard for how serious the psychiatric harm must be in order to become legally compensable, proposing that the new lower standard required a showing only that the psychological disturbance be 'serious and prolonged' [62]. To date, Canadian courts have continued to apply the more rigorous requirement [63, 64].

Given this relatively stringent standard, it may be difficult for claimants to prove legally compensable damages. Nonetheless, courts have recognized that family members may suffer serious mental harm, over and above that of routine grief, as a result of the negligence of HCP in the end-of-life context [40, 42, 52]. For example, "the courts have appreciated the emotional stress caused from grieving a loved one's death and have viewed the grieving family members as emotionally susceptible with regards to their feelings towards the decedent's remains" [45, p. 29].

The final element the claimant must prove is that the HCP's failure to meet the standard of care *caused* the mental harm to the plaintiff. This may be challenging for families to establish, given that they will have to show that it was the treatment of the family, and not just the bereavement, that caused or contributed to the psychiatric injury. Put another way, if the psychiatric injury would have happened in any event, the HCP's careless actions will not be viewed as the cause of the injury. However, the courts have been willing to draw the conclusion that HCP behaviour has caused suffering in some cases. For example, where a decedent's organs are donated without consent, the courts have concluded that "it is foreseeable that the immediate family will likely experience serious emotional distress from the mishandling of a loved one's dead body" [45, p. 28, 48]. Further, in *A B and Others*, the court stated that "most of the clinicians and experts who gave evidence said that they could have foreseen that the subsequent discovery that the organ of a deceased child had been retained after burial would have caused great distress and some that it would have led to psychiatric harm" [52, para 239].

4.4.3 Application of the Legal Principles to the Treatment of Families During the Organ Donation Request

In this section, we consider what the cases mentioned above suggest about ethical obligations to families in the context of attempts to persuade them to donate the organs of deceased loved ones. We do not offer firm legal conclusions given variation amongst jurisdictions on the applicable laws, the limited number of cases available, and the fact that many of the available cases involve misconduct such as removing tissue without any consent rather than alleged harm arising from the consent process. Although legal liability may be unlikely in the absence of bad faith, the cases illustrate the kinds of HCP behaviour that families perceive to have caused them psychological harm, and so are important as they provide some guidance on the ethical standards that should govern the conduct of HCP vis-à-vis the family.

First, the cases support the proposition that HCP owe a duty of care to families in the requesting process. Duties are owed where it is reasonably foreseeable that the HCP actions would foreseeably cause psychiatric injury to the average family member. Courts have recognized that families are emotionally susceptible during the critical illness and death of their loved ones in several cases involving the mishandling of their loved ones' remains. The qualitative studies outlined in the first half of this article also suggest that certain populations may be at increased risk of psychiatric complications if they are pressured to consent. In particular, families who have doubts about brain death, families in which there is dissension about donation, and families whose fears about donation are religiously motivated (e.g. about a loved one's disfigurement in the afterlife) may be at increased risk. This knowledge increases the ethical duty of care to these families.

The cases offer some guidance on the type of behavior that may be viewed as harmful and so possibly unethical in the context of requesting families to consent to donation. Families have complained about persistent requests that are perceived as harassment, attempts to persuade using misleading or inaccurate information, and persuasive attempts to pressure families by emphasizing the need of recipients or the potential for a deceased loved one to be a "hero." These cases suggest that communications that deliberately or carelessly induce feelings of guilt or shame in reluctant families should be avoided. They may also call into question the circumstances under which refusing families are re-approached. For example, where a family's refusal is due to a lack of information, or the absence of an important decision-maker, or to some other factor that can be reasonably modified to enable the family to make the decision under optimal conditions, re-approaching families seems justifiable. Where refusal flows from strong psychological aversions or religious objections, re-approaching families or applying persuasive techniques intended to overcome these objections seems risky.

On the other hand, the adoption of a positive and confident approach to donation does not seem likely to increase the risk of harm to families, and, on the contrary as illustrated by the empirical studies described above, may help some families at a time of great distress and uncertainty to become comfortable with a decision to donate that may ultimately bring some measure of comfort to the family. Furthermore, caring and compassionate treatment of families is justified by the benefits for family well-being, even if it is known that this may also increase consent to donation. Of course, all families, whether or not a loved one is a potential organ donor, should receive caring and compassionate treatment when a loved one is dying. We mention this here in order to make the point that caring and compassionate conduct does not become ethically suspect (within the utilitarian consequentialist framework we have chosen) just because it also tends to encourage consent and so could be regarded as a form of persuasion.

As for the use of persuasive "techniques" that operate in the background of discussions to channel family decisions (such as "source homophily," the "foot in the door" strategy, or other deliberately manipulative techniques) the ethics are more unclear in our view. In the broader context of the ethics of "nudging" and paternalism in healthcare decisions generally, Quigley [65] has noted that our decisions are shaped by a myriad of surrounding influences, some of which are unintended by anyone in particular but nonetheless malign. For example, a family may reflexively refuse to donate under the stress of the moment even if they might have consented under different conditions and they stand to gain psychologically later from the donation. In such cases, deliberate and admittedly manipulative attempts to assist them to reach this decision that (we presume) they would have made if possible seem more acceptable. These persuasive techniques seem unethical in cases where families are "nudged" toward decisions that they are unlikely to later be comfortable with having made.

4.5 Conclusion

Our impression is that the vast majority of medical practitioners and experts from organ donation organizations have family well-being in mind in the context of asking for family consent to organ donation. This concern for family well-being may be motivated purely by respect for family members, while in other cases it may be strategic or instrumental (i.e., in order to encourage consent to donation), and some HCP may act with a mixture of both motivations. However, some aspects of the organ donation system discourage attention to family interests, such as a focus on "conversion" statistics (success as a measure solely of how many consents can be obtained, rather than as a measure of how many families reach a decision that is ultimately satisfying for them), or a privileging of the interests of potential recipients. Clearly, given the life-or-death stakes for potential recipients, the focus on increasing donation is understandable, but the ethical pursuit of increased donations should include some constraints set by care for family well-being.

We have not attempted an ethical argument about how to balance the interests of all of those involved in organ donation and transplantation (families, deceased patients, potential recipients of donated organs, and the public) where those interests conflict. It is possible that some risk to families is justified by the importance of donation to others. In addition, it seems ethically justifiable to take the risk of some increased psychological distress for families by asking them about donation because the decision to donate seems ultimately to offer comfort to many. However, where there is reason to suspect that donation might cause increased and longer term psychological harm, attempts to persuade families seem unwise. Such reasons are present where families seem reluctant to donate due to confusion about brain death, family disagreement, or strong psychological or religiously-motivated aversion. Efforts to overcome reluctance due to a lack of information or inaccurate information, or other factors that can be reasonably modified, seem both acceptable and advisable as they may help families to reach informed decisions and to have a chance at drawing psychological benefits from having donated.

Acknowledgements The authors acknowledge the financial contribution of Health Canada, and the Provinces and Territories, as well as the financial support of the Canadian National Transplant Research Program (CNTRP) and the Canadian Institutes of Health Research (CIHR). The views expressed in this chapter do not necessarily represent the views of the Federal, Provincial or Territorial governments, the CNTRP or CIHR. We also acknowledge the excellent research assistance of Nir Harrel and Matthew Connors.

References

1. Barnieh L, Gill JS, Klarenbach S, Manns BJ. The cost-effectiveness of using payment to increase living donor kidneys for transplantation. Clin J Am Soc Nephrol. 2013;8(12):2165–73.
2. Rodrigue JR, Cornell DL, Howard RJ. The instability of organ donation decisions by next-of-kin and factors that predict it. Am J Transplant. 2008;8(12):2661–7.
3. Rodrigue JR, Scott MP, Oppenheim AR. The tissue donation experience: a comparison of donor and nondonor families. Prog Transplant. 2003;13(4):258–64.
4. Jacoby L, Jaccard J. Perceived support among families deciding about organ donation for their loved ones: donor vs nondonor next of kin. Am J Crit Care. 2010;19(5):e52–61.
5. Manzari ZS, Mohammadi E, Heydari A. Aghamohammadian, Modabber, Khaleghi E. Exploring families' experiences of an organ donation request after brain death. Nursing Ethics. 2012;19(5):654–65.
6. Siminoff LA, Traino HM, Gordon N. Determinants of family consent to tissue donation. J Trauma—Injury, Infect Crit Care. 2010;69(4):956–63.
7. Carey I, Forbes K. The experiences of donor families in the hospice. Palliat Med. 2003;17 (3):241–7.
8. Blok GA. The impact of changes in practice in organ procurement on the satisfaction of donor relatives. Patient Educ Couns. 2005;58(1):104–13.
9. Merchant SJ, Yoshida EM, Lee TK, Richardson P, Karlsbjerg KM, Cheung E. Exploring the psychological effects of deceased organ donation on the families of the organ donors. Clin Transplant. 2008;22(3):341–7.
10. Jansen NE, Van Leiden HA, Haase-Kromwijk BJJM, Van Der Meer NJM, Kruijff EV, Van Der Lely N, et al. Appointing 'trained donation practitioners' results in a higher family consent rate in the Netherlands: a multicenter study. Transpl Int. 2011;24(12):1189–97.
11. Bellali T, Papadatou D. Parental grief following the brain death of a child: does consent or refusal to organ donation affect their grief? Death Stud. 2006;30(10):883–917.
12. Siminoff LA, Mercer MB. Public policy, public opinion, and consent for organ donation. Camb Q Healthcare Ethics: CQ: Int J Healthcare Ethics Committees. 2001;10(4):377–86.
13. Cleiren MPHD, Van Zoelen AAJ. Post-mortem organ donation and grief: a study of consent, refusal and well-being in bereavement. Death Stud. 2002;26(10):837–49.
14. Tavakoli SAH, Shabanzadeh AP, Arjmand B, Aghayan SHR, Nozary Heshmati B, Emami Razavi SH, et al. Comparative study of depression and consent among brain death families in donor and nondonor groups from march 2001 to december 2002 in Tehran. Transpl Proc. 2008;40(10):3299–302.

15. Manzari ZS, Mohammadi E, Heydari A, Aghamohammadian M, Khaleghi E. Exploring families' experiences of an organ donation request after brain death. Nurs Ethics. 2012;19 (5):654–65.
16. Tong YF, Holroyd EA, Cheng C. Needs and experiences of Hong Kong Chinese cadaveric organ donor families. Hong Kong J. Nephrol. 2006;8(1):24–32.
17. Shih FJ, Lai MK, Lin MH, Lin HY, Tsao CI, Chou LL, et al. Impact of cadaveric organ donation on Taiwanese donor families during the first 6 months after donation. Psychosom Med. 2001;63(1):69–78.
18. Smudla A, Hegedus K, Mihaly S, Szabo G, Fazakas J. The HELLP concept—relatives of deceased donors need the help earlier in parallel with loss of a loved person. Ann Transplant. 2012;17(2):18–28.
19. Exley M, White N, Martin JH. Why families say no to organ donation. Critical Care Nurse. 2002;22(6):44–51.
20. Kesselring A, Kainz M, Kiss A. Traumatic memories of relatives regarding brain death, request for organ donation and interactions with professionals in the ICU. Am J Transplant. 2007;7(1):211–7.
21. Sque M, Long T, Payne S, Allardyce D. Why relatives do not donate organs for transplants: 'sacrifice' or 'gift of life'? J Adv Nurs. 2008;61(2):134–44.
22. Siminoff LA, Agyemang AA, Traino HM. Consent to organ donation: a review. Prog Transplant. 2013;23(1):99–104.
23. Metzger RA, Taylor GJ, McGaw LJ, Weber PG, Delmonico FL, Prottas JM. Research to practice: a national consensus conference. Prog Transplant (Aliso Viejo, Calif.). 2005;15 (4):379–84.
24. Zink S, Wertlieb S. A study of the presumptive approach to consent for organ donation: a new solution to an old problem. [Review] [8 refs]. Crit Care Nurse. 2006;26(2):129–36.
25. Williams MA, Lipsett PA, Rushton CH, Grochowski EC, Berkowitz ID, Mann SL, et al. The physician's role in discussing organ donation with families. Crit Care Med. 2003;31 (5):1568–73.
26. Streat S. Clinical review: moral assumptions and the process of organ donation in the intensive care unit. Crit Care. 2004;8(5):382–8.
27. Shih F, Lai M, Lin M, Lin H, Tsao C, Duh B, et al. The dilemma of "to-be or not-to-be": needs and expectations of the Taiwanese cadaveric organ donor families during the pre-donation transition. Soc Sci Med. 2001;53(6):693–706.
28. Whyte KP, Selinger E, Caplan AL, Sadowski J. Nudge, nudge or shove, shove—the right way for nudges to increase the supply of donated cadaver organs. Am J Bioeth. 2012;12(2):32–9.
29. Truog RD. When does a nudge become a shove in seeking consent for organ donation? Am J Bioeth. 2012;12(2):42–4.
30. Rodrigue JR, Cornell DL, Howard RJ. Organ donation decision: comparison of donor and nondonor families. Am J Transplant. 2006;6(1):190–8.
31. Siminoff LA, Gordon N, Hewlett J, Arnold RM. Factors influencing families' consent for donation of solid organs for transplantation. J Am Med Assoc. 2001;286(1):71–7.
32. Rodrigue JR, Cornell DL, Howard RJ. Pediatric organ donation: what factors most influence parents' donation decisions? Pediatr Crit Care Med. 2008;9(2):180–5.
33. Siminoff LA, Arnold RM, Hewlett J. The process of organ donation and its effect on consent. Clin Transplant. 2001;15(1):39–47.
34. Sanner MA. Two perspectives on organ donation: experiences of potential donor families and intensive care physicians of the same event. J Crit Care. 2007;22(4):296–304.
35. Joule RV, Bernard F, Geissler A, Girandola F, Halimi-Falkowicz S. Binding communication at the service of organ donations. Revue Int De Psychol Sociale-Int Rev Social Psychol. 2010;23(2–3):211–38.
36. Siminoff LA, Traino HM, Gordon NH. An exploratory study of relational, persuasive, and nonverbal communication in requests for tissue donation. J Health Commun. 2011;16 (9):955–75.

37. Anker AE, Feeley TH. Difficult communication: compliance-gaining strategies of organ procurement coordinators. J Health Commun. 2011;16(4):372–92.
38. Anker AE, Feeley TH. Asking the difficult questions: message strategies used by organ procurement coordinators in requesting familial consent to organ donation. J Health Commun. 2011;16(6):643–59.
39. Trillium Gift of Life Act, R.S.O. 1990, c. H.20.
40. Washington v. Knight, Collom & Carney Clinic Association and Texarkana Memorial Hospital, D/B/A/ Wadley Regional Medical Center, 887 S.W.2d 211 (Tex. App 1994).
41. Strachan v. John F. Kennedy memorial hospital and A.R. Pirolli, 507 A.2d 718 (N.J. Super. 1986).
42. Bastien et al v Ottawa Hospital (General Campus), (2001) 56 O.R. (3d) 397.
43. Devji v Burnaby (District), (1998) 158 D.L.R. (4th) 747.
44. Lew v Mount Saint Joseph Hospital Society, (1997) 44 B.C.L.R. (3d) 84, [1997] B.C. J. No. 2461.
45. Perry v. Saint Francis Hospital and Medical Center Inc and American National Red Cross, 886 F. Supp 1551 (U.S.Dist. 1995).
46. Mansaw v. Midwest Organ Bank and Truman Medical Center West, No. 97-0271-CV-W-6 (U.S. Dist. 1998).
47. George H. Lanier Memorial Hospital and Jason Ivey v. Andrews et al, 901 So. 2d 714 (Ala, 2004).
48. Adams et al v. King County et al, 164 Wn.2d 640 (Wash 2008).
49. Drever v. State of New York, 948 N.Y.S.2d 550 (N.Y. 2014).
50. Burgess v. Perdue and State of Kansas, 721 P.2d 239 (Kan. 1986).
51. Siegel v. LifeCenter Organ Donor Network, et al, 969 N.E.2d 1271 (Ohio 2011).
52. A B and Others v. Leeds Teaching Hospital NHS Trust and Cardiff and Vale HNS Trust, [2004] EWHC 644 (QB).
53. Joyner v. The State of New York, (2001), #2001-013-518, Claim No. 88536.
54. Carey v. New England Organ Bank, 446 Mass. 270 (Mass. 2006).
55. Bushlow v. INOVA Health Systems Hospitals, t/a Fairfax Hospital, 40 Va. Cir. 121 (Va. 1996).
56. Klar L. Tort law. 5th ed. Toronto: Carswell; 2012.
57. Osborne PH. The law of torts. 4th ed. Toronto: Irwin Law; 2011.
58. Mustapha v Culligan of Canada Ltd., 2008 SCC 27, [2008] 2 SCR 114.
59. Lodge v. Fraser Health Authority 2009 BCCA 108.
60. Vanek v. Great Atlantic & Pacific Co. of Canada (1999) 48 O.R. (3d) (Ont. C.A.).
61. Malette v Shulman, 2006 BCSC 1295, [2006] B.C.J. No. 1976.
62. Belanger-Hardy L. Thresholds of actionable mental harm in negligence: a policy-based analysis. Dalhousie LJ 2013; 36:103.
63. Healey v. Lakeridge Health Corporation, 2011 ONCA 55, 103 OR (3rd) 401.
64. Kotai v Queen of the North, 2009 BCSC 1604, [2009] BCJ No 2332.
65. Quigley M. Nudging for health: on public policy and designing for choice architecture. Med Law Rev. 2013;21(4):588–621.

Chapter 5
Death Determination in Pediatric Organ Donation

Sam D. Shemie, Ivan Ortega-Deballon and Sonny Dhanani

Abstract Complexities and misunderstanding related to the determination of death continue to evolve in response to remarkable technological advances in the ability to support, replace or transplant failing organs. This paper reviews the principles of management of life-threatening critical illness and the corresponding biological aspects of life, death and organ donation which are both informed and complicated by these technological and scientific achievements. Clarity in lexicon should be established, including movement toward functional definitions and away from anatomically based terms such as cardiac and brain death that erroneously imply death of the organ. Medicine is evolving towards a single unified determination of death, based on the cessation of brain function subsequent to catastrophic brain injury or circulatory arrest. In the presence of clear cause and in the absence of reversible or confounding conditions, the cessation of clinical functions of the brain that will not resume is determined by the absence of capacity for consciousness, centrally mediated motor responses, brainstem reflexes and capacity to breathe.

S.D. Shemie (✉)
Division of Critical Care, Montreal Children's Hospital, McGill University Health Centre, McGill University, Montreal, QC, Canada
e-mail: sam.shemie@mcgill.ca

S.D. Shemie
Canadian Blood Services, Ottawa, ON, Canada

I. Ortega-Deballon
Deceased Organ Donation Research program, Critical Care Division, Montreal Children's Hospital, Montréal, QC, Canada
e-mail: iviortega@gmail.com

S. Dhanani
Critical Care, Children's Hospital of Eastern Ontario, University of Ottawa, Ottawa, ON, Canada
e-mail: sdhanani@cheo.on.ca

S. Dhanani
Organ Donation, Trillium Gift of Life, Toronto, ON, Canada

R.A. Greenberg et al. (eds.), *Ethical Issues in Pediatric Organ Transplantation*, International Library of Ethics, Law, and the New Medicine 66,
DOI 10.1007/978-3-319-29185-7_5

Keywords Determination of death · Organ donation · Pediatric ethics · Brain death · Circulatory death · Cardiac death

5.1 Introduction

Humanity has thoughtfully struggled with the concept of and criteria for death for millennia and the 'line' between life and death continues to be debated [1]. The profound changes brought about by technological advances in organ failure support, organ replacement and transplantation continue to challenge our notions of life and death.

Discussions about death are challenging and complex, resulting from philosophical, religious, and cultural differences in the concept and definitions. The subject matter is emotionally charged. While controversies play out regularly in academia and the media, the discourse suffers from well know deficits in understanding and/or awareness of the issues surrounding death determination by the public [2] and health professionals [3]. There are difficulties in performing research in this field. As a result, there are deficits in high levels of evidence on a number of aspects of the dying process, a challenge particularly exaggerated in children.

Over the years, alarming language has emerged in the media and academic literature around death determination and deceased donation practices [1]. Emotive and contentious terminology such as 'not quite dead, nearly dead, dead enough' inflame rather than advance knowledge at the delicate interface of death and donation. It is difficult to distinguish valid scientific critique from those criticisms based on ideological perspectives and fear. Responsible scholarship and the best available, evidence-informed policies are required to address these well-publicized concerns about death determination and organ donation.

In any discussion about death, the profound emotional, psychological and spiritual impact that the loss of a loved one has on family and friends cannot be overstated. Discussions about death are deeply sensitive. Depending on circumstance, the death of a child may have overwhelming impact on families. There are a number of ethical issues related to death determination, including but not limited to, differences in concept, definition, criteria, meaning of the term 'irreversible', and the inexorable relationship between death determination and deceased organ donation. The ethics of death determination in children is largely informed by adult experience, as the availability of pediatric-specific research in death is limited. Despite these formidable challenges and limitations, it is important to understand how dying remains, first and foremost, a biological process. Death is an event in this biological process. In order to inform these discussions, this chapter will focus on the general biological aspects of organ support, replacement and transplantation that may both inform and complicate any discussion regarding the ethics of death determination.

5.2 The Biology of Dying and Death

Over the last 50 years, the advances in medicine, biology, and technology have been remarkable and helped us to save patients and better understand the biology of life and death. The specialties that have led to improving our understanding include: cardiopulmonary resuscitation and physiology; mechanical ventilation; cardiac surgery and cardiopulmonary bypass; Intensive Care Unit (ICU)-based life support; extracorporeal support and extracorporeal membrane oxygenation (ECMO); cell biology; and organ donation, preservation and transplantation. These advances have been truly astonishing in the collective effort to save lives. They have also informed, and complicated, how medicine and modern society understands what it means to be alive, dying or dead.

In the ICU, during the treatment of life-threatening illness, sustaining life is based on delivering oxygen and nutrients to cells, specifically, to the mitochondria of the cells. This process provides energy for metabolic processes required for life. Trillions of cells are grouped together and make up our organs and tissues—all distinct structures with very distinct functions. Vital organs have basic functions: for example, the lungs provide oxygen to the blood, the heart is the pump that circulates the blood containing oxygen, the liver metabolizes, and the kidney filters. The role of acute care and ICU professionals is to treat organ failure by recognizing life-threatening conditions and to intervene with life-sustaining treatments to maintain the delivery of oxygen to prevent death.

Technologies that support vital organs can sustain life in order for *time or treatment* to reverse the life-threatening condition. These complex, arduous and resource intensive treatments are extraordinarily successful, with survival rates ranging from 95 to 98 % in children and 80–90 % in adults. These treatments are fundamentally directed to provide oxygen delivery to the body. Without oxygen delivery, cells and organs stop working.

The dying process, which can be interrupted by life-sustaining treatment, is sequential and predictable. In general, death occurs by one of three physiological mechanisms [4]:

1. Primary respiratory illness/event causes breathing to stop, resulting in a fall in oxygen levels in the blood, which finally causes the heart to stop pumping which results in cessation of blood flow and oxygen to, and function of, all organs including the brain;
2. Primary heart disease such as a myocardial infarction—the heart arrests and cannot pump, resulting in cessation of blood flow and oxygen to all organs and especially the brain; or
3. Primary brain injury—the brain stops working and the brain's control of breathing is lost, breathing stops, oxygen drops and then the heart stops beating.

Remarkable advancements in life-sustaining technologies and transplantation permit the interruption of these dying processes by supporting or replacing failing organs, with the assumption that time and/or treatment (medical and/or surgical) will reverse the disease. Organs can now be supported by machines such as artificial hearts (ventricular assist devices), artificial kidneys (dialysis machines or blood filtration systems), breathing machines that effectively push oxygen into the blood stream, or artificial lungs that completely replace lung function. These treatments and technologies can be used inside the body or deployed external to the body. Examples of extracorporeal, or outside of the body technologies, include extra-corporeal membrane oxygenation (ECMO) [5] for respiratory failure or cardiac arrest, and heart-lung bypass machines used for open heart surgery. It is an incredible achievement to provide access to these complex heart/lung/kidney machines that can pump, circulate, oxygenate and filter blood. They can *completely* replace the total arrest of heart/lung/kidney and sustain function, potentially indefinitely. If that is the case, then how does one die?

These technologies serve as so-called '*bridges*' [6–8]. If the underlying life-threatening organ failure can improve with time or treatment, these technologies are '*bridges to recovery*'. If the failing organ cannot recover, they may become '*bridges to transplant*', but only if an organ transplant becomes available in time. In many unfortunate cases, when recovery is not possible and transplant is not indicated or is unavailable, these technologies effectively become onerous '*bridges to death*'. In this case, the technologies allow us to keep organs of the body working artificially, even when all effective treatment options are exhausted. Unfortunately, this is a circumstance many families find themselves in when a loved one has a non-recoverable illness and, based on the expert opinions of the health care team, must choose if it is time to stop life-sustaining treatment. In ICU's worldwide, a decision to withhold and withdraw life-support is the most common event preceding death [9]. These decisions regarding levels of life-sustaining treatment provided may vary and can include:

1. Withholding life-sustaining treatments. The following interventions may not be provided.
   a. Cardiopulmonary resuscitation.
   b. Initiating or escalating non-invasive or invasive mechanical ventilation.
   c. Initiating or escalating hemodynamic supports.
   d. Renal replacement therapies including dialysis.
   e. ECMO and/or mechanical circulatory assist.
   f. Antibiotics.
   g. Hydration and/or nutrition.
2. Withdrawing life-sustaining treatments. Interventions that have been initiated are withdrawn. The withdrawal of those interventions may occur gradually, sequentially or all at one time, depending on the circumstances and approach of the health care team.

In cases of non-recoverable illness, the goals of care may change from life sustaining to comfort measures, or a form of acute palliative care. In many countries, including Canada, the United States of America (USA) and most Western nations, the decision to limit or remove life-sustaining treatments can only be made with agreement by the family or surrogate decision-maker in accordance with the law and representative of the wishes and values of the patient.

5.3 Dead Donor Rule

While the advances in modern intensive care and life support technologies have both informed and complicated the line between life and death, clarity has been demanded by the concurrent success of organ transplantation. Through organ donation, transplantation saves lives. The majority of life-saving and life-preserving organ transplants occur through the process of deceased donation, whereby organs are removed after death has been determined. Patients must, by law, be declared dead prior to the removal of vital organs for the purposes of transplantation. This *dead donor rule* has been the central ethical, moral and legal requirement that has historically guided transplantation. Deceased organ donation can therefore occur when a person has been declared dead because either their heart beat (circulatory death) or brain function (brain death) has completely and permanently ceased. Calls to abandon the dead-donor rule [10] are related to the ethical, philosophical and scientific debates regarding the complexities of defining the moment of death. Proposals to redefine organ donation eligibility to a point before death may have some theoretical merit. However, they ignore the inaccuracies of prognostication and predicting those who inevitably will die, pose troubling clinical challenges and risk of abuse. In a voluntary system of organ donation based on public trust, the dead donor rule remains an indispensable ethical protection for dying patients [11].

5.4 Definitions of Death

The definitions of death proposed in the academic literature over the last several decades have focused on the irreversible loss of different capacities considered essential for life. Approaches to defining death can be grouped into three main categories as follows:

1. *Irreversible loss of functioning of the entire brain, including the brainstem.* This reflects the cessation of function of the organism as a "whole" (but not the whole organism) [12]. These are often cited in support of whole brain and lower brain (or brain-stem) criteria for death and examples of the underlying definition include "the irreversible cessation of the integrated functioning of an organism as a whole" [13] and "the permanent cessation of the critical functions of the organism as a whole" including the ability to breathe on one's own unassisted [14]. This is the most established definition and accounts for the most supported approach in the western hemisphere.

2. *Irreversible loss of the capacity for consciousness.* These are often cited in support of "higher brain" criteria for death (also known as cortical death) and definitions supporting such criterion may include: the loss of that which is considered to be essentially significant to the nature of humans where consciousness is the essential characteristic of human beings [15]; the irreversible loss of capacity for social interaction [16]; the total and irreversible extinction of consciousness and sensation, including discontinuation of actual survival of the individual personality [17]; or the irreversible loss of awareness that represents the loss of the person and signals human death [18].
3. *A combination of the previous two.* Examples may include: the irreversible loss of all capacity to integrate and to coordinate the physical and mental functions of the body; irreversible loss of the capacity for consciousness, combined with the irreversible loss of the capacity to breathe [19]; irreversible loss of the capacity for integrating human attributes with an integrative functioning of the body [20]; the loss of ability of the organism to work as expressed in its commerce with the surrounding world through receptivity to stimuli and signals, the drive to act and the ability to act [21].

To complicate these three proposed definitions, other authors conclude that it may not be possible to provide a definition of death and therefore definitions should be used only with the possibility to proceed with organ donation. Examples of these definitions include the currently applied criterion of irreversible (however defined) asystole, with the open admission that it does not define *death* but only a moment in the process of dying where organ retrieval can be allowed [22]. Some authors suggest abandoning the search for a definition of death [23] and focus on a pragmatic concept of *explantability window*, which allows us to discuss when explantation can be realized without committing ourselves to strong metaphysical and/or religious beliefs [24]. Lastly, others choose not to define death but to request that death should be properly established by formalized and set criteria [25].

Recent international work has advocated for focusing on an operational definition of human death that refers to a practical and concrete definition that describes the state of human death based on measureable and observable biomedical standards [26]. This operational definition does not attempt to explain death in religious, spiritual, philosophical or abstract terms. This definition includes movement away from anatomically-based terms such as brain death or cardiac death, which may erroneously imply the death of that organ. We argue that emphasis should be placed on the cessation of brain or heart function and the predominance of brain function for determination of death.

5.5 Death and Cessation of Circulation

Historically, there has been little need for precision in regard to death. Early medicine associated living with breathing and cessation of life was marked by unresponsiveness and the absence of respiration. The discovery of blood circulation by William Harvey in 1628 and the stethoscope in 1816 allowed the absence of heartbeat to be included in the determination. In the absence of organ donation, determination of death after cardiac arrest had remained rudimentary and of low rigour. Death has been declared upon a doctor's or coroner's determination. The criteria used were not articulated and remained untaught in training, ranging from absence of movement, breathing, heart sounds, pulse or electrocardiogram activity, applied at discretion of the attending physician. Observation and confirmation was not required and the irreversibility of death was not a practical concern, although diagnostic errors were made. Organ donation in general, and the practice of donation after circulatory death in particular, has by necessity enhanced the rigour of the determination of death [4].

An international literature review demonstrated that variability exists in the guidelines and statements for the determination of death amongst selected countries where donation after circulatory death (DCD) occurs, particularly in regard to the criteria, diagnostic procedures and wait times [27]. Where testing was specifically addressed, unresponsiveness, absent arterial pulse and apnea were most consistently recommended. Waiting periods ranged from 2 to 10 min, with a 5-min interval the most frequent.

5.5.1 Controversies in Cessation of Circulatory Function

A major criticism of DCD protocols has been the concern regarding the time of observation to determine death and the possibility that the loss of circulatory function is not 'irreversible' within the time limits proposed. There is ongoing academic debate between proponents of defining the term irreversible as meaning '*cannot be reversed under any circumstances*' versus permanent where circulation '*will not reverse under existing circumstances*' [28, 29].

While the irreversibility versus permanence debate continues, it is generally confined to academia. Most clinicians responsible for the declaration of death have accepted the permanence concept in death determination after circulatory arrest [29]. For the purposes of DCD, the circulatory arrest leads to absent brain blood flow. In humans [30–32] and animal studies [33], cortical brain function takes less than 20 s to stop after circulatory arrest. If the brain blood flow is quickly re-established, then cortical brain function may resume. It is not clear how long brain blood flow must be arrested to uniformly preclude reanimation of neurological function. However, any permanent absence of brain blood flow beyond 20 s will lead to permanent absence of brain function. This physiology has been accepted in the context of controlled DCD, where there has been a consensual decision to withdraw life-sustaining therapies (WLST) and to not provide cardiopulmonary resuscitation (CPR). As a note, uncontrolled DCD occurs in situations where CPR is initiated but after resuscitation efforts have failed. The physiologic changes to cortical brain function after circulatory arrest in the context of uncontrolled DCD is less clear. Fundamentally, and most relevant to DCD, the issue is not whether the body or brain circulation and function can be resumed (because it can), but rather, whether it will be. The Institute of Medicine [34] and the ethics committee of the American College of Critical Care Medicine [35] have addressed the ambiguity surrounding the term “irreversible” similarly.

As the permanent cessation of circulation has been accepted as fulfilling death determination standards, then the question of *autoresuscitation* (AR) becomes relevant. There have been case reports of spontaneous, unassisted resumption of heart function after cardiac arrest, ranging from seconds to minutes. Of importance is the distinction between the electrical activity versus the mechanical function of the heart, which generates circulation. In comatose patients who die after withdrawal of life-sustaining treatments, electrocardiogram activity can persist for many minutes after terminal cardiac arrest without resumption of circulation [36, 37]. Thus it is mechanical function, or the heart’s ability to generate blood flow and circulation that is of key importance to declaration of death for the purposes of organ donation.

The true incidence of AR and conditions that increase the potential for such an event are unclear and many reports are hampered by inadequate monitoring [31, 32]. Some cases are related to mistaken diagnosis [33, 38, 39]. In a retrospective study of 73 DCD patients (including eight children), no patients exhibited AR during the five-min observation period following asystole [40]. A systematic review of published studies [41] found no reported cases of spontaneous resumption of circulation after withdrawal of life-sustaining treatment (absence of CPR), as applies to controlled DCD practices. In eight studies that reported continuing electrocardiogram monitoring and exact times, AR did not occur beyond seven min after discontinued CPR, as may apply to uncontrolled DCD practices where donation is considered after unsuccessful CPR. Two subsequent case reports of AR in children were similar in that it occurred only after unsuccessful CPR. A recent prospective pilot study monitoring vital signs in pediatric and adult ICU deaths after withdrawal of life-sustaining treatments (absence of CPR) showed no evidence of AR beyond 89 s, well within the minimum of two-min observation period [37]. A larger prospective study is underway to confirm these preliminary results.

It appears that the chance of AR may be related to the provision of CPR. At minimum, the current discussion and criticisms of death determination related to DCD should distinguish between controlled (absence of CPR) and uncontrolled DCD (after termination of unsuccessful CPR). The vast majority of DCD worldwide is controlled, in situations where life support is withdrawn and CPR is not provided. To our knowledge, uncontrolled DCD has never been performed in children aged <14 years.

5.6 Death and the Cessation of Brain Function

Brain death is a concept that has been accepted as a medical and legal criterion for declaring death in many countries with advanced health care systems. While the incidence or prevalence of brain death worldwide is not known, it is estimated to constitute 1–2 % of all deaths in countries with advanced health care systems. The actual number will be affected by the incidence of catastrophic brain injury, availability of and access to intensive care services, neurosurgical and neuroprotective interventions, professional practice patterns and the development of deceased donation and transplant systems. A surrogate measure of brain death may be tracked through deceased donation rates. Worldwide, there are approximately 27,000 deceased donors per year in 70 countries with deceased donation programs [42]. This is a substantial underestimate of brain death as it excludes those catastrophically brain injured patients who did not have diagnostic criteria applied or have not consented organ donation. Twenty seven thousand brain death determinations annually is a substantive global clinical experience.

The concept of brain death (the idea that the cessation of brain function is an instance of death) arose as a result of two major advances in health care in the 1960s: firstly, the development of intensive care units with patients support with artificial airways and mechanical ventilators that treated irreversible apnea, thus interrupting the natural evolution from brain failure to cardio-circulatory death; and, secondly, to address the ethical concerns associated with organ donation arising from the then-new discipline of transplant surgery. Prior to the introduction of mechanical ventilators in the mid-20th century and the evolution of resuscitative measures, a non-brain or circulation formulation was used to determine death. Historical records indicate that 12th century rabbi and physician scholar Moses Maimonides suggested that the brain through control of breathing was of primary importance in sustaining life, when he noted that decapitated individuals would invariably die. The clinical appearance of brain death was first described in a seminal work by the French physicians Mollaret and Goulon in 1959 and termed *coma dépassé* meaning '*a state beyond coma*', which described 23 cases in which loss of consciousness, brain stem reflexes, and spontaneous respiration were associated with absent encephalographic activity [43]. In 1968, the Ad Hoc Committee of the Harvard Medical School, lead by neurologists Schwab and Adams, undertook to define irreversible coma and brain death [44]. They established a new, neurologically-based definition of death, defined as "*unresponsiveness and lack of receptivity, the absence of movement and breathing, the absence of brain-stem reflexes and coma whose cause had been identified*". Readers are urged to review the original text, from which the following selective quotes are instructive and relevant to this discussion: "from ancient times, when the respiration and heart stopped, the brain would die in a few minutes"; "in those times, the heart was considered to be central organ of the body"; "this is no longer valid when modern resuscitative and supportive measures are used"; "characteristics of a *permanently* nonfunctioning brain; no discernable central nervous system activity".

Subsequently, Mohandas and Chou emphasized the crucial role of the brainstem in the functioning of the entire brain [45]. This was then the focus of a published statement by the Conference of Medical Royal Colleges and Their Faculties in the United Kingdom (U.K.) in 1976 [46]. Championed by Pallis and Harley, the brainstem formulation of brain death was formally adopted in the U.K. in 1995 [19]. In the United States, however, the Uniform Determination of Death Act [13] codified the whole-brain formulation in stating that "*an individual who has sustained irreversible cessation of all functions of the entire brain, including the brainstem, is dead*". This formulation is the one most commonly applied worldwide and forms the foundation for legal codification in many Western nations.

The brain is responsible for our ability to breathe independently. A mechanical ventilator, the mechanism to sustain life after catastrophic brain injury, is based on replacing the lost ability to breathe. However, unlike all other organs, the vital and fundamental clinical functions of the brain, responsible for who we are, cannot be replaced or supported. The brain controls consciousness, awareness, sensation, movement, thinking, feeling and acting intentionally—as well as brainstem reflexes. Interaction and conducting exchange of information with our environment depends on brain function. This commercial exchange of information [21] may be as neurologically complex as a verbal response to a posed question or as rudimentary as the pupil of the eye constricting to light.

The most extreme form of brain injury is brain death. It is better characterized as *brain arrest* [47] which is the complete and permanent cessation of all clinical functions of the brain. *All clinical functions of the brain have been lost and they will never resume*—no ability to breathe independently, no capacity for any consciousness, no awareness, no sensations (no sight, smell, hearing, taste, pain, feeling), no thinking, no emotions, no acting, no brainstem reflexes and no interaction/exchange of information with the environment .... This can be better described as a dead person with a body/organs that still function. Brain death remains fundamentally based on the seminal work from the Harvard Ad Hoc Committee [44]. Death that is correctly diagnosed by neurological criteria [48] is not reversible—there is no chance of recovery of brain function [49].

5.6.1 Controversies in Cessation of Brain Function

There are well described variabilities in brain death diagnostic practices [50–52]. The differences in practice are generally related to procedural aspects of apnea testing, observation time for confirmation, age-specific criteria and the use of confirmatory, supplemental or ancillary testing. This variability may damage the credibility of the determination and adversely influence public and professional trust. It also serves as motivation to improve consistency and standardize practices [47, 48].

A deeper analysis reveals a remarkable and laudable uniformity of the fundamental clinical criteria for brain death [47, 48, 53–55]. These fundamentals of practice based on the Harvard criteria are remarkably consistent across the world and include:

1. Unresponsive coma with an established etiology.
2. Absence of reversible conditions.
3. Absence of cortical or brainstem mediated motor responses.

4. Absent brainstem reflexes.
5. Loss of the capacity to breathe.

Thus the apparent discrepancy between the U.S.A., based on so called '*whole brain death*' criteria and the U.K., based on '*brainstem death*' criteria, misrepresent the fact that both are based on these fundamental elements signaling the cessation of brain function. These are not divergent views or concepts especially when both countries understood brainstem death as the infratentorial consequence of a supratentorial catastrophe [53]. Work is in process to develop international consensus in this regard [26].

These are the clinical requirements to confirm the absence of whole brain function. Depending on the country, different medical, legal, cultural or religious factors may oblige fulfilling requirements beyond this consistent minimum standard. Conceptually, this means that the person can be declared dead as a result of complete and irreversible cessation of all brain function. However, their organs, which remain inside their bodies, may be kept alive. This is the most important premise for most deceased donation practices, where the patient is declared dead by neurological criteria and only then can the supported organs be eligible for donation.

In pediatrics, there may be unique requirements due to the relatively lower clinical experience, particularly in infants and newborns. These vary according to guidelines and may include extended waiting periods prior to the determination in newborns, extended time intervals between exams [55]. While these pediatric specific considerations may be cautious enhancements to the determination, they have been criticized as unnecessarily complicated [56].

Given these complexities, it is easy to comprehend the source of public's and non-specialist health professionals' ongoing misunderstanding [3] regarding brain death. Confusion has arisen from recent media cases of brain death in pregnancy [57]. After brain death in a pregnant mother, it is possible to sustain organ function with mechanical breathing, infection control, hormone replacement and diligent ICU care for long periods of time to allow for fetal development to mature birth [58]. Effectively, these pregnant but brain dead mothers serve as life support systems for the baby, similar to extracorporeal life support systems such as ECMO, until the fetus is viable. It does not change the medical and legal fact that the mother remains a dead person with an artificially sustained body to allow the baby to be delivered, after which organ support systems are terminated.

5.7 The Unification of Circulatory and Brain Based Determinations of Death

The emergence of modern intensive care, in the face of advancing organ support and replacement technologies has challenged our ability to identify the margins of human physiology. Organs may be alive or dead in a living person, alive or dead in a dead person, non-functioning in a living or dead person, or resuscitated outside of the body of a dead person. For example, the reversibility of circulatory arrest is now a function of the availability and application of support systems such as prolonged CPR and ECMO to maintain oxygenated circulation [59, 60]. Complete and irreversible arrest of the heart may no longer be death, as long as oxygenated circulation to the body can be provided mechanically. In addition, ex vivo or external to the body organ support systems are based on the same physiological principles of providing oxygenated circulation to the organ. These technologies now permit the salvage of hearts [61] and lungs [62] that may no longer function well inside the body of a dead person, but can be reanimated in a laboratory outside of the body prior to transplantation.

These technological advances are fundamentally changing the prevailing historical understanding of the essential aspect of personhood from the heart to the brain. It has become brain function—not heart or circulatory function—as defining the critical margin between life and death [26, 63]. There is an inextricable link between circulation and brain function. Similar to all vital organs, brain function depends on the provision of oxygenated circulation generated by heart and lung function. Most cases of brain death are also associated with an absence of blood flow to the injured brain due to severe elevation of intracranial pressure. Similarly, after circulatory arrest, the brain has lost its blood flow. Following cardiac arrest in humans, loss of consciousness occurs in 5–6 s [64]. In human studies of brief cardiac arrest, the electroencephalogram becomes isoelectric within 7–20 s [31–33]. Thus cortical brain function stops working within seconds after circulation to the brain ceases.

As long as circulation will not resume, i.e. is permanently lost, and the period of time for autoresuscitation has passed, then brain function has permanently ceased and will not resume. The longer the time period without oxygenated circulation to the brain, the progressively higher likelihood that the cessation of brain function is irreversible, even if oxygenated circulation can be re-established (either spontaneously or through intervention). The precise time period for the complete cessation of brain function to be non-reversible and non-resuscitatable (through intervention) after circulatory arrest is unresolved but may vary depending on factors such as baseline neurological function, duration of anoxia, temperature and the effectiveness of restoring oxygenated circulation to the brain if attempted. If circulation will not resume spontaneously and will not be restored by intervention, brain function cannot resume and the final end point reached is the permanent cessation of brain function.

Public and professional misunderstanding is exacerbated by imprecise terminology and inflammatory language used in academic discourse [1]. Simplifying the concepts involved, and making the definitions accessible and comprehensible to the lay and professional public, while accounting for evolving knowledge and technological capacity, is essential. An emerging definition of death should be based on the cessation of function (the primary and fundamental purpose of an organ that can be assessed by observation and examination and is necessary for sustained life) rather than activities (physiologic properties of cells and groups of cells that can be measured by laboratory means) [65]. Emphasis is increasingly being placed on the predominance of brain function for determination of death, even after circulatory arrest. Death is becoming a single phenomenon based on the permanent cessation of brain function (loss of capacity for consciousness and brainstem reflexes) with two mechanisms to reach that point: (i) permanent absence of circulation or (ii) subsequent to a catastrophic brain injury—two entrances, one exit [26].

5.8 Conclusion

While the dead donor rule remains an indispensable ethical foundation for deceased donation practices, there will continue to be ongoing debate about the line between life and death. The interface of life saving interventions, life preserving interventions, organ failure support, death determination and organ donation/transplantation continues to challenge and force the evolution of our notions and concepts. In order to better inform ethical debate, this chapter reviewed the biology of the dying process in the face of these remarkable advances in science and medicine. It is anticipated that the central and primary role of brain function in life and death will gain even greater emphasis over time. There is a need to enhance pediatric specific research in regard to the biology of death.

References

1. Naffine N, Richards B, Rogers W. Scaring us all to death: the need for responsible legal scholarship on post-mortem organ donation. J. Law and Med. 2009;16:696–707.
2. Shah SK, Kasper K, Miller FG. A narrative review of the empirical evidence on public attitudes on brain death and vital organ transplantation: the need for better data to inform policy. J Med Ethics. 2014. doi:10.1136/medethics-2013-101930. [Epub ahead of print].
3. DuBois JM, Anderson EE. Attitudes toward death criteria and organ donation among healthcare personnel and the general public. Prog Transplant. 2006;16(1):65–73.
4. Shemie SD. Clarifying the paradigm for the ethics of donation and transplantation: was 'dead' really so clear before organ donation? Philos Ethics Humanit Med. 2007;24:2–18.
5. Gray BW. Extracorporeal life support: experience with 2000 patients. ASAIO J. 2014. [Epub ahead of print].
6. Ashraf O, Sharif H. Cardiac failure, transplantation and donation: current perspectives. J Cardiovasc Surg 2013. [Epub ahead of print].
7. Ibrahim M, Yacoub MH. Bridge to recovery and weaning protocols. Heart Fail Clin. 2014; 10(1 Suppl):S47–55. doi:10.1016/j.hfc.2013.08.004.
8. Holley CT, Harvey L, John R. Left ventricular assist devices as a bridge to cardiac transplantation. J Thorac Dis. 2014;6(8):1110–9. doi:10.3978/j.issn.2072-1439.
9. Sprung CL. End-of-life practices in European intensive care units: the ethicus study. JAMA. 2003.13;290(6):790–7.
10. Truog RD, Miller FG, Halpern SD. The dead-donor rule and the future of organ donation. N Engl J Med. 2013;369(14):1287–9.
11. Bernat JL. Life or death for the dead-donor rule? N Engl J Med. 2013; 369(14):1289–91.

12. Bernat JL, Culver CM, Gert B. On the definition and criterion of death. Ann Intern Med. 1981;94(3):389–94.
13. President's Commission for the Study of Ethical Problems in Medicine and Biomedical and Behavioral Research. Defining death: a report on the medical, legal and ethical issues in the determination of death. 1981. Accessed 16 Feb 2012. http://bioethics.georgetown.edu/pcbe/reports/past_commissions/defining_death.pdf.
14. Bernat JL. A defense of the whole-brain concept of death. Hastings Cent Rep. 1998;28:14–23.
15. Veatch R. Death, dying and the biological revolution: our last quest for responsibility: new haven: Yale University Press; 1976.
16. Veatch RM. Defining death: the role of brain function. JAMA. 1979;242:2001–2.
17. Walton DN. On defining death. An analytical study of the concept of death in philosophy and medical ethics: montreal: McGill-Queens University Press; 1979.
18. Green MB, Wikler D. Brain death and personal identity. Philos Public Aff. 1980;9:105–33.
19. Pallis C. Brainstem death. In: Braakman R, editors. Handbook of clinical neurology: head injury. BV. Amsterdam: Elsevier Science Publisher, 1990;13 (57):441–496.
20. Machado C. Consciousness as a definition of death: its appeal and complexity. Clin Electroencephalogr. 1999;30:156–64.
21. Controversies in the Determination of Death: a white paper by the president's council on bioethics. 2008.http://www.bioethics.gov:80/reports/death/index.html. Accessed Feb 2012.
22. Zamperetti N. Defining death in non-heart beating organ donors. J Med Ethics. 2003;29:185.
23. Chiong W. Brain death without definitions. Hastings Cent Rep. 2005;35(6):20–30.
24. Boniolo G. Death and transplantation: let's try to get things methodologically straight. Bioethics. 2007;21:32–40.
25. Byk C. The European protocol on organ transplant: key issues. J Int Bioethique. 2009;20: 119–33.
26. Shemie SD, Hornby L, Baker A, Teitelbaum J, Torrance S, Young K, Capron AM, Bernat JL, Noel L. The international guidelines for determination of death phase 1 participants, in collaboration with the World Health Organization. International guideline development for the determination of death. Intensive Care Med. 2014;40(6):788–97. doi:10.1007/s00134-014-3242-7 Epub 2014 Mar 25.
27. Dhanani S, Hornby L, Ward R, Shemie S. Variability in the determination of death after cardiac arrest: a review of guidelines and statements. J Intensive Care Med. 2012;27:238–52.
28. Bernat JL, Capron AM, Bleck TP, Blosser S, Bratton SL, Childress JF, DeVita MA, Fulda GJ, Gries CJ, Mathur M, Nakagawa TA, Rushton CH, Shemie SD, White DB. The circulatory-respiratory determination of death in organ donation. Crit Care Med. 2010;38 (3):963–70.
29. Bernat JL. How the distinction between "irreversible" and "permanent" illuminates circulatory-respiratory death determination. J Med Philos. 2010;35(3):242–55.
30. Clute HL, Levy WJ. Electroencephalographic changes during brief cardiac arrest in humans. Anesthesiology. 1990;73:821–5.
31. Losasso TJ, Muzzi DA, Meyer FB, Sharbrough FW. Electroencephalographic monitoring of cerebral function during asystole and successful cardiopulmonary resuscitation. Anesth Analg. 1992;75:1021–4.
32. Moss J, Rockoff M. EEG monitoring during cardiac arrest and resuscitation. JAMA. 1980;244:2750–1.
33. Bircher N, Safar P, Eshel G, Stezoski W. Cerebral and hemodynamic variables during cough-induced CPR in dogs. Crit Care Med. 1982;10:104–7.
34. Institute of Medicine. Non heart beating organ transplantation: practice and protocols. http://www.iom.edu/reports/2000/non-heart-beating-organ-transplantation-practice-and-protocols.aspx. Accessed 13 Oct 2014.
35. Recommendations for nonheartbeating organ donation. A position paper by the ethics committee, american college of critical care medicine, society of critical care medicine. Crit Care Med. 2001;29:1826–31.

36. Wijdicks EF, Diringer MN. Electrocardiographic activity after terminal cardiac arrest in neurocatastrophes. Neurology. 2004;62:673–4.
37. Dhanani S, Hornby L, Ward R, Baker A, Dodek P, Chamber-Evans J, et al. vital signs after cardiac arrest following withdrawal of life-sustaining therapy: a multicenter prospective observational study. Crit Care Med. 2014 May 7.
38. Hossmann KA, Kleihues P. Reversibility of ischemic brain damage. Arch Neurol. 1973;29:375–84.
39. Steen PA, Gisvold SE, Milde JH, et al. Nimodipine improves outcome when given after complete cerebral ischemia in primates. Anesthesiology. 1985;62:406–14.
40. Sheth KN, Nutter T, Stein DM, Scalea TM, Bernat JL. Autoresuscitation after asystole in patients being considered for organ donation. Crit Care Med. 2012;40:158–61.
41. Hornby K, Hornby L, Shemie SD. A systematic review of autoresuscitation after cardiac arrest. Crit Care Med. 2010;38:1246–53.
42. Global Observatory on Donation and Transplantation. 2012. http://www.transplant-observatory.org/Pages/Facts.aspx. Accessed 12 Aug 2014.
43. Mollaret P, Goulon M. Le coma dépassé. Rev Neurol. 1959;101:3–15.
44. A definition of irreversible coma. Report of the ad hoc committee of the harvard medical school to examine the definition of brain death. JAMA. 1968;205(6):337–40.
45. Mohandas A, Chou SN. Brain death. A clinical and pathological study. J Neurosurg. 1971; 35(2):211–8.
46. Diagnosis of brain death: Statement issued by the honorary secretary of the conference of Medical Royal Colleges and their Faculties in the United Kingdom on 11th October 1976. Ann R Coll Surg Engl. 1977;59(2):170–2.
47. Shemie SD, Doig C, Dickens B, et al. Severe brain injury to neurological determination of death: Canadian forum recommendations. CMAJ. 2006;174:S1–13.
48. Wijdicks EF. Evidence-based guideline update: determining brain death in adults: report of the quality standards subcommittee of the American academy of neurology. Neurology. 2010;74:1911–8.
49. Wijdicks EF. There is no reversible brain death. Crit Care Med. 2011;39(9):2204–5. author reply 2206.
50. Wijdicks EF. Brain death worldwide: accepted fact but no global consensus in diagnostic criteria. Neurology. 2002;58:20–5.
51. Hornby K, Shemie SD, Teitelbaum J, Doig C. Variability in hospital-based brain death guidelines in Canada. Can J Anaesth. 2006;53:613–9.
52. Powner DJ, Hernandez M, Rives TE. Variability among hospital policies for determining brain death in adults. Crit Care Med. 2004;32:1284–8.
53. Wijdicks EF. The transatlantic divide over brain death determination and the debate. Brain. 2012;135(4):1321–31. doi:10.1093/brain/awr282 Epub 2011 Dec 24.
54. The ANZICS statement on death and organ donation. 2008. http://www.anzics.com.au/uploads/ANZICSstatementfinal26sept08.pdf Accessed Feb 2012.
55. Nakagawa TA, Ashwal S, Mathur M, Mysore M. Clinical report-guidelines for the determination of brain death in infants and children: an update of the 1987 task force recommendations. Pediatrics. 2011;128:e720–40.
56. Wijdicks EF, Smith WS. Brain death in children: why does it have to be so complicated? Ann Neurol. 2012;71(4):442–3.
57. ABC news. Brain dead and pregnant. Why one baby was born and one wasn't. http://abcnews.go.com/blogs/health/2014/02/12/brain-dead-and-pregnant-why-one-baby-was-born-and-one-wasnt/. Accessed 25th Sept 2014.
58. Powner DJ, Bernstein IM. Extended somatic support for pregnant women after brain death. Crit Care Med. 2003;31(4):1241–9.
59. Massetti M., et al. Back from irreversibility: extracorporeal life support for prolonged cardiac arrest. Ann Thorac Surg. 2005;79(1):178–83; discussion 183–4.
60. Stub D, Bernard S, Pellegrino V, Smith K, Walker T, Sheldrake J et al. Refractory cardiac arrest treated with mechanical CPR, hypothermia, ECMO and early reperfusion (the CHEER

trial). Resuscitation. 2014. doi: http://dx.doi.org/10.1016/j.resuscitation.2014.09.010. [E-pub ahead print.].
61. Iyer A, Gao L, Doyle A, Rao P, Jayewardene D, Wan B, et al. Increasing the tolerance of DCD hearts to warm ischemia by pharmacological postconditioning. Am J Transplant. 2014 Aug. 14(8):1744-52. Ex vivo heart (pending).
62. Van Raemdonck D, Neyrinck A, Cypel M, Keshavjee S. Ex-vivo lung perfusion. Transpl Int. 2014 Mar 15. doi:10.1111/tri.12317 [Epub ahead of print].
63. Baker A, Shemie SD. Biophilosophical basis for identifying the death of a person. J Crit Care. 2014;29(4):687–9. doi:10.1016/j.jcrc.2014.04.013 [Epub 29 Apr 2014.].
64. Rossen R. Acute arrest of cerebral circulation. Arch Neurol Psychiatry. 1943;50:510–28.
65. Bernat, JL. Refinements in the definition and criterion of death. The definition of death: contemporary controversies. In: Youngner SJ, Arnold SJ, Schapiro R, Baltimore R, editors. The Johns Hopkins University Press.1999. 83–92.

Chapter 6
Controlled Donation After Cardiac Death in Pediatrics

Erin Paquette and Joel Frader

Abstract Donation after circulatory death (DCD) is a relatively new topic in pediatric ethics. This chapter explores the ethical dimensions of controlled DCD, first examining the distinction between donation following brain death and donation following controlled circulatory death. While uncontrolled donation following circulatory death increases, a detailed consideration of the unique issues in that setting is beyond the scope of this chapter. We describe the background of and challenges to the dead donor rule, including debate over whether it comprises the appropriate ethical or legal basis for organ donation. With respect to donation in pediatrics, we examine the DCD process from the standpoint of multiple stakeholders, including parents/family members, clinicians, and organ procurement organizations. Finally, we explore the role of national and organizational regulations and guidelines in systematically addressing ethical issues with DCD practices. This discussion gives special attention to parental involvement and consent. Lastly, we consider the role for DCD protocols as additional mechanisms to manage the many ethical issues that arise with pediatric DCD.

Keywords Organ donation · Donation after circulatory death · Pediatric ethics

6.1 Introduction—Use and Outcomes

The practice of organ donation has evolved substantially since the 1960s. At that time, the concept of brain death emerged, and while early attempts at organ donation involved organ retrieval after circulatory death, the growing acceptance of the concept of brain death dominated discussions about the source of organs for transplantation over the next several decades. In the 1990s, at least partly in response to a growing shortage of available organs, renewed interest emerged in

E. Paquette (✉) · J. Frader
Northwestern University, Ann & Robert H. Lurie Children's Hospital of Chicago, Chicago, IL, USA
e-mail: epaquette@luriechildrens.org

R.A. Greenberg et al. (eds.), *Ethical Issues in Pediatric Organ Transplantation*, International Library of Ethics, Law, and the New Medicine 66,
DOI 10.1007/978-3-319-29185-7_6

obtaining organs after circulatory death. The terms used to describe organ donation following circulatory determinations of death have varied. Non-heart-beating organ donation (NHBD), donation after cardiac death/donation after cardiocirculatory death/donation after circulatory death (DCD), and donation after circulatory determination of death (DCDD) all refer to the same practice of organ removal from non-brain dead donors. Most commonly referred to as DCD or DCDD, organ donation in this setting follows the "irreversible" cessation of circulatory and respiratory functions [1].

In the last decade, numerous centers have reported their experiences with DCD. These include experiences with liver, kidney, pancreas, lung, and heart transplantation. Individual center experiences in the adult population suggest similar organ survival and post-transplant complication rates to donation after brain death for transplanted kidneys, pancreases, and lungs [2–6]. Success rates for transplantation of livers following DCD vary more, with early positive reports and more recent concern about increased complication rates and more graft failure [2–4, 6].

The renewed attention to DCD using adult donors soon sparked interest in the practice in pediatrics, given the growing gap between need for and availability of organs for pediatric transplantation. This gap presents a particular challenge for some candidates due to their small size and the need for size-appropriate organs [7]. Advocates for pediatric DCD suggest developing policies and protocols to guide hospitals in implementing DCD programs for children [8]. Some institutional experiences with pediatric DCD indicate a large potential pediatric donor pool, while other studies indicate that a small percentage of dying pediatric patients could become donors [9]. Similar to the experience in adult medicine, reports generally note successful pediatric kidney transplantation with organs from DCD donors [7, 10]. Other pediatric studies also suggest similar success using DCD cardiac, pancreas, and liver donors when compared to donation following brain death donors (DBD) [7]. DCD in some pediatric populations has raised greater controversy, including transplantation of organs from neonates and transplantation of hearts from pediatric patients following cardiocirculatory death [11–15]. Single institution studies report a significant potential donor pool for liver, kidney, and cardiac transplant from deaths in neonatal intensive care units, ranging from 4 to 8 % of deaths during study periods [11, 12]. Importantly, a large proportion of pediatric organs obtained following DCD processes, and kidneys in particular, are transplanted into adult recipients [16].

While some argue that the use of pediatric organs in adult recipients disfavours children, others point out that any increase in organ donation shortens waitlists for pediatric recipients awaiting organs [17]. Moreover, these authors advise caution in perceiving organs as "belonging" to a certain subgroup of recipients [17]. Additionally, there are separate lists for certain organs for adults and children, and, in other situations, children may be given priority for organs. Importantly, allocation of organs from pediatric donors is only one of myriad issues that arise in the use of DCD in the pediatrics [18]. The most discussed ethical issues involve defining death; debate over the appropriate interval between the determination of death and procurement of organs (from 1.25 min to up to 20 min) [15, 19]; the use of pre-mortem interventions, including anticoagulation and insertion of vascular catheters; and changes in categories of DCD. Additional concerns have included the appropriateness of DCD versus other donation mechanisms given the age of the recipient, the adequacy of parental consent for pediatric organ donation, the impact of DCD on the dying process and family grief, and public perceptions of the use of organs from children [20, 21]. The next section addresses these ethical issues in detail; subsequent sections consider the practice of DCD from the perspectives of families, health care providers, and society, as well as the impact of regulatory and legal frameworks on the practice of DCD. Although protocols guiding DCD in pediatrics are widespread, we give only limited attention to these, as the matter receives extensive attention elsewhere in this volume.

6.2 Ethical Issues in DCD

Before thinking about the ethical issues surrounding the practice of organ donation after cardiac death, it seems reasonable to consider the appropriateness of organ donation for a given patient. Although many children wait for organs and risk both additional morbidity and death with every day on a waiting list, it is important to consider the benefits and burdens of transplantation for each patient and the timing of transplantation, as typically the children themselves do not choose to undertake the responsibilities and burdens that transplantation involves. Listing clinicians expect the potential benefits of transplantation to outweigh these burdens, but responsible organ stewardship requires thought about maximizing the use of (scarce) available organs [22]. DCD raises questions from the standpoint of the donors, health care team members, families, and society. Chronologically, these issues involve how to determine death once a decision has been made to proceed with DCD, followed by the impact of DCD on the dying process, and conclude with consideration of the suitability of DCD organs for subsequent transplantation.

6.2.1 The Dead Donor Rule

The Hippocratic tradition informs subsequent codes of medical ethics for physicians, imparting on physicians a duty to avoid harm to their patients, and, a corresponding duty not to participate in killing of patients [23]. These are the core values that underlie the dead donor rule [23]. The dead donor rule (DDR) requires that clinicians caring for a patient (i.e., not clinicians involved with caring for potential transplant recipients) should, prior to organ procurement, determine a potential donor to be dead. In the United States, the President's Commission on Bioethics addressed the definition of death in the 1980s [24]. The Commission ultimately endorsed the Uniform Determination of Death Act (UDDA), which describes two approaches for determining death using accepted medical standards.

According to the UDDA, "[a]n individual who has sustained either (1) irreversible cessation of circulatory and respiratory functions, or (2) irreversible cessation of all functions of the entire brain, including the brain stem, is dead" [24]. Under the UDDA, organ procurement is possible after an individual has met, at least, one of these criteria for death. For DCD, this requires a determination of irreversible cessation of circulatory and respiratory functions. While other chapters in this volume focus on issues in defining death, how clinicians and the public view what constitutes death is a core matter for the acceptability of donation following circulatory or brain death, and the following aims to highlight the principal areas of conflict.

Advocates for the DDR point out normative, legal, and practical justifications for its existence. These justifications support the DDR using either neurologic (brain death) or circulatory criteria to determine death, ultimately advocating that the declaration of death must precede the procurement of organs for transplantation [25–27]. Two deontological principles underlie the DDR. The first stems from the moral imperative that doctors should not kill [28, 29]. Removal of organs from an individual before the individual is declared dead would result in the individual's death, thereby violating the moral imperative that prohibits doctors from killing. The DDR avoids this problem by ensuring the patient has been declared dead prior to organ procurement. A second deontological justification for the DDR relates to worry about exploitation of vulnerable populations. Deontological principles teach that an individual should be seen as an end in herself and never as merely a means to some other aim. Adhering to the DDR guards against the possibility of using an individual as a means only (i.e., a source of organs) by preventing the procurement of organs until there has already been a determination of death [28]. Others have argued that the DDR's prohibition on exploitation of others has consequentialist underpinnings in that the rule avoids a slippery slope towards allowing killing for social purposes [30].

The DDR also holds legal weight. This rule was implicitly incorporated into the Uniform Anatomical Gift Act, which most states have subsequently made law [31]. Adoption of the DDR bars the procurement of organs prior to determination of an individual's death, and puts ethical and medical questions about determining death at the center of controversies over donation following either brain death or cardiac death. Practically, advocates of the DDR argue that restricting organ procurement until after a proper determination of death has occurred assures individuals, and the public generally, that clinicians will not withdraw life-sustaining therapies simply to obtain organs. This assurance may be false, however, as it could be possible that a withdrawal is intended to allow a timely determination of death in order to make possible organ procurement. Thus, it is not clear that the dead donor rule practically functions as advocates hope it does.

6.2.2 Problems with the Dead Donor Rule

The DDR has been criticized because, some claim, it actually fails to define death by either circulatory or neurological criteria. With respect to cardiac death, numerous scholars point out that the requirement for "irreversible cessation" of circulatory and respiratory functions cannot be met [32]. This is because circulatory and respiratory functions can often be restored with resuscitative efforts, even up to twenty minutes following cessation of these functions [25, 32]. Some countries have developed practices based on this possible resuscitation period, not allowing organ procurement for up to 20 min following the cessation of circulatory and respiratory functions [33]. In rare circumstances, circulatory function may return spontaneously without resuscitation, a concept known as autoresuscitation [25]. There are variable reports of time to autoresuscitation, with the majority of reports in adults, raising debate over the acceptable interval between declaring death by circulatory criteria and procuring organs [25, 34, 35]. Additionally, critics point out that cardiac transplantation following DCD is inherently inconsistent with irreversible cessation of the heart, meaning that both maintenance of the DDR and current definitions of death require either reliance on (i) cessation of *circulatory* rather than *cardiac* functions *or* (ii) limiting cardiac donation to deaths only by neurological criteria [36].

In attempting to defend the determination of death by circulatory criteria against the challenge to irreversibility, and to preserve the DDR, two arguments have emerged. The first is that determining death by circulatory criteria remains valid because the cessation of circulatory functions is permanent after a window for possible autoresucitation (even if not fully irreversible) if the individual does not intend resuscitation [32]. Advocates for this view argue that permanence is a perfect surrogate for irreversibility once the timeframe for autoresuscitation can no longer occur [37]. The second argument appeals to the physiology of the dying and dead person. From this perspective, while the cessation of circulatory functions may be reversible after a determination of death by circulatory criteria, shortly after circulatory death the interruption of circulation to the brain results in an *irreversible* cessation of brain functions, thereby rendering the individual dead by neurologic criteria, even if circulation were to be restored [25]. This claim that death by circulatory criteria leads to death by neurologic criteria requires the fulfillment of two conditions. First, enough time must elapse from the cessation of circulation and prior to organ removal such that complete cessation of brain function can occur [38]. Second, any resuscitative efforts that might restore blood flow must not be permitted, even efforts to perfuse the organs intended for donation, because they may undermine progression towards neurological death [25].

6.2.3 *Abandoning the Dead Donor Rule?*

Many critics of the DDR argue that abandoning the rule will not require changes in current organ donation practices [32]. Challengers of the DDR make arguments on moral, legal, and practical grounds. Moral challenges include the view that removal of organs prior to death, when this act will lead to an individual's death in the setting of intended organ donation, is not morally different from withdrawal of life-sustaining medical treatment from individuals who refuse such treatment [28, 39]. These critics challenge the validity of the moral distinction between killing and allowing to die that underpins traditional arguments in favour of the DDR. Supporters of the DDR respond that when clinicians withdraw life-sustaining treatment based on the wishes of a patient or family, the cause of the individual's death is the underlying condition rather than the withdrawal of treatment. Some critics of the DDR, however, point out that from the standpoint of legal causation, withdrawal of life-sustaining treatment, for *whatever* reason, leads to one's death, and that therefore, this act causes death, rejecting the distinction between killing and allowing to die [39]. They go on to argue that while withdrawal of life support is not morally different from killing, legal justifications for permitting withdrawal of life support do not hinge on whether this practice causes death, but rather on the right of the individual person to accept or refuse medical treatment [39]. They conclude that a decision regarding organ donation following a decision to withdraw life-sustaining treatment would be justified based on this reasoning, even without the DDR, and thus, argue that the DDR does not serve a legitimate purpose [28, 39]. Critics also challenge the argument that abandoning the DDR would lead to exploitation of persons, treating them as a merely a means to others' ends. According to this view, the concept of informed consent transforms a wrongful use of an individual as a means into a morally permissible action, because the consenting donor desires the death to accomplish donation [28]. By limiting organ donation to patients who have previously planned to withdraw life-sustaining treatment, proponents of this view suggest that exploitation is avoided.

Those advocating for abandoning the DDR also argue that there are significant legal implications of maintaining the DDR and permitting organ donation practices to continue under current definitions of death. Some argue that current DCD practices violate criminal homicide statutes [40]. As noted above, these scholars argue that determining death by circulatory criteria does not assure irreversibility and, therefore, such individuals are dying, but not yet dead. Thus, removing organs from individuals who are not yet dead, equates to killing under the law. Those who reject the dead donor rule, however, argue that even though such life-ending practices are justified, it remains illegal to remove organs from these individuals if the DDR remains a normative requirement for vital organ procurement [41]. For legal reasons, proponents of the view that removal of organs from an individual who intends to donate them is justified (irrespective of a prior declaration of death) argue for abandoning the DDR, noting the practical value of such a change in the law and the potential implications for public trust of current organ procurement practices [28].

6.2.4 *Legal Fictions and the Dead Donor Rule*

Some legal scholars contend that an alternative to abandonment of the DDR would be to accept that although the legal definitions of death may not conform to biological conceptions of death, the use of legal fictions allows maintenance of the status quo in organ donation. Proponents of this view explain that legal fictions can be helpful in allowing the law to deal with difficult situations by accepting as true in law something that is in fact not true. An often cited example of a useful legal fiction is the idea that corporations are persons [31]. In the context of defining death, this view suggests that an individual is dead for legal purposes if he or she meets medically acceptable standards for defining death, even if not meeting the biological definition of death. Under this framework, even if there are legitimate challenges to the idea that an individual who is declared dead by neurologic or cardiac criteria is in fact dead, the legal fiction allows the individual to be treated as if dead. This framework allows the DDR to remain in place, while acknowledging that current definitions of death do not fully comport with generally accepted biological definitions of death [31]. In the United States, the most recent national consideration of the ethics of organ donation supports maintenance of the DDR despite its legal and ethical challenges [42].

6.2.5 *Variability in Declaring Death Under the Dead Donor Rule*

With the DDR remaining the status quo in many jurisdictions, practical issues related to declaring death carry potentially important ethical implications. There is an inherent tension between ethical obligations to respect the potential donor by declaring death with certainty prior to organ retrieval and the ethical obligation to optimize the chances for successful organ transplantation. This tension often reduces to debate over the appropriate time interval between pulselessness and initiating organ procurement. There is considerable variability in how cessation of circulatory function is actually assessed, prompting many to advocate for objective criteria for determining circulatory death. Empirically, there is considerable variation in the interval institutions use from asystole to organ procurement, with some programs accepting less than two minutes and others requiring 10–20 min before declaring death and allowing organ retrieval to proceed [43, 44]. Also, the inability to accurately predict whether an individual will have cessation of circulatory functions within a reasonable period following withdrawal of life-sustaining treatment adds a layer of complexity [45]. Bernat et al., who hold one of the most recognized views on the irreversibility requirement under the DDR, suggest a minimum of 2 min and preferably greater than 5 min waiting period, arguing that no known cases of autoresuscitation occur after this point [37]. In children, the data on autoresuscitation are less well reported [35]. Others criticize strict time limits, maintaining that data on autoresuscitation is lacking and, furthermore, that strict adherence to this view may undermine the desires of those wishing to become organ donors, as longer waits many render organs unsuitable for transplantation. These critics advocate for transparency in the process of consent for donation, rather than strict adherence to any particular timeframe [46].

6.2.6 Beyond Definitions—Additional Ethical Concerns

In addition to defining death, concerns related to DCD include how the process impacts the experience of death. Some scholars contend that adherence to the DDR compromises the fidelity of the physician-patient-family relationship at a time when trust matters most, i.e., when the physician must support the dying (but not-yet-dead) patient and the patient's loved ones [47]. Others have noted that the practice of DCD could alter the experience of death by (i) removing the family from the patient's side at the time of death (remembering that common practice in pediatrics involves allowing parents to hold and remain with the child's body for as long as they wish), (ii) rushing a declaration of death for the purposes of optimizing organ procurement and subsequent organ function, and (iii) possibly shortening the interval between consideration of DCD and the subsequent time of death [42]. At least one study suggests that there is no difference between donors and non-donors in survival time, although this remains an empirical concern with ethically relevant consequences [48].

A separate concern relates to using organ preservation measures (anticoagulation, insertion of catheters, and/or use of vasoactive medications) after a decision to attempt DCD but before the declaration of death and the beginning of organ procurement. Some worry that a decision to pursue DCD may compromise comfort care in favour of measures to either hasten death or ensure organ perfusion and subsequent organ function [49]. While some studies describing organ preservation measures describe specific consent for using organ preservation measures, others are less clear about whether and how consent was obtained, heightening ethical concerns with such practices [50].

One final ethical issue related to DCD is the emerging practice of uncontrolled donation after cardiac death. DCD usually takes place in a controlled setting, where the timing of the withdrawal of life-sustaining treatments is known and, therefore, the process leading up to organ procurement is orderly. Uncontrolled DCD refers to the process of procuring organs for transplantation following unsuccessful resuscitation for out-of-hospital cardiac arrest. The practice has largely been reported in adults, although, it is also permitted in children in Spain [51]. This practice raises considerations of conflicts of interest for the resuscitation team, the ability to obtain informed consent in the midst of crisis, and the use of organ preservation measures prior to obtaining consent for organ donation [52, 53]. Both supporters and challengers of uncontrolled DCD point to the well-accepted separation of decisions regarding resuscitation from decisions regarding organ donation and approaches to optimizing organ perfusion for donation [52, 54]. Challengers contend, however, that in some uncontrolled DCD protocols, a resuscitation team could decide to stop resuscitative measures and begin efforts to preserve organs before seeking and obtaining consent for donation (or for preservation measures) [52]. These concerns carry greater weight in countries where presumed consent for organ donation is permitted [52].

6.2.7 *Problems with Pediatric DCD—A Call for a Moratorium*

Donation after cardiac death has continued despite ongoing debate about definitions of death and the ethics of organ donation from non-heart-beating donors. However, some have called for complete cessation of the practice (particularly in children) until there is greater transparency in the process of obtaining parental or other surrogate consent for DCD [55–57]. Those calling for a moratorium cite the ethical issues discussed above that arise in DCD generally and they have heightened worries about DCD in pediatrics. The dissenters contend that (1) children are more likely to autoresuscitate, challenging the DDR's requirement for irreversibility; (2) the requirement for irreversibility should not depend on whether or not one *chooses* to attempt resuscitation but whether a child *could be* resuscitated; (3) the brain is not convincingly dead at the time of organ procurement in DCD and therefore death cannot have truly occurred; (4) pre-mortem interventions further challenge the definition of irreversibility and cannot be ethically justified; and (5) clinicians in institutions performing DCD may be biased (and less forthcoming) in their presentation of DCD to families because the clinicians want to increase available organs for procurement [56, 57]. Advocates for a moratorium argue that until full public disclosure and debate occur on these issues, informed consent for DCD-based organ donation is not possible, and, more importantly, not ethically permissible [56, 57].

6.3 Perspectives on DCD

As discussed, abstract ethical concerns related to DCD include worries about the DDR, about the adequacy of consent for donation, and about potential conflicts of interest. At the same time, a number of studies have examined whether those actually involved in the DCD process have perceived problems that match the theoretical concerns.

6.3.1 Parents/Family Members

Parents are important stakeholders in pediatric DCD. In fact, DCD partially emerged from family desires to donate organs from a loved one who did not yet meet criteria for brain death. Studies considering organ donation in the setting of brain death have generally demonstrated parental willingness to donate, motivated by a desire to help others [58]. One of the most important factors for parents considering donation following brain death is their certainty or uncertainty about the diagnosis of brain death [58]. Recent literature describing motivations for organ donation, irrespective of criteria for death, highlight families' desires to uphold the wishes of their loved ones regarding donation and to make a positive impact through donation [59]. Willingness to donate also relates to families' previous ideas and beliefs about donation, their perceptions of the quality of care their loved one received before donation became an issue, and family members' trust in, or acceptance of, their loved one's prognosis [59]. Although the literature specifically describing parental experience with DCD is limited both in sample size and sample diversity, common themes emerge from parents who have considered DCD [60, 61]. Parents commonly reported a motivation to donate so that something positive could come from the loss of a child [60]. Other parents are motivated by what they believed their child would have wanted [60]. Parents who trusted both the institution where their child received care and the child's health care professionals generally relied on the providers' communications about the child's poor prognosis in order to make donation decisions [61]. Parents expressed confusion over what organs could and could not be donated, as well as a desire to protect their child's body [61]. A community study of viewpoints on the acceptable time between declaration of death and organ removal suggests that family members recognize the tension between ensuring organ viability and the need to respect the death of the individual prior to donation [62]. The literature, however, has not specifically described parent perceptions about controversial issues in DCD, including the DDR, time from declaration of death to the recovery of organs, or the appropriateness of antemortem interventions [61]. The themes of trust in the medical team and the need to accept diagnosis/prognosis highlight a need to consider how controversies in diagnosing death and use of antemortem interventions could affect family-clinician relationships and families' willingness to participate in organ donation.

6.3.2 Providers

Various providers participate in the care of pediatric DCD donors. Most studies describing provider perspectives focus on critical care, anesthesia, and surgical clinicians including nurses, respiratory therapists, clergy, and physicians. At least one study assessed organ procurement organization perspectives on consent for

DCD, reporting that many organ procurement organizations (OPOs) treat DCD similarly to donation following brain death, while the hospital clinicians tend to focus on differences between the processes involving DCD and donation following brain death [63]. In multidisciplinary studies, common themes across provider types include worries about the certainty of futility and prognosis for a given patient, the need for consistent protocols for declaration of cardiopulmonary death, and the questionable quality of organs procured from for DCD [64]. Specific provider groups, however, also express variable concerns.

Pediatric Intensive Care Unit (PICU) nurses generally support organ donation and endorse the view that donation gives meaning to death [65]. Nurses, however, self-report knowing less about organ donation following circulatory death than following brain death, and some perceive that DCD may be more problematic than donation following brain death [65, 66]. Nurses also express additional concerns and barriers related to DCD including that (1) donors can feel pain and should receive pain medication; (2) unlike donation following brain death, DCD requires an active role in the dying process; and (3) there is ambiguity about whether or when death occurs and when transition begins from care provider to donation-process participant [65, 66]. Many nurses disagreed that a 5-min interval after pulselessness is sufficient for a declaration of death [65]. They generally agree that organ procurement organizations perform their roles ethically [66]. Like nurses, respiratory therapists raise concerns about ensuring that the process of death goes as well as it can for the family [21]. Clergy members express a tension between making families aware of the potential to donate and the impact that even asking the question might have on a family who declines [21].

Pediatric physicians generally support organ donation following cardiac death and specifically the DDR [67]. Physicians working in adult and pediatric intensive care units, however, worry about perceived or actual conflicts of interest in introducing DCD to families; the appropriateness of the time interval between death declaration and organ procurement; negative effects that the DCD process may have on end-of-life care, including patient pain and suffering; and maintaining the dignity of the dying child [21, 67–69]. Furthermore, there is disagreement amongst physicians over whether an asystolic period of five minutes is sufficient for declaring death [70]. Some physicians feel DCD should only be offered when the physician perceives that discussing donation might provide a benefit to families, while others believe that every family should know of the option if their child is a potential DCD donor [21, 71]. Acting in ways consistent with family desires is paramount for clinicians, although physicians recognize that both families and society may benefit from DCD and that there is tension in managing obligations to both groups [68]. Anesthesiologists express tension between their traditional role of resuscitation in an operating room, including pain management, and their role in a DCD procurement case, where (1) some argue pain control is not necessary but others worry there is still capacity for pain, and (2) the emphasis of resuscitative measures is on organ preservation rather than resuscitation of the patient [21]. It has been suggested that anesthesiologists play a unique role and should contribute a separate voice to discussions about the role of DCD in hospitals [72]. Clinicians suggest that collaboration and consensus between providers can mitigate uncertainty about prognosis, the structure of the DCD process, and perceived conflicts of interest [21, 71].

In addition to variability among providers in acceptability of DCD, significant social, cultural and religious variability exists regarding DCD in different countries. While some of this variability flows from distinct conceptions of death and care of the dead body, in some cases existing laws call DCD into question given the uncertainties surrounding death by circulatory criteria. In Arab countries, donation after brain death is carried out, but DCD raises difficulties with the legal definition of death and acceptability of the practice of DCD from various Islamic traditions [73]. In Italy, proceeding with DCD requires a 20-min period of electrocardiographic silence to comply with legal definitions of death [19]. Lower donation rates in China are at least partially attributed to the belief that the body should be protected and respected following death [74]. In Canada, where the DDR is also generally accepted, uncertainty regarding the definition and determination of death and confusion about accepted medical practice create legal challenges in determining eligibility for organ donation following circulatory death [75].

6.4 Systematic Approaches to Pediatric DCD: Processes and Protocols

6.4.1 Impact of Organizational Guidance and Regulatory Frameworks

As most countries currently accept the DDR, approaches to DCD must consider local definitions of death prior to ascertaining eligibility for organ donation [75]. In addition to laws defining death, presumed consent legislation also plays a role in DCD practices. In comparison to opt-in informed consent approaches, some argue that presumed consent legislation favorably impacts rates of organ donation, although other studies have not shown increased organ donation in presumed consent regions [76]. As described below, informed consent legislation often relies on the process of consent to address many of the controversial issues in DCD. It is less clear how such issues are negotiated in presumed consent regions.

Consensus-building within hospitals, regions, national organizations, and even entire countries has shaped the landscape for DCD in different settings. In some countries, the need to develop universal principles on which DCD could rely led to delay in implementing DCD [77]. Indeed, pragmatic policy approaches to DCD practices require consideration of its ethical dimensions [78]. Professional societies in different countries have contributed to consensus statements on the ethics of DCD. In 2005, a national conference of key stakeholders in DCD convened in the U.S. to develop consensus about the ethical issues related to DCD [79]. Representation came from the American Medical Association, the Society of

Critical Care Medicine, the American Association of Critical Care Nurses, the American Society of Anesthesiologists, the American Society of Transplant Surgeons, Eurotransplant, the Association of Organ Procurement Organizations, the United Network for Organ Sharing, and the World Health Organization and other groups [79]. Six working groups considered many of the issues we have reviewed regarding organ donation following circulatory death. The group defined irreversibility to mean that function would not spontaneously resume. They stressed the importance of maintaining quality end-of-life care, separate from the decision to donate. In considering appropriate measures for organ preservation and successful transplantation, the group supported requiring explicit informed consent for any premortem interventions. Other considerations included the need for a robust informed consent process, discussions regarding organ allocation, and optimizing coordination between hospitals, OPOs, potential donors, other interested parties, and the public [79].

Also in 2005, national guidance was formulated in Canada, considering many of the same issues raised in the US consensus group [26]. The Canadian group considered issues specific to uncontrolled organ donation, advocating that this process only be considered when controlled DCD is not possible [26]. Consensus groups in France and the United Kingdom provide similar guidance [80, 81]. In the most recent statement of the National Institute for Health Care Excellence (NICE) in the U.K., a best interests standard was proposed to evaluate interventions to sustain an individual's life prior to knowing the individual's wishes regarding organ donation. While some have questioned whether any measures to prolong a dying individual's life can ever be in his or her interests, the NICE guidance contends that best interests include not only the individual's medical well-being, but also social, emotional, cultural and religious interests, which may require temporizing measures to ascertain [81]. Although these consensus and guidance statements contain little specific to pediatric donors, the American Academy of Pediatrics recently issued revised guidance on the ethics of DCD [43]. This guidance document reiterates concerns expressed by others related to determining death and potential conflicts of interest. The statement acknowledges the special role of parents as decision-makers for children and that the framework for decisions about children may differ from that used for decisions for previously competent adults who may have had express preferences [43]. Additionally, the statement recognizes the need for additional data on the limits and likelihood of autoresuscitation in children to help frame discussions on appropriate waiting intervals. Importantly, the statement highlights the need for comprehensive palliative care in the pediatric population, and the goal to minimize differences between the dying process in DCD versus non-DCD patients [43]. Clinicians should minimize location, environment and care-team differences for potential DCD donors. Additionally, care teams must consistently defer to the parents as decision-makers and need to obtain specific parental consent for any interventions, including premortem preservation measures performed on a dying child [43].

6.4.2 Consent to Donation

Given the ethical challenges of DCD, many contend that a robust consent process is essential, and that the process of consent holds more importance than the definition of death [82]. Even within the process of consent, however, multiple questions arise, including: who should obtain consent, how should families be approached, how should consent for donation be distinguished from consent for organ preservation measures, and whether opt-in or opt-out (presumed) consent models provide the best framework for DCD consent [83]. As previously reported with respect to brain death, research on consent for DCD underscores the importance of the initial presentation of the option to donate, including timing, creating a supportive environment, and the need for trained communicators to obtain consent [84–86].

6.4.3 Development of Pediatric DCD Protocols

In response to national and organizational guidance on the need for systematic approaches to DCD, the introduction of federal regulations requiring hospital protocols for DCD, and the myriad issues that arise in the consent process, a growing number of children's hospitals have developed policies and protocols for proceeding with organ donation after cardiac death [87]. A fuller discussion of the development of such protocols is covered by Harrison in Chap. 9 of this book, although a few common themes deserve mention here. Some hospitals have used a step-wise approach to consensus-building in formulating ethical foundations for DCD programs [88]. Consistent with reports of variability among adult DCD protocols, in the United States and Europe considerable variation exists between children's hospitals' policies on organ donation following circulatory death [89–91]. There is disagreement over criteria for eligibility for DCD; variability in the standards used for the determination of death and waiting periods prior to organ procurement; the separation of decisions to withdraw life-sustaining treatment and donation; required disclosures in the consent process; and the use of palliative care as death approaches [90].

6.5 Conclusion

Pediatric DCD remains a relatively new area in organ transplantation ethics. Important ethical considerations related to pediatric DCD include whether current practices for declaring death simultaneously facilitate organ donation and comply with the DDR; whether the DDR respects potential donor (or donor family) wishes; what the appropriate asystolic interval is prior to organ procurement; whether and what kind of organ preservation measures are permissible; how conflicts of interest can be avoided between caring for a dying patient and optimizing chances for successful organ procurement and transplant; and the role for informed consent in mitigating much of the uncertainty around concerns with DCD. From this list it seems evident that the most prominent ethical concerns in pediatric DCD are not unique to pediatrics; some are not even unique to DCD, although the unique role of parents as decision-makers often heightens reactions in this special population. This chapter cannot settle the myriad ethical dilemmas in pediatric DCD; our obligation to our youngest patients demands as much thoughtful attention to the needs and goals of dying children and their families as to settling these debates. In the meantime, ongoing and honest dialogue with families about options and uncertainty seems the only legitimate way forward.

References

1. Detry O, Le Dinh H, Noterdaeme T, De Roover A, Honore P, Squifflet JP, et al. Categories of donation after cardiocirculatory death. Transpl Proc. 2012;44(5):1189–95 Epub 2012/06/06.
2. Merion RM, Pelletier SJ, Goodrich N, Englesbe MJ, Delmonico FL. Donation after cardiac death as a strategy to increase deceased donor liver availability. Ann Surg. 2006;244(4):555–62.
3. Reich DJ, Munoz SJ, Rothstein KD, Nathan HM, Edwards JM, Hasz RD, et al. Controlled non-heart-beating donor liver transplantation: a successful single center experience, with topic update. Transplantation. 2000;70(8):1159–66 Epub 2000/11/04.
4. Neyrinck A, Van Raemdonck D, Monbaliu D. Donation after circulatory death: current status. Current Opin Anaesthesiol. 2013;26(3):382–90 Epub 2013/05/02.
5. Cooper JT, Chin LT, Krieger NR, Fernandez LA, Foley DP, Becker YT, et al. Donation after cardiac death: the university of wisconsin experience with renal transplantation. Am J Transplant: Official J Am Soc Transplant Am Soc Transplant Surg. 2004;4(9):1490–4 Epub 2004/08/17.
6. D'Alessandro AM, Fernandez LA, Chin LT, Shames BD, Turgeon NA, Scott DL, et al. Donation after cardiac death: the University of Wisconsin experience. Ann Transplant: Q Polish Transplant Soc. 2004;9(1):68–71 Epub 2004/10/14.
7. Yoo PS, Olthoff KM, Abt PL. Donation after cardiac death in pediatric organ transplantation. Current opin Organ Transplant. 2011;16(5):483–8 Epub 2011/08/23.
8. Kolovos NS, Webster P, Bratton SL. Donation after cardiac death in pediatric critical care. Pediatric Crit Care Med: J Soc Crit Care Med World Fed Pediatr Intensive Crit Care Soc. 2007;8(1):47–9 Epub 2007/01/26.
9. Pleacher KM, Roach ES, Van der Werf W, Antommaria AH, Bratton SL. Impact of a pediatric donation after cardiac death program. Pediatr Crit Care Med: J Soc Crit Care Med World Fed Pediatr Intensive Crit Care Soc. 2009;10(2):166–70 Epub 2009/02/04.
10. Naim MY, Hoehn KS, Hasz RD, White LS, Helfaer MA, Nelson RM. The children's hospital of Philadelphia's experience with donation after cardiac death. Crit Care Med. 2008;36 (6):1729–33 Epub 2008/06/04.
11. Labrecque M, Parad R, Gupta M, Hansen A. Donation after cardiac death: the potential contribution of an infant organ donor population. J Pediatr. 2011;158(1):31–6 Epub 2010/08/25.
12. Mathur M, Castleberry D, Job L. Identifying potential heart donors among newborns undergoing circulatory determination of death. J Heart Lung Transplant: Official Publ Int Soc Heart Transplant. 2011;30(4):389–94 Epub 2010/12/28.

13. Potts M, Byrne PA, Evans DW. Infant heart transplantation after cardiac death: ethical and legal problems. J Clin Ethics. 2010;21(3):224–8 Epub 2010/11/26.
14. Ross LF, Frader J. Are we ready to expand donation after cardiac death to the newborn population? J Pediatr. 2011;158(1):6–8 Epub 2010/09/28.
15. Boucek MM, Mashburn C, Dunn SM, Frizell R, Edwards L, Pietra B, et al. Pediatric heart transplantation after declaration of cardiocirculatory death. New Engl J Med. 2008;359 (7):709–14.
16. Mazor R, Baden HP. Trends in pediatric organ donation after cardiac death. Pediatrics. 2007;120(4):e960–6 Epub 2007/10/03.
17. Kim HB, Lillehei CW. Organ donation for children: the road ahead. Pediatrics. 2013;131(6): e1946–7 Epub 2013/05/22.
18. Blackstock MJ, Ray DC. Organ donation after circulatory death: an update. Eur J Emerg Med: Official J Eur Soc Emerg Med. 2013. Epub 2013/11/13.
19. Geraci PM, Sepe V. Non-heart-beating organ donation in Italy. Minerva Anestesiol. 2011;77 (6):613–23 Epub 2011/05/28.
20. Hardart G. Pediatric organ donation after cardiac death–still in its infancy. Crit Care Med. 2008;36(6):1956–7 Epub 2008/06/04.
21. Curley MA, Harrison CH, Craig N, Lillehei CW, Micheli A, Laussen PC. Pediatric staff perspectives on organ donation after cardiac death in children. Pediatr Crit Care Med: J Soc Crit Care Med World Fed Pediatr Intensive Crit Care Soc. 2007;8(3):212–9 Epub 2007/04/10.
22. Frader JE. Deconstructing donation after cardiac death. Pediatr Crit Care Med: J Soc Crit Care Med World Fed Pediatr Intensive Crit Care Soc. 2007;8(1):76–7 Epub 2007/01/26.
23. Robertson JA. The dead donor rule. Hastings Center Rep. 1999;29(6):6–14.
24. PsCftSoEPiMaBaB Research. Defining death: medical, legal, and ethical issues in the determination of death. Washington, D.C.: Library of Congress; 1981.
25. Shemie SD. Clarifying the paradigm for the ethics of donation and transplantation: was 'dead' really so clear before organ donation? Philos Ethics Humanit Med: PEHM. 2007;2:18.
26. Shemie SD, Baker AJ, Knoll G, Wall W, Rocker G, Howes D, et al. National recommendations for donation after cardiocirculatory death in Canada: donation after cardiocirculatory death in Canada. CMAJ: Can Med Assoc J=journal de l'Association medicale canadienne. 2006;175(8):S1.
27. Bernat JL. Life or death for the dead-donor rule? New Engl J Med. 2013;369(14):1289–91 Epub 2013/10/04.
28. Miller FG, Truog RD, Brock DW. The dead donor rule: can it withstand critical scrutiny? J Med Philos. 2010;35(3):299–312.
29. Birch SC. The dead donor rule: a defense. J Med Philos. 2013;38(4):426–40 Epub 2013/07/17.
30. Rodriguez-Arias D, Smith MJ, Lazar NM. Donation after circulatory death: burying the dead donor rule. Am J Bioeth: AJOB. 2011;11(8):36–43 Epub 2011/08/03.
31. Shah SMF. Can we handle the truth? Legal fictions in the determination of death. Am J Law Med: Copyright (c) 2010 American Society of Law, Medicine & Ethics, Inc., and Boston University Am J Law amp; Med; 2010. p. 540–85.
32. Sade RM. Brain death, cardiac death, and the dead donor rule. JSC Med Assoc. 2011;107 (4):146–9 Epub 2011/11/08.
33. Fanelli V, Geraci PM, Mascia L. Donation after cardiac death: is a "paradigm shift" feasible in Italy? Minerva Anestesiol. 2013;79(5):534–40 Epub 2013/03/02.
34. Joffe AR. The ethics of donation and transplantation: are definitions of death being distorted for organ transplantation? Philos Ethics Humanit Med: PEHM. 2007;2:28 Epub 2007/11/27.
35. Duff JP, Joffe AR, Sevcik W, deCaen A. Autoresuscitation after pediatric cardiac arrest: is hyperventilation a cause? Pediatr Emerg Care. 2011;27(3):208–9 Epub 2011/03/08.
36. Veatch RM. Transplanting hearts after death measured by cardiac criteria: the challenge to the dead donor rule. J Med Philos. 2010;35(3):313–29.
37. Bernat JL. How the distinction between "irreversible" and "permanent" illuminates circulatory-respiratory death determination. J Med Philos. 2010;35(3):242–55.

38. Harrington MM. The thin flat line: redefining who is legally dead in organ donation after cardiac death. Issues Law Med. 2009;25(2):95–143 Epub 2010/01/28.
39. Miller FG, Truog RD. Rethinking the ethics of vital organ donations. Hastings Center Rep. 2008;38(6):38–46.
40. McGregor J, Verheijde JL, Rady MY. Do donation after cardiac death protocols violate criminal homicide statutes? Med Law. 2008;27(2):241–57 Epub 2008/08/13.
41. Truog RD, Miller FG, Halpern SD. The dead-donor rule and the future of organ donation. New Engl J Med. 2013;369(14):1287–9 Epub 2013/10/04.
42. Issues PsCftSoB. Controversies in the determination of death: a white paper by the president's council on bioethics 2008.
43. Committee on Bioethics AAoP. Ethical controversies in organ donation after circulatory death. Pediatrics. 2013;131:1021–6. Epub April 29, 2013.
44. Daar AS. Non-heart-beating donation: ten evidence-based ethical recommendations. Transpl Proc. 2004;36(7):1885–7.
45. Rabinstein AA, Yee AH, Mandrekar J, Fugate JE, de Groot YJ, Kompanje EJ, et al. Prediction of potential for organ donation after cardiac death in patients in neurocritical state: a prospective observational study. Lancet Neurol. 2012;11(5):414–9 Epub 2012/04/13.
46. Halpern SD. Donation after circulatory determination of death: time for transparency. Ann Emerg Med. 2014;63(4):401–3 Epub 2013/11/12.
47. Chaten FC. The dead donor rule: effect on the virtuous practice of medicine. J Med Ethics. 2013. Epub 2013/06/14.
48. Ledoux D, Delbouille MH, Deroover A, Lambermont B, Meurisse M, Damas P, et al. Does comfort therapy during controlled donation after circulatory death shorten the life of potential donors? Clin Transplant. 2014;28(1):47–51 Epub 2013/11/23.
49. Rady MY, Verheijde JL, McGregor J. Organ donation after circulatory death: the forgotten donor? Crit Care. 2006;10(5):166. Epub 2006/10/06.
50. Bastami S, Krones T, Biller-Andorno N. Whose consent matters? Controlled donation after cardiac death and premortem organ-preserving measures. Transplantation. 2012;93(10):965–9 Epub 2012/05/12.
51. Organizacion Nacional de Trasplantes. Donacion en asistolia en Espana: Situacion Actual Y Recomendaciones. 2012. Available from: http://www.ont.es/infesp/DocumentosDeConsenso/DONACI%C3%93N%20EN%20ASISTOLIA%20EN%20ESPA%C3%91A.%20SITUACI%C3%93N%20ACTUAL%20Y%20RECOMENDACIONES.pdf.
52. Ortega-Deballon I, Vailhen DR, Smith MJ. Uncontrolled donation after circulatory determination of death protocols: ethical challenges and suggestions for improvement. Ann Emerg Med. 2013;61(2):255–6 Epub 2013/01/22.
53. Simon JR, Schears RM, Padela AI. Donation after cardiac death and the emergency department: ethical issues. Acad Emerg Med: Official J Soc Acad Emerg Med. 2014;21(1):79–86 Epub 2014/02/21.
54. Snoeijs MG, Wind T, van Heurn E. Protocols for uncontrolled donation after circulatory death. Lancet. 2012;380(9846):974–5; author reply 5. Epub 2012/09/18.
55. Potts M. Truthfulness in transplantation: non-heart-beating organ donation. Philos Ethics Humanit Med: PEHM. 2007;2:17.
56. Carcillo JA, Orr R, Bell M, Joffe A, Maffei FA, Sullivan J, et al. A call for full public disclosure and moratorium on donation after cardiac death in children. Pediatr Crit Care Med: J Soc Crit Care Med World Fed Pediatr Intensive Crit Care Soc. 2010;11(5):641–3; author reply 3–5. Epub 2010/09/09.
57. Joffe AR, Carcillo J, Anton N, deCaen A, Han YY, Bell MJ, et al. Donation after cardiocirculatory death: a call for a moratorium pending full public disclosure and fully informed consent. Philos Ethics Humanit Med: PEHM. 2011;6:17 Epub 2011/12/31.
58. Walker JA, McGrath PJ, MacDonald NE, Wells G, Petrusic W, Nolan BE. Parental attitudes toward pediatric organ donation: a survey. CMAJ: Can Med Assoc J = journal de l'Association medicale canadienne. 1990;142(12):1383–7. Epub 1990/06/15.

59. Walker W, Broderick A, Sque M. Factors influencing bereaved families' decisions about organ donation: an integrative literature review. West J Nurs Res. 2013;35(10):1339–59 Epub 2013/04/27.
60. Hoover SM, Bratton SL, Roach E, Olson LM. Parental experiences and recommendations in donation after circulatory determination of death*. Pediatr Crit Care Med: J Soc Crit Care Med World Fed Pediatr Intensive Crit Care Soc. 2014;15(2):105–11 Epub 2013/12/18.
61. Bauer PN, Han YY. Parental perspectives of donation after circulatory determination of death in children: have we really investigated the heart of the matter?*. Pediatr Crit Care Med: J Soc Crit Care Med World Fed Pediatr Intensive Crit Care Soc. 2014;15(2):171–3 Epub 2014/02/05.
62. Rogers WA, de Lacey S, Avery JC. Donation after cardiac death: community views about 'decent' intervals. Am J Transplant: Official J Am Soc Transplant Am Soc Transplant Surg. 2011;11(3):583–90 Epub 2011/02/09.
63. Kalkbrenner KJ, Hardart GE. Consent for donation after cardiac death: a survey of organ procurement organizations. J Intensive care Med. 2012;27(4):253–63 Epub 2011/05/18.
64. Mandell MS, Zamudio S, Seem D, McGaw LJ, Wood G, Liehr P, et al. National evaluation of healthcare provider attitudes toward organ donation after cardiac death. Crit Care Med. 2006;34(12):2952–8 Epub 2006/11/01.
65. Mathur M, Taylor S, Tiras K, Wilson M, Abd-Allah S. Pediatric critical care nurses' perceptions, knowledge, and attitudes regarding organ donation after cardiac death. Pediatr Crit Care Med: J Soc Crit Care Med World Fed Pediatr Intensive Crit Care Soc. 2008;9 (3):261–9.
66. D'Alessandro AM, Peltier JW, Phelps JE. An empirical examination of the antecedents of the acceptance of donation after cardiac death by health care professionals. Am J Transplant: Official J Am Soc Transplant Am Soc Transplant Surg. 2008;8(1):193–200 Epub 2007/11/02.
67. Stavel M, Akinsoji Y, Latifi S. PICU staff attitude towards organ donor identification and donation after circulatory death. Arch Dis Child. 2011;96(7):700. Epub 2011/02/11.
68. Sarnaik AA, Clark JA, Meert KL, Sarnaik AP. Views of pediatric intensive care physicians on the ethics of organ donation after cardiac death. Crit Care Med. 2013;41(7):1733–44 Epub 2013/05/11.
69. Hart JL, Kohn R, Halpern SD. Perceptions of organ donation after circulatory determination of death among critical care physicians and nurses: a national survey. Crit Care Med. 2012;40 (9):2595–600 Epub 2012/06/27.
70. Joffe AR, Anton NR, deCaen AR. Survey of pediatricians' opinions on donation after cardiac death: are the donors dead? Pediatrics. 2008;122(5):e967–74 Epub 2008/11/04.
71. Coleman NL, Bonner A. Exploring Australian intensive care physicians clinical judgement during donation after cardiac death: an exploratory qualitative study. Aus Crit Care: Official J Confed Aus Crit Care Nurses. 2014. Epub 2014/05/28.
72. Truog RD. Organ donation after cardiac death: what role for anesthesiologists? Anesthesiology. 2003;98(3):599–600.
73. Faraj W, Fakih H, Mukherji D, Khalife M. Organ donation after cardiac death in the Middle East. Transpl Proc. 2010;42(3):713–5 Epub 2010/05/01.
74. Zheng S, Shen Z. Current status of donation after cardiac death in children in China. Pediatr Transpl. 2012;16(8):813–4 Epub 2012/11/08.
75. Downie J, Kutcher M, Rajotte C, Shea A. Eligibility for organ donation: a medico-legal perspective on defining and determining death. Can J Anaesth. 2009;56(11):851–63 Epub 2009/07/09.
76. Abadie A, Gay S. The impact of presumed consent legislation on cadaveric organ donation: a cross-country study. J Health Econ. 2006;25(4):599–620.
77. Squifflet JP. Why did it take so long to start a non-heart-beating donor program in Belgium? Acta Chir Belg. 2006;106(5):485–8.
78. Kirby J. Ethically informed pragmatic conditions for organ donation after cardiocirculatory death: could they assist in policy development? J Clin Ethics. 2013;24(4):373–80 Epub 2013/01/01.

79. Bernat JL, D'Alessandro AM, Port FK, Bleck TP, Heard SO, Medina J, et al. Report of a national conference on donation after cardiac death. Am J Transplant: Official J Am Soc Transplant Am Soc Transplant Surg. 2006;6(2):281–91.
80. Graftieaux JP, Bollaert PE, Haddad L, Kentish-Barnes N, Nitenberg G, Robert R, et al. Contribution of the ethics committee of the French intensive care society to describing a scenario for implementing organ donation after Maastricht type III cardiocirculatory death in France. Ann Intensive Care. 2012;2(1):23. Epub 2012/07/04.
81. Littlejohns S, Bontoft H, Littlejohns P, Richardson J, Robertson A. Legal and ethical implications of NICE guidance aimed at optimising organ transplantation after circulatory death. Clin Med (London, England). 2013;13(4):340–3. Epub 2013/08/03.
82. Valapour M. Donation after cardiac death: consent is the issue, not death. J Clin Ethics. 2006;17(2):137–8 Epub 2006/08/18.
83. Childress JF. Organ donation after circulatory determination of death: lessons and unresolved controversies. J Law, Med Ethics: J Am Soc Law Med Ethics. 2008;36(4):766–71, 610.
84. Siminoff LA, Agyemang AA, Traino HM. Consent to organ donation: a review. Prog Transplant. 2013;23(1):99–104 Epub 2013/03/02.
85. Siminoff LA, Arnold RM, Hewlett J. The process of organ donation and its effect on consent. Clin Transplant. 2001;15(1):39–47 Epub 2001/02/13.
86. Browne A. The ethics of organ donation after cardiocirculatory death: do the guidelines of the Canadian Council for donation and transplantation measure up? 2010.
87. Rady MY, Verheijde JL, McGregor J. Organ procurement after cardiocirculatory death: a critical analysis. J Intensive Care Med. 2008;23(5):303–12.
88. Harrison CH, Laussen PC. Controversy and consensus on pediatric donation after cardiac death: ethical issues and institutional process. Transpl Proc. 2008;40(4):1044–7 Epub 2008/06/17.
89. Fugate JE, Stadtler M, Rabinstein AA, Wijdicks EF. Variability in donation after cardiac death protocols: a national survey. Transplantation. 2011;91(4):386–9 Epub 2010/12/04.
90. Antommaria AH, Trotochaud K, Kinlaw K, Hopkins PN, Frader J. Policies on donation after cardiac death at children's hospitals: a mixed-methods analysis of variation. JAMA J Am Med Assoc. 2009;301(18):1902–8 Epub 2009/05/14.
91. Wind J, Faut M, van Smaalen TC, van Heurn EL. Variability in protocols on donation after circulatory death in Europe. Crit Care (London, England). 2013;17(5):R217. Epub 2013/10/05.

Chapter 7
The Newborn as Organ Donor

Christopher Tomlinson and Jonathan Hellmann

Abstract From a medical perspective, organ donation from newborn infants is fraught with difficulty: the diagnosis of brain death is difficult to make in this age group and, by current definition, exceedingly rare. In addition, severely brain-injured newborns often retain sufficient respiratory ability to maintain circulation long past the time when organs may be suitable for donation after circulatory death (DCD) following the withdrawal of life-sustaining treatment. The issue of donor suitability in terms of organ size and the frequent absence of a clear diagnosis compound the complexity. These medical issues lead to ethical challenges for caregivers and the families of newborn infants when the potential for organ donation is raised. In this chapter, we will address the issues relating to the diagnosis of brain death in the newborn; with regard to DCD we will discuss the timing of death and how the dying and grieving process for families and the team may be altered by the prospect of donation. While organ donation may salvage some good for the family and broader society, caution is required not only for practical and medical reasons but also in view of the ethical uncertainty inherent in the concept of the newborn as an organ donor.

Keywords End-of-life · Neonatal intensive care unit · Best interests · Withdrawal of life sustaining treatments

---

C. Tomlinson (✉)
Departments of Paediatrics and Nutritional Sciences, The Hospital for Sick Children, University of Toronto, Toronto, Canada
e-mail: Christopher.tomlinson@sickkids.ca

J. Hellmann
Department of Bioethics, The Hospital for Sick Children, University of Toronto, Toronto, Canada
e-mail: Jonathan.hellmann@sickkids.ca

R.A. Greenberg et al. (eds.), *Ethical Issues in Pediatric Organ Transplantation*, International Library of Ethics, Law, and the New Medicine 66,
DOI 10.1007/978-3-319-29185-7_7

7.1 Introduction

Despite the high mortality that occurs in the newborn period, this group of patients has not been viewed, until very recently, as potential organ donors. In this chapter we review the ethical dilemmas that are faced by caregivers and families when a newborn infant has the potential to be an organ donor. In many ways these dilemmas are not dissimilar to those faced in the older paediatric patient; however, the precipitating diagnoses and physiology of the newborn introduce a level of medical uncertainty over and above those of the older child. This in turn adds greater complexity to the challenges inherent in end-of-life (EOL) decision-making and potential organ donation in the neonatal intensive care unit (NICU).

Organs for transplantation in older paediatric patients and adults are retrieved from three groups of donors (i) living (often related), (ii) after neurological death and (iii) after circulatory death. Though living related donation in paediatric practice is accepted as ethically justified, the unique considerations and medical risks surrounding infancy are such that the same arguments made in favour of newborn infants do not apply [1]. The potential for the newborn to be an organ donor thus falls into the category of donation after neurological or circulatory death.

7.2 Neurological Death in Newborns

The concept of neurological death is accepted today in both adult medicine and paediatrics by both physicians and the general public [2]. However, neurological death is difficult to diagnose in newborns and, although possibly under-diagnosed, is still likely to be extremely rare [3, 4]. Indeed in the United Kingdom (UK), organ donations do not occur in infants under two months of age due to this difficulty [2]. UK guidelines state: “Between thirty-seven weeks of gestation and two months of age, it is rarely possible confidently to diagnose death as a result of cessation of brain-stem reflexes and below thirty-seven weeks of gestation the criteria to establish this cannot be applied” [5, p. 93]. This has not been updated since 1991 and, in the light of the acceptance of the diagnosis of brain death in the US, Canada and elsewhere, brain death determination in newborns in the UK is currently under review [6].

The reason that brain death occurs rarely in the newborn and is difficult to diagnose relate to the newborn’s neurological immaturity, the ‘open fontanelle’ which makes irreversible brain-stem injury from raised intracranial pressure extremely uncommon, and the apparent ability of the brain-stem to recover, even after significant insult. In Canada and the United States, diagnostic criteria for brain death in newborns have been developed. They differ from that in paediatric practice with the main difference being a minimum of 24 h observation between the two determinations of neurological death in neonatal patients [3, 4].

At the Hospital for Sick Children, Toronto, a major tertiary referral centre there are approximately 50–70 deaths in the NICU each year; there have been only two confirmed cases of brain death over the past 20 years. It is likely there have been missed cases that were classified as severe neurological injury without an attempt to formally diagnose neurological death. However, it is unlikely there were substantial numbers of cases where potential organ donation from brain-dead newborns could have occurred. That being said, brain death does occur, and there has been one successful solid organ donation from such a donor at this institution. In the second instance of brain death, the family of the deceased newborn declined consent to transplant.

7.3 Circulatory Death in Newborns

Given the rarity of the determination of neurological death in newborns, attention has shifted to the potential for the newborn to donate after a diagnosis of circulatory death [2]. A policy for donation after circulatory death (DCD) has been in existence at The Hospital for Sick Children, Toronto for ten years which nominally includes newborns. Notwithstanding this policy, there has not been a case of organ retrieval after circulatory death from a newborn, despite this occurring in the paediatric age group.

The reasons for the rarity of DCD in the newborn are multifactorial. The first relates to the uncertainty of diagnosis. The most common underlying diagnosis for potential donation in newborn patients is hypoxic-ischaemic encephalopathy secondary to complications around the time of delivery [7, 8]. However, the cause of the encephalopathy is not always clear and there is a risk, however small, of an undiagnosed disorder giving rise to the clinical picture of brain injury. The concern is that an undiagnosed or unknown metabolic disorder, for example, would affect organ function in the recipient.

Second, as discussed elsewhere in this book, successful donation requires a minimal period of warm ischaemia after withdrawal of life-sustaining treatment (WLST). Timing of death after discontinuation of respiratory support in an otherwise stable newborn patient with severe neurological injury is difficult to predict. It often does not occur within the defined, acceptable time frame for donation. There are also concerns about the different criteria for the "hands off" observation period after the first determination of death which ranges from 5 min to 75 s [6].

Studies from different settings do indicate that a small but potentially significant number of newborn patients dying in NICUs *could* be donors. Saha and Kent, in a retrospective review of deaths in an NICU over a ten-year period, suggested that 6 % of their patients could have been organ donors after circulatory death [9]. Two other studies from different institutions report similar figures. Labrecque et al. estimated that ~8 % of neonatal deaths at a tertiary institution were potential donors and, looking specifically at heart donors after DCD [8], Mathur et al. determined that 4 % of patients would, theoretically, have been eligible and Charles et al. reported 34 potential DCD donors over a six year period in a UK quaternary centre [10, 11]. Data from the Canadian Neonatal Network suggests similar numbers of patients from Canadian NICUs (Hellmann et al., submitted for publication). In this latter study, which collected data on all deaths in 75 % of Canadian NICUs, in full-term infants there were 61 deaths from hypoxic-ischaemic encephalopathy; of these, 25 % died after WLST within a time frame that might have allowed DCD to proceed. It must be noted that all these studies assessed only *potential* donors from the point of view of gestational age at birth, weight, cause of death and time after WLST. They did not take into account aspects such as blood grouping, detailed organ function and, most importantly, whether families would have consented. These studies suggest that at a given institution the likelihood of a successful DCD would only occur at most once per year, even in the largest of institutions.

In addition to the underlying diagnostic uncertainty and the issue of timing of death after WLST, a compelling obstacle to donation in the newborn is that attempting DCD dramatically alters the process of dying in the NICU. There is the fear that changing the dynamic from one focused on the suffering of the newborn and the grieving family to one in which technology and the altruistic desire of the family and care team will interfere with an NICU's established EOL care process. Curley et al. showed that clinicians at the bedside find the issues raised with DCD ethically challenging, yet despite these concerns, staff did want to "make it happen" for the families that are motivated [12, p. 216].

Death is not uncommon in the NICU and most units have procedures and routines that enable families to be actively involved in EOL care and WLST. This usually takes place in a quiet room in which family members gather. This is then followed by an opportunity for parents and family to spend time with the child before and after the death, facilitating memory and legacy creation through, for example, the creation of hand and foot moulds and the opportunity for photographs to be taken [13]. The suggestion by Labrecque et al. that potential organ donors be transferred to specialized institutions primarily for the purposes of organ retrieval would certainly be considered contentious by some [8].

Furthermore, when DCD is considered there is significant 'medicalisation' of the EOL process. By necessity, the removal of respiratory support must take place in or next to the operating room, and, rather than disconnecting monitors, the level of invasive monitoring may actually increase with, for example, the siting of an arterial line or other interventions. This disruption of a unit's normal process around NICU death may be disconcerting for staff members; those in the operating room do not normally deal with dying patients, while those in the NICU see this as a disturbance of a usually sensitive and effective process. As Larcher suggests, "All these measures (to enable DCD) potentially maximise the benefits that can be derived from donor organ but may be ethically problematic for families of the dying child and those caring for them" [14, p. 108].

Two cases are presented to highlight these medical and practical uncertainties when considering organ donation in the newborn.

Case study 1

Baby G was admitted to the NICU for therapeutic hypothermia following perinatal asphyxia. The pregnancy had been uncomplicated until 24 h prior to delivery when the mother noted a reduction in fetal movements. Labour was induced with difficulty and was complicated by sudden loss of the fetal heart rate. Delivery was by emergency Cesarean section under general anaesthesia. The infant was born without vital signs and resuscitation was immediately commenced. There was recovery of the heart rate after 9 min; the first respiratory gasp was recorded at 28 min after birth. Blood gases demonstrated extreme metabolic acidosis. Therapeutic passive hypothermia was commenced at the community hospital and the infant was transferred to a tertiary centre.

Clinically the infant demonstrated signs of extreme neurological injury: there were no spontaneous movements or primitive reflexes, EEG showed minimal cerebral activity, and there was evidence of multiorgan involvement with elevated liver enzymes and oliguria, some spontaneous respiratory effort was present, indicating the infant was not brain dead.

After discussion with the family regarding the poor prognosis for long-term neurological outcome, care was redirected to a palliative care pathway. As the mother was still an inpatient at the hospital of birth the infant was transferred back to the referral centre ($\sim$30 min by ambulance) where the infant was extubated and palliative care initiated. Given this scenario and the impression that death would not occur within a time frame to consider donation after circulatory death, a decision was made not to approach the family about organ donation. Despite some initial respiratory effort the infant died six hours after WLST.

This case highlights many of the issues discussed above, even though there was no diagnostic dilemma or uncertainty over the predicted long-term neurological prognosis. In this not uncommon scenario in the NICU the infant had severe neurological injury yet was quite clearly not brain dead. If care were continued the development of cerebral edema might compromise brain-stem function but, likely due to the open fontanelle, would not lead to brain death, as is more common in older paediatric patients. It is evident that organ donation after neurological death would not be possible in such cases: donation after circulatory death could occur in principle and would appear to be the only option for organ donation from similarly afflicted newborns.

In attempting to pursue DCD, however, significant problems arise: despite the severe neurological injury the infant remained haemodynamically stable and, while dependent on the ventilator, would likely breathe for an unpredictable period of time after WLST. To add to the complexity, the mother is unable to be discharged from the hospital of birth. The practice in many neonatal outborn centres is to minimise social disruption for families who live some distance from the tertiary centre, and attempts are made to retro-transfer babies and assist the local team in providing bereavement support for the family. It is well shown that perinatal asphyxia is far less common in tertiary maternity and perinatal centres, whereas unexpected and unanticipated perinatal asphyxia in full-term infants is more commonly seen in community or outlying hospitals, necessitating transfer to specialized centres in children hospitals for ongoing management [15].

Case study 2

Baby F was an infant born at term after an uncomplicated pregnancy. Soon after birth it was noted that he had generalized hypotonia and poor respiratory effort requiring ventilatory support and continuous positive airway pressure. He was transferred to a tertiary centre for ongoing investigation.

After a prolonged admission for investigation, during which it was clear he was dependent on respiratory assistance, he was diagnosed as having a congenital disorder of skeletal muscle function which precluded long-term survival without ongoing respiratory support. Other vital organ function was not affected. Rather than pursue lifelong ventilation the family wished to initiate palliative care. They met with the organ donation coordinator after making this decision and decided to proceed with an attempt to donate after their son's circulatory death. The medical care team felt there was a reasonable expectation that he would die within 45 min after removal of respiratory support, the maximum time recommended by the receiving institution for harvesting kidneys. Prior to the withdrawal of respiratory assistance an arterial line was placed to assist with blood sampling and monitoring during the dying process. He was transferred (with several family members present) to an anesthetic area in the operating room. Heparin was administered to facilitate organ retrieval and, in accordance with the palliative care plan, IV morphine and midazolam were given to alleviate anxiety and discomfort. Respiratory support was withdrawn. Despite minimal respiratory effort Baby F's heart rate, blood pressure and pulse oximetry were maintained, and at 45 min after withdrawal of ventilatory support the attempt at DCD was abandoned. Baby F and his family were transferred back to the palliative care room in the NICU. Death was certified 20 min later, approximately 1 h 10 min after removal of the respiratory support. At follow up the family stated that they felt grateful to have had the opportunity to pursue the potential for donation.

In the first case the expectation was that there would be sufficient respiratory effort that would preclude DCD; this was borne out. In the second case, an infant who appeared dependent on respiratory support was thought to have a reasonable likelihood of succumbing within an acceptable time frame to allow retrieval of organs. Despite minimal respiratory effort he survived for over an hour.

Accurate prediction of time of death after WLST is poor. There are no validated tools in paediatrics and none in newborns. It is important that as neonatal practitioners we routinely convey to families that death after withdrawal of respiratory assistance may take minutes, hours or days. Given the uncertainty around the timing of death it is pertinent to question whether it is appropriate to approach families for DCD and risk disrupting and medicalising an already fraught situation when it is known that the chance of a successful donation is low. Or is it unfair to deny families, such as that of Baby F, the opportunity to at least consider donation?

7.4 The Ethical Dimensions of Organ Donation in the Newborn

The fundamental ethical principle in decision-making in neonatal practice is to act in the "best interests" of the newborn. This moral and legal standard of judgment helps to establish the primacy of duties to infants, ensuring they be regarded as fully human persons with interests, even when clearly unable to express their own autonomous value system. Pursuing the best interests of an infant implies determining what treatment course has a more favorable benefit/harm ratio than other possible options; whether the treatment is likely to improve the newborn's condition or well-being, prevent it from deteriorating, or reduce the extent or rate at which it is likely to deteriorate; and whether the condition is likely to improve, remain the same, or deteriorate without the treatment.

The concept of best interests engenders much debate, particularly around who should be making the assessment and on what grounds the assessment is being made. As to who are the rightful decision-makers, in most jurisdictions the state confers the responsibility for health care decisions of young children on the parents. They have a unique relationship with their children, one of concern, obligation, and intimacy. Recognition of parents' moral authority does presume they will act to promote their children's best interests and endeavour to make decisions in that light.

With regard to decision making in the newborn a process of 'shared' decision-making is undertaken: by first defining the rightful decision-makers as those who bear the greatest burden of care and conscience (the parents), those with special knowledge (the appropriate medical team), and those with the most continuous, committed and trusting relationship directly involved in the care of the newborn (the care team) [16]. The decision-makers then both individually and collectively determine what they believe to be the best interests of the newborn infant. The unique circumstances of each situation will affect the moral weight given to each of these perspectives, particularly when there may be disagreement about what is in the child's best interest. A shared decision-making process, in which *adult* patients are educated about their role and given effective tools and opportunity to understand options and consequences as well as emotional support, is regarded by Barry and Edgman-Levitan as "the pinnacle of patient centered care" [17]. Extrapolation of this model of decision making to the parents of the newborn infant is considered in most circumstances to be the preferred mode of deriving decisions in the best interest of the newborn.

Critiques of the best interest standard largely focus on its use by *surrogate* decision-makers (i.e., those making decisions for the infant in the absence of the infant making their own decisions) and the subjectivity of the surrogates criteria for any best interest assessment. In addition, defining "best" may be unattainable [18], and is culture specific. It is also considered by some to be too individualistic and ignoring of the interests of others, which may be the parents, family or society in general. Furthermore, Truog and Sayeed feel that the best interest standard can distort the process of shared decision-making for parents when any factors that do not bear *directly* on the infant's interests are rejected [19]. Rhodes and Holzman recently suggested that instead of the best interest standard parents should be held to a standard they term "the not unreasonable standard," that allows a greater range for parental discretion [20, p. S127]. Despite these and other critiques of the best interest standard, it is generally accepted as a guiding principle for decision-makers as it requires "acceptable and reasonable choices grounded in a web of established duties to incompetent persons" [21]. Therefore, in line with the current prevailing principle in newborn decision-making, it is appropriate to ask, as Brierley has done: "Could it be in a dying baby's best interests to donate organs so that others may survive?" [2, p. F82].

While parents and the health care team normally act as surrogates in the assessment and decision-making for the newborn infant, as described above, in the rare instance when potential organ donation is raised it is conceivable that parents may consider acting as *substitute* decision-makers for their child, i.e., parents might ask themselves "what would our child have wanted to do in this situation?" They may, for example, consider that their child would have wanted them to pursue an altruistic pathway and consent to the retrieval of his/her organs. We fully acknowledge that this pattern of thinking is counter to the usual 'parent-as-surrogate' in the decision-making process. While it may be inadvisable for caregivers to raise this pattern of rationalization with parents, it is worth considering whether parents might view this as a possible expression of their child's desire that some good ensue from a tragic situation. It would be a courageous perspective for parents to contemplate that their dying child might see organ donation as an altruistic last gesture.

It is important to emphasize that normally a determination of best interests entails consideration of what treatment (or non-treatment) pathway to pursue for the sake of a child's life or lived experience only. In potential organ donation situations the death of the child is inevitable and no option for active treatment is being considered. Therefore, does the concept of best interests still apply? Once death has been diagnosed it is generally accepted that parents have the moral authority for decision-making over their child's body. This would be true for whichever pathway is used for determining death, whether neurological or cardiac. What is less clear, however, is how interventions that must take place ante-mortem to optimise organ retrieval and function—e.g., siting of femoral lines, heparin infusion, even transfer of the child to a transplant centre—fit within the best interests of the child, when clearly they are targeted at the best interests of another, remote patient. It is our contention that justification for these antemortem interventions to accomplish DCD are coherent with respect for the families' desire to pursue organ donation from their newborn infant. However, the discomfort of staff during this period of 'organ management' has been well documented [12].

7.5 Family Centred Care

The shift towards a more participatory process of decision-making in health care in general and the recognition of the stresses parents experience with a child in an NICU have significantly advanced the concept of family centred care (FCC) in this environment. FCC shapes the underlying framework for discussion between parents and caregivers and serves a particular role in facilitating and ensuring the parental role in EOL decisions with physicians and the team. It is fundamental to consider family interests when making health care decisions for sick newborns [22]. Indeed, the majority of neonatologists ascribe to an incorporation of family interests into decision-making for their incompetent newborn patients [23]. Neonatal health care professionals demonstrate respect for parental authority because of the presumption that parents will make decisions that are in their child's best interests. Respect for parental authority incorporates the notion that parents act intentionally, with understanding and without external controlling influences, and that they choose voluntarily and intelligently from among various options whose relative risks and benefits have been fully explained to them by their physicians.

7.6 From End-of-Life Decision-Making to Organ Donation: Changing the Conversation

Introducing parents to the concept of organ donation after brain death moves the EOL conversation from a discussion of the inevitability of death or a predicted poor quality of life to a discussion about the criteria for confirmation of whole brain silence. A possible negative consequence of this shift is that parents may mistakenly believe that withdrawal of life-sustaining medical treatment should be undertaken *only* when electrocerebral silence is confirmed. This could result in an infant being kept alive longer than might otherwise be the case, solely to determine if the criteria for brain death *can* be met. This raises the potential for the infant's condition to change during the evaluation period such that, for example, an infant with severe hypoxic-ischemic encephalopathy who would have succumbed after withdrawal of assisted ventilation may survive in a severely compromised state.

For neonatal teams, finding the right balance between EOL decision-making and broaching organ donation requires sincere empathy, insight and excellent communication skills. Neonatologists need to be conscious of the impact of *their* authority and the power that derives from expertise or that follows from information given in a convincing manner. They also have to recognize that the ability of parents to reason and to act in a self-directed way may be diminished at the time when death appears imminent. Even when both the parents and the caregivers wish to salvage some good from a tragic neonatal death, extreme caution should be exercised in discussions about potential organ donation and even more so when responding to parents' requests to pursue donation in situations where success is unlikely.

As raising the possibility of organ donation changes the dynamic of neonatal physician-parent interaction it becomes important to question who should broach the idea of organ donation with the parents. Should there be a clear separation between the NICU care team and the organ procurement team? Truog has suggested that the desire for organ procurement has become excessive [19]. He describes the tension when organ procurement organizations adopt a strategy known as the "presumptive approach for organ donation" and contrasts this with the standard approach. For example, in the former approach the language used would be—"Most people, if given the chance to save a life will do it"—as compared with the standard "We will support you whatever choice you make." He considers that "the pendulum has swung too far in the direction of procuring organs at the expense of commitments fundamental to the patient-physician relationship" [24, p. 1210]. This concern regarding the potential conflicting obligations for the neonatal team is worth noting. Integrating neonatal care teams' communication and end-of-life practices with an organ procurement spokesperson or team is essential to ensure harmonious transition at this critical juncture [24].

A further issue that may emerge is how neonatal organs are allocated. Children, and especially newborns, have a higher waitlist mortality than other patient groups [7]. It is assumed that increasing donations from newborns will increase the rate of transplantations to children. This may not be the case, however, and data suggest that organs from newborns often go to adults instead [25]. Whether this is important to families of neonatal donors is unclear and will need to be elucidated.

7.7 Conclusion

Solid organ transplantation is becoming an increasingly effective treatment for a range of disorders affecting newborns [6]. Currently, one of the main limitations in advancing the field is the shortage of donor organs. There are many practical and ethical reasons why organ donation from newborns has lagged behind that of older infants and children. Current evidence does suggest there is a small but defined population of neonatal patients who could fulfill criteria for neurological or circulatory death, despite the considerable medical and ethical challenges that accompany consideration of the newborn as organ donor. Nevertheless, through education, exploration, reflection, and public discourse, it is likely that medical, nursing and transplantation teams will find these challenges to be surmountable, and so enable families to consider organ donation from their newborn infants.

References

1. Sheldon M. Children as organ donors: a persistent ethical issue. Camb Q Healthc Ethics. 2004;13(2):119–22.
2. Brierley J. Neonatal organ donation: has the time come? Arch Dis Child Fetal Neonatal Ed. 2011;96(2):F80–3.
3. Nakagawa TA, Ashwal S, Mathur M, Mysore M. Clinical report-Guidelines for the determination of brain death in infants and children: an update of the 1987 task force recommendations. Pediatrics. 2011;128(3):e720–40.
4. Shemie SD, Doig C, Dickens B, Byrne P, Wheelock B, Rocker G, et al. Severe brain injury to neurological determination of death: Canadian forum recommendations. CMAJ. 2006;174(6): S1–13.
5. Brierley J, Hasan A. Aspects of deceased organ donation in paediatrics. Br J Anaesth. 2012;108(Suppl 1):i92–5.
6. Wright JC, Barlow AD. The current status of neonatal organ donation in the UK. Arch Dis Child Fetal Neonatal Ed. 2015;100(1):F6–7.
7. Boucek MM, Mashburn C, Dunn SM, Frizell R, Edwards L, Pietra B, et al. Pediatric heart transplantation after declaration of cardiocirculatory death. N Engl J Med. 2008;359(7): 709–14.
8. Labrecque M, Parad R, Gupta M, Hansen A. Donation after cardiac death: the potential contribution of an infant organ donor population. J Pediatr. 2010;158(1):31–6.
9. Saha S, Kent AL. Length of time from extubation to cardiorespiratory death in neonatal intensive care patients and assessment of suitability for organ donation. Arch Dis Child Fetal Neonatal Ed. 2013 Oct 8.
10. Charles E, Scales A, Brierley J. The potential for neonatal organ donation in a children's hospital. Arch Dis Child Fetal Neonatal Ed. 2014;99(3):F225–9.
11. Mathur M, Castleberry D, Job L. Identifying potential heart donors among newborns undergoing circulatory determination of death. J Heart Lung Transplant. 2011;30(4):389–94.
12. Curley MA, Harrison CH, Craig N, Lillehei CW, Micheli A, Laussen PC. Pediatric staff perspectives on organ donation after cardiac death in children. Pediatr Crit Care Med. 2007; 8(3):212–9.
13. Ives-Baine L, Faust J, Drewry J. When birth and death collide: best practices in end-of-life at the beginning. In: Cox GR, editor. Amityville, NY: Baywood Publishing Co; 2013.
14. Larcher V. Ethical considerations in neonatal end-of-life care. Semin Fetal Neonatal Med. 2012;18(2):105–10.
15. Berry MA, Shah PS, Brouillette RT, Hellmann J. Predictors of mortality and length of stay for neonates admitted to children's hospital neonatal intensive care units. J Perinatol. 2008;28 (4):297–302.
16. Mitchell C. Care of severely impaired infant raises ethical issues. Am Nurse. 1984;16(3):9–10.
17. Barry MJ, Edgman-Levitan S. Shared decision making–pinnacle of patient-centered care. N Engl J Med. 2012;366(9):780–1.
18. Holm S, Edgar A. Best interest: a philosophical critique. Health Care Anal. 2008;16(3): 197–207.
19. Truog RD, Sayeed SA. Neonatal decision-making: beyond the standard of best interests. Am J Bioeth. 2011;11(2):44–5.
20. Rhodes R, Holzman IR. Is the best interest standard good for pediatrics? Pediatrics. 2014;134 (Suppl 2):S121–9.
21. Kopelman LM. The best interests standard for incompetent or incapacitated persons of all ages. J Law Med Ethics. 2007;35(1):187–96.
22. Hardwig J. What about the family? Hastings Cent Rep. 1990;20(2):5–10.
23. Hardart GE, Truog RD. Attitudes and preferences of intensivists regarding the role of family interests in medical decision making for incompetent patients. Crit Care Med. 2003;31 (7):1895–900.

24. Truog RD. Consent for organ donation–balancing conflicting ethical obligations. N Engl J Med. 2008;358(12):1209–11.
25. Ross LF, Frader J. Are we ready to expand donation after cardiac death to the newborn population? J Pediatr. 2011;158(1):6–8.

Chapter 8
Ethical and Organizational Issues in Adopting a Pediatric Protocol for Controlled Donation After Circulatory Determination of Death

Charlotte H. Harrison

Abstract Despite two decades of ethical and clinical debate, controlled donation after circulatory determination of death (cDCDD) remains a controversial practice, especially in pediatrics. While debates continue both nationally and internationally, it can be challenging for pediatric hospitals to build agreement on a cDCDD policy among their own constituents. After briefly reviewing the state of the larger debates, this chapter describes organizational steps involved in developing a cDCDD policy for one deeply divided pediatric hospital community. Tracing the experience of a large, tertiary-care teaching hospital and transplant center, the chapter reports the work of a multidisciplinary task force that developed an empirical evidence base and an ethical "constitution" for a cDCDD program, as well as a protocol for pediatric cDCDD. The chapter highlights a number of philosophical and practical issues considered in this process, including questions about the compatibility of a cDCDD program with the overall mission of the hospital. Also described are organizational measures aimed at forging a transparent, legitimate policy in a context of partial stakeholder agreement and persistent controversy.

Keywords Organ donation after cardiac death · Organ donation after circulatory determination of death · Pediatric ethics · Organizational ethics · Transplantation ethics

Controlled donation after circulatory determination of death (cDCDD) has long been practiced in an environment of ethical controversy. There are many points of contention, including several issues that are made more complex when the prospective donor is a child. When disagreement exists within a single institution, it can be difficult to develop an institutional policy regarding cDCDD that is widely accepted as ethical among the staff involved.

C.H. Harrison (✉)
Boston Children's Hospital, Boston, MA, USA
e-mail: charlotte.harrison@childrens.harvard.edu

R.A. Greenberg et al. (eds.), *Ethical Issues in Pediatric Organ Transplantation*,
International Library of Ethics, Law, and the New Medicine 66,
DOI 10.1007/978-3-319-29185-7_8

This chapter will describe a multistage organizational ethics process that was undertaken by Boston Children's Hospital (BCH) with the aim of addressing the deep divisions within its staff with regard to offering cDCDD. BCH is a 395-bed freestanding pediatric tertiary-care hospital and transplant center, a teaching hospital of the Harvard Medical School. In describing the BCH experience, we will trace organizational steps involved in debating and developing a policy for a community of staff and families with a range of views on the ethics of pediatric cDCDD. We will highlight a number of philosophical and practical issues considered in this process, including questions about the compatibility of a cDCDD program with the overall mission of the hospital. We will also summarize some key elements of the policy and protocol that have resulted. First, by way of context, we will briefly note the state of debate on some of the more contentious issues that were faced by the hospital when this initiative began, in 2005, and that remain controversial in the larger society today.

8.1 National and Global Issues

In national and international surveys, ethically-significant differences in policy and protocol for cDCDD have been identified both within and among societies [1–4]. Perhaps most prominent among the points of disagreement is the theoretical definition of circulatory death (in particular, the meaning of "irreversibility" in the cessation of circulation), along with the clinical standards for determining that the relevant point of irreversibility has been reached in a patient's dying process. (Please see Chap. 5 *Death Determination in Pediatric Organ Donation* by Shemie et al. for a more thorough discussion of death determination and "irreversibility.") Some also contest whether death should continue to be a prerequisite for the donation of vital organs [5–7]. At present, however, the dominant legal and ethical standard is the "Dead Donor Rule," which requires that "donors of vital organs must be declared dead before organ removal rather than dying as a result of donation" [8, p. 973].

Death determination in cDCDD remains controversial with respect to patients of any age, though some convergence in standards has been reported. At an international level, Manara et al. [9, p. i115] have described a "growing global consensus that a minimum of 5 min of continuously observed and appropriately monitored absence of the circulation, apnoea, and coma will define the point at which death can be diagnosed"; however, some continuing divergence in both theoretical and practical standards has also been observed [10]. Among some groups, particularly within the US, there is a trend toward acceptance of a shorter observation period than that suggested by Manara et al. The Pittsburgh protocol of the 1990s had set that period at 2 min [11]; however, in an influential 1997 report, the Institute of Medicine recommended a 5-min standard [12]. More recent guidelines from several professional panels and groups have suggested an "emerging … consensus" [13, p. 146] on a more flexible metric—a range of 2–5 min of a circulation [14–16]. A recent statement from an international multistakeholder group has advocated a 2-min standard as a legal matter [17]. The statement acknowledges, however, that this recommendation represents a pragmatic stipulation rather than a theoretical consensus, as the signatories were unable to agree on a "philosophical or conceptual" [17] foundation to support their recommendation. Still more recently, an international consensus process, convened in collaboration with the World Health Organization, has achieved greater agreement on the conceptual understanding of the moment of death than on the precise time period required to reach that moment, once certain criteria have been observed [18].

Several sources have recognized greater doubt with respect to the time of death in children as compared with adults. In its 2001 guidance, the ethics committee of the American College of Critical Care Medicine and Society of Critical Care Medicine (ACCM/SCCM) spoke in somewhat tentative terms regarding the appropriate pediatric standard [14].[1] More than a decade later, the online supplement to the 2013 international multistakeholder statement also highlighted special concerns about children [17]. A 2012 report on autoresuscitation, on which the multistakeholder group appeared to rely most heavily, itself urged further study of autoresuscitation in younger patients [19].

While it has received the most attention in the ethics literature, the determination of the prospective donor's circulatory death is far from being the only ethically controversial aspect of cDCDD, especially in pediatrics. Other issues include the relative uncertainty of neurological prognosis for children [20, 21], the absence of first person consent to donation in most pediatric cases [20], and the designation of appropriate agents (whether clinical team or organ bank) to initiate discussion of cDCDD with parents [21, 22]. Some clinicians have called for a moratorium on pediatric cDCDD [23], as well as adult cDCDD [24], until there is greater public understanding and resolution of a number of issues, including death criteria, informed consent, and conflicts of interest regarding the decision to withdraw the prospective donor's life sustaining treatment [23, 24]. In response, others have contended that these issues can be addressed with sufficient safeguards to make the offering of cDCDD a morally acceptable and even desirable option that families have the right to consider and choose [25, 26].

[1]"Based on anecdotal reports, … basic science reports [studies of laboratory rats and pigs], *and existing guidelines at pediatric transplant centers*, it appears that a period of not less than 2 min of asystole, apnea and unresponsiveness is sufficient to declare death in a pediatric population. Therefore, although the IOM did not specifically evaluate the appropriate interval for death determination in a pediatric population, we believe their *recommendation for 5 min in adults can reasonably apply to pediatric death determination, although a somewhat shorter time interval may be acceptable as well*" (emphasis added) [15, p. 1829]. In the first italicized phrase, note the curious use of de facto hospital guidelines as evidence of a scientific basis for pediatric standards.

In this debate, pediatric health care providers have been found to have substantially divergent views [21, 27, 28]. A recent statement from the American Academy of Pediatrics (AAP) declared that, in light of "legitimate ethical disagreement" [16, p. 1025] regarding cDCDD death criteria, "individual physicians should not be required to participate" in cDCDD [16, p. 1025]. While asserting that hospitals should "provide access to DCD,"[2] the statement also reports that the AAP "could not reach consensus on whether institutions may refuse to participate" [16, p. 1023] in cDCDD.

8.2 The Context at Boston Children's Hospital

Given this diversity in views within the pediatrics community, similar divisions may be expected to exist within the constituencies that are served by a single program or hospital [29]. That was certainly the case for BCH in the early to mid-2000s, when there was a persistent impasse among the staff regarding whether the hospital should offer the option of cDCDD for donation by children. While there was strong support for cDCDD, particularly among renal and transplant teams, serious ethical concerns were expressed by some other staff members, especially within the intensive care units where prospective donors would likely be cared for at the end of their lives. Such a division was observed in a number of pediatric hospitals at the time, due in part to the fact that "staff responsible for implementing [DCD] protocols are not the transplant team members" [30, p. 4].

Along with the concerns we have just reviewed from the literature, there were additional worries among BCH staff about practical intrusions on patient and family experience at the end-of-life, as a result of the necessity to prepare the dying child for organ donation. It was understood that the protocols of some other hospitals called for prepping and draping the prospective donor's body while the patient was still alive [1], allowing at most for family members to hold the hand of the dying donor-to-be. Further, the need to begin procurement within minutes of acirculation, [3]immediately after the declaration of death, appeared to require a quick and abrupt separation of parents from children at a moment when many families would ordinarily choose to linger with their children in the ICU. Some staff were concerned that the dying child might have residual awareness of a loved one's presence—a comfort that should not be lost in the final minutes or seconds of life. Others worried that parents would suffer from having to leave their child at the moment when circulation ceased, or would be unable to bring themselves to leave the room and allow donation to proceed.

[2]The abbreviation "DCD" is retained in direct quotations from sources in which the practice is referred to as "donation after cardiac death".

[3]acirculation—absence of circulation.

8.3 The Organizational Ethics Task Force

With these kinds of controversy in mind, the senior leadership at BCH appointed a 17-member hospital-wide task force to consider "the appropriateness of a pediatric DCD protocol at [BCH][4] in light of the values and mission of the Hospital" [31, p. 5]. The Task Force charge identified two phases of activity. Phase I would comprise a general assessment of the ethical acceptability of pediatric cDCDD. Then, if the group concluded that offering DCDD would be compatible with the hospital's values and mission, a specific protocol would be developed in Phase II.[5]

Task Force membership included representatives from a range of stakeholder services that could be expected to have differing perspectives on the desirability of offering cDCDD. There were members from critical care medicine, critical care nursing, ethics and legal services, nephrology, neurology, operating room nursing, palliative care, pastoral care, pulmonary care, respiratory care, transplant surgery and social work, as well as two parent members. The group was co-chaired by the present author, then a BCH Clinical Ethicist, together with the hospital's Chief of Cardiac Critical Care. While not a voting member, the executive director of the regional organ bank participated as a consultant in a number of meetings.

8.3.1 Initial Research and Evidence Base

Phase I took place from February to June of 2005. Members conducted a thorough literature review, finding that most of the literature dealt primarily or exclusively with adult patients. The Task Force also heard from an invited speaker whose institution was offering pediatric cDCDD. In addition, Task Force subcommittees conducted original research on the following topics.

Numbers of BCH children and families likely to benefit from cDCDD A review of ICU deaths at BCH for the years 2002–2004 indicated that, if cDCDD were offered here, the number of eligible patients whose families choose cDCDD would likely to be 1 or 2 per year, out of 254 total deaths in the relevant ICUs during the 3-year period reviewed [32].[6] Also, it was predicted that most cDCDD organs donated by BCH patients would probably go to adults rather than to children [30].

---

[4]At the time, the Hospital was known as Children's Hospital Boston (CHB), rather than the current BCH.

[5]Unless otherwise specifically noted, the referenced Task Force proceedings and findings are documented in the Task Force Final Report or in other unpublished documents, including protocols and consent forms. All are available from this author at charlotte.harrison@childrens.harvard.edu.

[6]This research excluded patients under 3 months of age. A subsequent BCH study focused on the potential scope of donation among infants [33].

Attitudes and concerns about cDCDD among BCH staff Focus group discussions were conducted with 8 groups of BCH staff, including nurses from the OR and 2 ICUs, members of chaplaincy and clergy, respiratory therapists, surgeons, anesthesiologists and intensivists. Participants' primary reason for supporting DCDD was to make donation possible for families who would find it meaningful. Most participants appeared to wish to help families by offering cDCDD if the issues that troubled them could be satisfactorily addressed. Each focus group identified a slightly different array of highest-priority concerns. These included questions about the dying child's best interests, issues in approaching and preparing families, and concerns about determining death and maintaining the quality and integrity of the cDCDD process. There were also worries about public understanding of cDCDD and trust in the hospital [22].

Policies, protocols, experience and views of cDCDD from other pediatric institutions Structured phone interviews were conducted with 19 selected pediatric institutions and centers across the US, using a standardized questionnaire. Findings indicated substantial differences in attitudes toward cDCDD, with a small majority of institutions surveyed adopting or leaning toward adopting a protocol and the others declining to do so or deferring a decision. Three institutions had developed protocols primarily for pediatric patients, while six others permitted pediatric cases but used adult protocols. One had performed pediatric cDCDD procedures but did not have a specific protocol. At one institution, all families of potential cDCDD candidates were approached about their interest in donation; at three others, cDCDD was undertaken only when the family initiated a request [31, p. 52].

In addition, Task Force members investigated family and public attitudes toward cDCDD, including a review of scholarly literature and popular media, as well as literature on parent and public attitudes toward organ donation in general. The hospital chaplaincy service provided a summary of relevant religious values and views of religious traditions regarding organ donation, while also reporting the concerns of BCH chaplains about cDCDD. A hospital attorney reviewed the relevant legal issues. Finally, in collaboration with the hospital finance team and the regional organ bank, the financial implications of cDCDD for the institution and donating families were assessed.

8.3.2 *"Foundation Conditions" for Pediatric cDCDD*

At the end of Phase I research and deliberation, there remained strong differences in judgment among the members as to the ethical acceptability of cDCDD for children at BCH. One of the 17 members felt unable to support the adoption of cDCDD because of such factors as the use of premortem interventions that were not for the patient's direct benefit, the need to retrieve organs within a fixed period of time after death, and the limitations placed on parental contact with the child. Rather than opposing the adoption of a protocol, however, this member chose to "stand aside"[7] from what was otherwise a group consensus: that the answer to the ultimate question—yes or no to cDCDD—depended on whether a protocol could be developed that was both clinically feasible and ethically justifiable. Thus a policy conclusion was reserved until attempts could be made to design a protocol.

At the same time, the Task Force agreed on fundamental ethical standards to be employed in the development and evaluation of the protocol. The group identified a set of eight "Foundation Conditions" as the ethical grounding on which any cDCDD program should be built. These conditions (which are condensed in the list below[8]) were aimed at ensuring the following protections for hospital patients, families and staff:

- Appropriate decisions to withdraw life support, independent of donation considerations.
- Thorough disclosure and genuine informed consent to changes in end-of-life care required for cDCDD.
- Compassionate end-of-life care, with no physical harm, suffering or hastening of death to the child.
- Consistent quality of ICU care, whether child is an organ donor or not, such that family can trust that the dying child's welfare is the caregivers' paramount concern.
- Justifiable criteria for death determination, by both medical and ethical standards, so that the child will clearly be dead before organ procurement begins.
- Independent oversight and monitoring of the cDCDD process.
- Respect for diverse values within the staff, including a right for staff to exercise conscientious objection.
- Protection of the family from any extra financial costs by virtue of their child's participation in cDCDD.

At the request of the Task Force, these conditions were approved unanimously by the hospital's Medical Staff Executive Committee as a core organizational commitment in the event that the hospital chose to adopt a cDCDD program.

[7]This member felt that these donation-related measures would "compromise the human dignity of the patient and therefore cause […] harm"; however, she determined that "these reservations, based on personal spiritual beliefs and values, should not prevent the Task Force as a whole from proceeding" to try to design a protocol that met the Foundation Conditions [31, p. 56–57]. In consensus decision-making, a member's decision to "stand aside" represents his or her determination that it is better for the group to proceed with its decision, without his or her support, than to be left without a decision [34]. After standing aside at the end of Phase I, this BCH Task Force member participated actively in Phase II. Her views are represented in the votes and comments reported below.

[8]The complete text of the Foundation Conditions is reproduced in a 2008 report [20].

8.3.3 *From Principle to Practice*

With official adoption of its guiding ethical principles, the Task Force turned to designing the specifics of a cDCDD program. The protocol was designed with kidney donation in mind, since most pediatric cDCDD donations were kidney donations at that time. (There was no determination that other organs should not be donated; they simply were not discussed.) The program was intended for candidates who were not brain dead but were severely and irreversibly brain damaged (e.g., from head trauma, anoxia, cerebral vascular accident). Contraindications included HIV/AIDS, an active malignancy (except primary brain tumor) or encephalopathy of unknown origin. Task Force and hospital leaders recognized the possibility that donation would be desired by conscious patients with degenerative neuromuscular disorders; however, they determined that planning for such donations would require special consideration beyond the scope of the initial Task Force process.

Four new subcommittees were formed to address key programmatic elements. Two dealt directly with program design; one focused on the premortem protocol and the second on the process and content of informed consent. Two others investigated underlying issues: the time of death in children and the meaning and relevance of the "best interests" standard for severely brain-damaged children.

These subcommittees extended their reach to engage additional stakeholders beyond the original task force members. For example, the protocol design subcommittee included additional ICU nurse and physician representatives, and the time of death group enlisted an additional neurologist. Both the protocol and informed consent groups consulted with the regional organ bank. The time of death subcommittee also sought to engage a larger group of colleagues, by publishing a letter highlighting concerns regarding children's possible residual awareness at the time of organ procurement [35]. The "best interests" group developed a question for later consideration in a community-engagement process.

The subcommittees met in parallel following adoption of the Foundation Conditions by the hospital's Medical Staff Executive Committee. Subcommittee findings, along with specific protocol and policy features, were considered in a rolling process of plenary meetings between September 2005 and February 2006. An informed consent procedure and form were also reviewed. All proposed provisions were tested against the Foundation Conditions.

8.3.3.1 Protocol

Key features of the resulting protocol included a range of safeguards addressing the Foundation Conditions. For example, in order to protect the dying child's comfort and avoid hastening his or her death, the protocol ruled out the use of premortem cannulation or administration of vasodilators. It also restricted donation-related interventions to minor measures, such as fluid administration, subject to the consent of the family, and forbade major interventions solely for the purpose of facilitating donation, such as blood transfusion, mechanical circulatory support or cardiopulmonary resuscitation. Premortem administration of heparin was permitted, but only in the last moments before death, in patients who met strict blood pressure guidelines [12].

Working from a legal definition of death as "irreversible cessation of circulatory and respiratory functions" [12, p. 40], the Task Force accepted a 5-min period of acirculation as sufficient to indicate that death had occurred, in circumstances in which there had been a morally justifiable decision that resuscitation would not be attempted. The group was aware of more liberal standards in the Pittsburgh protocol and the ACCM/SCCM position paper; however, it appeared that a 5-min period would be compatible with successful kidney donation.

Much attention was given to orchestrating the end-of-life process so as to maintain conditions as close as possible to those available when life support was withdrawn in the absence of plans for organ donation. The protocol called for the use of two adjacent operating rooms, one of which would function as an anteroom for extubation; there family members could be present and hold the child if they wished to do so. Arrangements would be made for furnishings not usually available in the OR, such as rocking chairs and a second bed next to the patient's. Music, rituals and other observances could be facilitated as if in an ICU room.

Extubation would be managed by the ICU team, while transplant-related monitoring would be conducted by OR nurses and transmitted to organ bank staff waiting outside the room. Once acirculation was observed and announced to the family, the patient would be wheeled to a second OR, where the ICU attending would remain with the patient until s/he had confirmed the absence of autoresuscitation after 5 min of acirculation and declared the patient's death. During the 5-min wait period, the patient would be prepped and draped for surgery by an OR nurse. After the death declaration, the family would be notified and the care of the patient's body would be transferred to the surgical team for donor surgery. Meanwhile, the family could remain in the first OR or return to the ICU, chapel or another space as they wished.

8.3.3.2 Informed Consent

Unlike donation after brain death, cDCDD requires consent not only for disposition of organs after death but also for changes in end-of-life care. The former is the province of the organ bank and the individual authorized by law to donate the child's organs. The latter is the concern of the patient's health care providers and the parents or other health care proxies.

With separation of these two domains in mind, an informed consent procedure was designed by the Task Force in collaboration with the organ bank. In conversations with the family, the ICU team was to be responsible for describing how the cDCDD process would change the usual delivery of end-of-life care, discussing any interventions that required specific consent, informing the family that organs might go to other children or to adults, and explaining the family's right to cancel the donation at any time, up to the moment when the patient was taken to the second operating room for organ procurement.

All of these points were documented in the ICU consent form for premortem care. Blank spaces were left in the form to describe unusual features of the expected protocol and those that required explicit permission. The ICU consent form was separate from the authorization form that was used by the organ bank to obtain permission from the family for disposition of organs and tissues after death.

Recalling the findings from the Phase I study of ICU deaths among BCH patients [32], the Task Force anticipated that eligible cDCDD donors would be few. Accordingly, it was expected that staff-family discussion of cDCDD would be a relatively rare event. Given the complexity of the relevant ethical considerations, it was recognized that staff could be aided by a just-in-time resource for conducting these sensitive discussions. An educational DVD was created by members of the Task Force in collaboration with the New England Organ Bank and the Institute for Professionalism and Ethical Practice at BCH. The DVD includes a simulated discussion by BCH and NEOB staff with professional actors taking the roles of the parents [36].

8.3.4 *Questions of Organizational Mission*

In early February of 2006, one year after the Task Force had first convened, the state of agreement within the membership was mixed. A detailed, confidential survey was conducted, in which each Task Force member was asked to indicate his or her level of satisfaction that each of the eight Foundation Conditions could be met and to identify any concerns or information needs. Responses were anonymized and shared with all members of the Task Force.

Task Force members generally agreed that the proposed informed consent process and most aspects of the protocol were the best blueprints the Task Force could design for cDCDD at BCH, if cDCDD were to be offered at all; however, some were not certain that all Foundation Conditions would be met in practice. For example, if too few staff members were willing to take part in cDCDD, there could be pressure on others to participate against their conscience. In addition, it could be hard to maintain a separation between the decision to withdraw life support and the decision to donate organs, if a particular family asked about donation before the clinical team raised the subject. There was also concern that details of the oversight process had not been established; it was emphasized that the integrity of the protocol would require the involvement of high-ranking staff and a commitment of hospital resources.

Most members had come to believe that it would be ethical for families to choose cDCDD for their own children under the protocol proposed by the Task Force. That is, they no longer saw the "best interests" standard for proxy consent as a bar to parental agreement with the changes in end-of-life care required for cDCDD. At the same time, opinion was divided as to whether the hospital itself could ethically undertake a cDCDD program. For example, some questioned whether it was right to make donation-driven changes to ICU procedures that would apply to many ICU patients, not just to those few who would turn out to be eligible cDCDD donors. A key point of contention was the addition of preliminary donor-screening for all potentially cDCDD-eligible children before their families were approached to discuss the process of withdrawal of life support. Would the addition of that screening process be consistent with the hospital's overall mission, including its responsibilities for the care of children and families who would not be eligible to participate in cDCDD, as well as its obligations to support the integrity of its staff?

The most contested issues included these:

Significance of the numbers; effects of cDCDD on overall ICU care The Phase I Durall data had shown that offering cDCDD would probably benefit only 1 or 2 BCH donor families and 2–4 (likely adult) kidney recipients each year [32]. To some Task Force members, these numbers were irrelevant to whether cDCDD should be offered: each patient or parent was to be treated as an individual rather than a number. It was important to offer the option of donation, regardless of how rarely that option could be exercised in practice.

For other members, the data suggested that the costs of cDCDD could be greater than the benefits. These members felt that greater priority should be given to the far larger number of children who would die in the relevant ICUs without being able to donate. They feared that organ bank requirements to screen most dying children for eligibility as cDCDD donors would dilute the child-focused ethos of the ICU. Some worried that dying children would begin to be regarded primarily as "vessels" for supplying organs to others. One saw a link to the history of abusive experimentation on disabled patients and wondered if cDCDD could be the first step down a slippery slope toward further abuse, such as using these children as subjects on whom trainees could practice procedures.

Obligations to families: choice, benefit and fairness Among Task Force members, the most widely-accepted reason for cDCDD was the desire to make donation possible for families (and older patients) who clearly wanted it. At stake were two immediate considerations: autonomy and benefit. First, offering the option for cDCDD would be a way to honor the preferences and values of some BCH patients and families. Also, many members believed that organ donation could be very comforting to families. Some who had worked with families of brain dead donors had been moved by the depth of their gratitude for the chance to donate. This view was supported by anecdotal reports from other pediatric institutions regarding their families who had chosen cDCDD. It was recognized that, in these cases, families might place a higher value on donating organs than they would on experiencing the kind of end-of-life care that had become standard in BCH ICUs. It was also noted that some older children might have indicated a desire to be an organ donor. In the view of many members, failing to offer this option would be unjustifiably paternalistic.

At the same time, other Task Force members believed that more research was necessary in order to understand the short- and long-term effects on families of participating in cDCDD [37]. What if families regretted their choice to attempt donation? There was also concern for families who would *not* choose cDCDD and who might be upset by being asked about organ donation. Some members worried that these families would perceive the staff to have a conflict of interest concerning their child's prognosis and care, and that these families would come to mistrust their ICU caregivers and the hospital.

A final aspect of this debate was a larger question of fairness or reasonableness in the choice of services offered at BCH, in comparison with those that might be available at some other (adult) hospitals in the Boston area. Was it unfair or unreasonable to deny the option of cDCDD to families at BCH, if they would have been able to donate at another local hospital? It was noted that BCH did not offer every service that could benefit its patients and families; for example, burn care and maternity care had to be obtained at other hospitals. On the other hand, patients could usually transfer to other hospitals if they needed services of that kind, whereas transfer would probably not be feasible or humane (or desired by families) when their children were being cared for in an ICU near the end of life. After recognition of this difficult situation, discussion focused on two questions. First, would cDCDD be so important to families that, if they had known at the time of admission that they could obtain it elsewhere but not at BCH, they would have preferred to take their child to another hospital? Would they consider themselves worse off, overall, because they had come to BCH? Second, was cDCDD such a standard, non-controversial, therapeutic option that it would be reasonable to expect every pediatric hospital to provide it?

Obligations to the common good Many members believed that BCH should generally aim to support transplantation as an element of the common good. Offering cDCDD was seen as a way of affirming the importance of the life-saving potential of transplantation. Some members emphasized the continuing need of BCH patients for donated organs. It was suggested that, as an institution that accepts organs from the national organ pool, BCH owed a duty of reciprocity to contribute organs in return. Others argued, however, that donated organs were allocated to individual recipients, not to institutions, and that the idea of direct reciprocity entailed an unacceptable way of looking at human organs—as commodities for exchange rather than "gifts of life." They also suggested that any institutional obligation to the transplant network should not be met at the expense of cDCDD donor-children if these children might suffer any harm in the process. It was suggested that the hospital could and should look for other ways to bolster its support for organ donation. In addition, there were questions as to whether increasing donations via cDCDD might result in a decrease in donations after brain death, with a net decrease in the number of organs transplanted successfully.

Obligations to support staff members' integrity; tensions between commitments to accommodate conscientious objection and provide continuity of care Some members questioned the justifiability and feasibility of a cDCDD program in the event of extensive conscientious objection by staff. Based on their own objections or on objections they had heard from others, they worried that too few ICU clinicians and psychosocial staff, especially chaplains, would be willing to participate in cDCDD. In that case, they feared, it would not be feasible for many staff to avoid participation without pressure or sanction. In addition, morale and patient care could be adversely affected, as staff members were required to choose between their objections to cDCDD and the need to provide continuity of care to patients and families.

One member's comments summarized many concerns:

> I am now convinced that the protocol is ethical, but I do not think it is appropriate for our population. The best scenario is that a family requests DCD, and we are able to respond to their deeply felt desire to make this happen and do it well. It is high risk for families and staff. The worst scenario is that we assume all potential DCD donor families should be approached, and in approaching all families cause them trauma at a time they are very vulnerable and [lose] the trust they have placed in [us]. Another bad case scenario is that we decide to proceed, but the staff that would be asked to facilitate this would be morally distressed or unwilling to participate.

A specific area of contention, suggested by this comment, was whether cDCDD should be offered to all families or only to those who asked about donation. The latter approach, which was dubbed "don't ask/do tell," was generally criticized because of perceived unfairness to families who might wish to donate but be unaware of the possibility. At the same time, it could be seen that waiting for family requests could greatly reduce two general objections to having a cDCDD program at all: first, that being approached for donation would traumatize some families, and second, that the necessity to screen every potentially-eligible patient would change the ethos of ICU care.

8.3.5 Final Views and Recommendations of the Task Force

As all these views were explored, it became clear that there was a central fault line among the members' positions on the ultimate acceptability of cDCDD protocol at BCH: the non-medical criteria for eligibility of individual patients to become prospective cDCDD donors. If patient eligibility were strictly limited, there was much greater support within the Task Force for offering cDCDD. The final agenda for Task Force deliberations was structured accordingly.

8.3.5.1 Policy Options

After circulating a comprehensive summary of the pros and cons of cDCDD as the Task Force had seen them, the co-chairs reconvened the group for a final discussion and vote. Members were asked to indicate their degree of agreement with offering cDCDD to two different populations, assuming BCH adoption of the Task Force protocol and consent procedure. They were also invited to explain their reasoning, in comments that would be shared in the final report to hospital leaders.

Eleven of the 17 Task Force members favored offering cDCDD to patients of any age, including small children and infants. Their comments included the following:

> I feel that parents can make this decision for minors. The decision to withdraw care is more weighty than DCD [31, p. 128].

Other members strongly disagreed with offering cDCDD unless it were restricted to "competent adults and mature or emancipated minors who have signed donor cards or entered their names in a donor registry." Among these members, three indicated complete agreement with cDCDD for this limited group and three others offered lesser support. In arguing for these limits to eligibility, one member stated:

> This policy – compared to offering DCD for all ages – reduces the effects of DCD on overall ICU care, since we would only have to consider DCD (i.e., screen patient, involve organ bank, work in the shadow of conflicts of interest around end-of-life decisions) for this small and clearly defined sub-population. Other patients and families wouldn't be affected [31, p. 127].

Another member explained affirmative reasons for supporting this option for older patients:

> A DCD protocol to advocate for the wishes of these patients is important and desirable. We should support the autonomy of our adult and mature patients [31, p. 127].

These results, including all members' final comments, were included in a 170-page final report [31].

8.3.5.2 Implementation Contingencies

Also included in the final report was a unanimous recommendation from the Task Force that three organizational steps should be taken as prerequisites for offering cDCDD:

An independent survey of staff members who would be expected to participate in the protocol. General concerns of staff, for and against cDCDD, had been brought to the surface in Phase I of the Task Force process, through the non-confidential focus groups whose views were about to be published [22]. The case scenario presented to those groups had been based on a protocol written by the organ bank for use with adult patient-donors. By contrast, the protocol developed by the Task Force had been written specifically for pediatric patients at BCH. Once the protocol was available for discussion, and in the context of an identity-protected survey, the

Task Force judged that it should be possible to gauge with more precision the acceptability of cDCDD to each staff member who might be asked to participate.

A process of community review. In its 1997 report on cDCDD, the Institute of Medicine had recommended that the adoption of protocols be characterized by "complete … public openness" [12, p. 4–5] and that community representatives, including "patient and donor families," be involved in the approval process [12, p. 5]. The BCH Task Force had included two parent members, but it was felt that a more independent public consultation would be a helpful next step.[9]

An undertaking by the hospital and the regional organ bank to build a relationship of greater trust and collaboration between them, with particular emphasis on working relationships in the ICUs. During the years of pre-Task Force debate, tensions had arisen between the organ bank and the ICUs. Some of these had been ameliorated in the Task Force process, through mutually respectful collaboration with the then-current Executive Director of the organ bank. The Task Force asked for a commitment from senior leaders of both institutions to direct and support a continuing effort at relationship-building.

All Task Force members reviewed the report in draft and agreed that it accurately reflected their deliberations. The full report was then submitted to hospital leadership. It was suggested that the hospital's senior leaders were in the best position to determine a policy, based on the work of the Task Force, the results of the recommended staff survey and community consultation, and the leaders' own understanding of the hospital's overall mission.

8.4 Evolution of Policy

As a matter of initial policy, hospital leaders affirmed the Foundation Conditions and approved the Task Force protocol and consent form. With respect to patient eligibility, leaders chose to begin with the group that had the broadest support from the Task Force: competent adults and mature or emancipated minors for whom there was documented first person consent to donate their organs for transplantation (e.g., a signed donor card or an entry in a donor registry). Some months later, the policy was expanded to allow the offering of cDCDD for clinically-eligible candidates of all ages, but only if the family had inquired about organ donation on its own initiative. This "don't ask/do tell" policy was envisioned as a temporary measure; it was determined that these limitations would be revisited after the staff survey and community consultation. Ultimately, with the strong support of both the staff survey respondents and the Community Ethics Committee, these categories

[9]The Task Force made no specific recommendation as to how or with whom the consultation should be conducted. In the year following issuance of the final report, the Harvard Community Ethics Committee was formed, as a policy-related resource for all of the Harvard-affiliated teaching hospitals (http://www.medicalethicsandme.org). The BCH consultation with regard to cDCDD would become the first hospital submission to this group.

were expanded. The opportunity for cDCDD was then offered for clinically-eligible candidates of all ages and their families, regardless of who had initiated the conversation about organ donation.

The protocol was initially limited to the procurement of kidneys, the only organ considered by the Task Force. Eventually, on the basis of clinical developments in transplantation, this policy was expanded. Any organ could be procured, as long as the premortem process could be conducted in conformity with the Foundation Conditions and without altering the approved protocol. For purposes of this requirement, organ-related variances in warm ischemic times considered compatible with successful donation have not been considered to be alterations in the protocol.

Pursuant to the Foundation Condition regarding oversight, the hospital established a cDCDD advisory committee to supervise the implementation of the protocol and policy, including staff education and survey, community involvement, and development of the relationship between the ICUs and the organ bank. The committee was also charged with advising participants about the protocol in each of the first several cDCDD cases, reviewing the conduct of each case and reporting that review to hospital leaders, and vetting any proposals to alter the protocol or policy, which would require senior leadership approval. After a number of years, these functions were transferred to an Organ Donation Oversight Committee that is responsible for policies, protocols and implementation of both cDCDD and donation after brain death.

Policy changes have continued to be proposed, both by internal stakeholders and by the organ bank. In these cases, reference to the Foundation Conditions has continued to be essential; there has been no proposal to alter those conditions. In recent years, when memories have faded and several leaders from the Task Force era have left the hospital, the detailed arguments articulated in the final report have been instrumental in helping to clarify what is at stake in considering particular proposals for change.

8.5 Lessons and Limitations

In the BCH experience, several organizational approaches proved to be useful for the development and implementation of a policy. These included the creation of a hospital-wide deliberative task force and the conduct of primary research on foundational questions, such as the likely incidence of cDCDD donation at the hospital and the views of staff focus groups as to the ethics of cDCDD. Since much disagreement within the Task Force concerned the likely long-term effects on the hospital's patients, families and staff, two other measures were helpful in soliciting the views of those constituents. First was the conduct of an extensive, identity-protected survey of all staff members who might be expected to participate in DCDD. Second was the involvement of an external community group to review targeted issues from the perspective of prospective donor-parents and the public.

This process involved significant costs and limitations, and some aspects were criticized. In general, the process and its outcomes reflected a particular organizational culture, where there was a substantial history of ethical objection to cDCDD.[10] Task Force members spent many hundreds of hours engaged in the process, over a period of nearly two years. Some particularly difficult ethical issues were not considered, since the charge to the Task Force did not include heart donation or donation from patients with cognitive capacity who had neurodegenerative conditions. While agreement was reached within the group on most features of a cDCDD protocol, there was far from a consensus on every ethics-related question. Some questions, especially regarding the effects on families, might have been better informed by an earlier effort at parent or community engagement [38], although the forum of the Community Ethics Committee did not exist at the time.

The broad inclusiveness of the Task Force membership was questioned by some, who felt that members involved in the care of prospective organ recipients had a conflict of interest in considering parameters for prospective donor care. From the opposite perspective, others were concerned that accommodation of strong objections from some in the ICU community was having the effect of denying donation to families who wanted to request cDCDD for their young children. In a hospital with a higher likely incidence of clinically-eligible cDCDD candidates, it might well have been more difficult to sustain the extended process followed at BCH.

In the staff survey that followed the Task Force process, respondents indicated significant levels of support for the protocol and for the inclusion of patients of all ages and all families; however, the response rate was only 39 %, leaving uncertainty as to the views of nonrespondents. There has been no difficulty maintaining teams for the few cDCDD cases actually undertaken since adoption of the protocol. At the same time, there remain a number of staff members who choose not to participate in cDCDD.

In this environment of persistent disagreement, three further measures have been valuable in supporting the legitimacy and acceptance of the hospital's cDCDD policy. Perhaps most important was the early adoption of a set of guiding ethical principles to which both Task Force members and senior leaders of the hospital were committed. In the Task Force process, this commitment laid a foundation for trust and provided a consistent guideline against which protocol proposals could be checked; when the same conditions were adopted into hospital policy, they established a continuing "constitution" to guide any future reexamination of policy

[10]The process also took place at a time when there was no national requirement in the US for transplant hospitals to offer cDCDD. That requirement was first published for public comment in August of 2006 [30]. During the comment period, the co-chairs of the BCH Task Force submitted a letter detailing specific pediatric concerns and urging that two measures be undertaken: first, a delay in implementation of the proposed requirement solely with respect to pediatric institutions, and, second, the convening of a national consensus conference on pediatric cDCDD [31]. A report recording the OPO Committee's discussion of these proposals, together with comments from other hospitals, highlights some of the struggles of pediatric institutions at the time in confronting the ethics of cDCDD [30].

and protocol. Also fundamental, at the end of the Task Force process, was the written articulation of both majority and minority opinions on a broad range of contentious issues. Not only did this Final Report provide transparency for colleagues and decision-makers outside the Task Force, but it also supported the integrity of dissenting members, who did not want to be associated with certain prevailing views. Both of these documents—the Foundation Conditions and the Final Report—have provided significant points of reference, both for the Task Force and for other members of the hospital community, in evaluating cDCDD policies over time. Finally, in recognition of persistent points of contention within the Task Force, there was a decision by hospital leadership to move conservatively in formulating a cDCDD policy. Initial policies were so limited as to meet with the approval of most skeptical Task Force members. With the passage of time and with more information about the views of other stakeholders, the hospital expanded its policy, while maintaining conformity with the Foundation Conditions.

8.6 Conclusion

Despite two decades of ethical and clinical debate, cDCDD remains a controversial practice [3, 39], especially among pediatric caregivers and programs [16, 40]. If hospitals are seeking broad staff support for a cDCDD donation protocol, it may be challenging to achieve that support within some morally-diverse populations of professionals [29, 30]. This article has described the policy-development efforts of one pediatric institution in which there were deep and longstanding divisions about the ethics of cDCDD. In that environment, there was greater agreement on underlying ethical principles than on the feasibility of devising a cDCDD protocol that could comport with them. There was also substantial uncertainty about the effects of a cDCDD program on other aspects of the hospital's mission. The organization was able to move forward through an extended, iterative process of fact-finding, deliberation and community consultation.

Debates about the ethics of pediatric cDCDD can be expected to continue. There are questions about death determination, in children and adults, that have yet to be resolved [17, 18]; meanwhile, interest in heart transplantation, particularly for children, will prolong debates about conceptual and clinical aspects of the determination of death [41]. In addition, since younger children will never have made a competent choice for themselves to be an organ donor, one basic ethical backbone of organ donation—the effectuation of the donor's wish—will never be available to justify possible risks to the youngest donors' welfare. Even family consent for children's donation may continue to be contested, because of the challenges to public understanding of cDCDD [42, 43]. New technological developments will continue to introduce new ethical questions [8, 43]. It will require the thoughtful engagement of pediatric professionals, hospitals, patients and families, as well as the larger community, in order to develop ethical responses to these persistent challenges.

References

1. Antommaria AH, Trotochaud K, Kinlaw K, Hopkins PN, Frader J. Policies on donation after cardiac death at children's hospitals: a mixed-methods analysis of variation. J Am Med Assoc. 2009;301(18):1902–8.
2. Fugate JE, Stadtler M, Rabinstein AA, Wijdicks ER. Variability in donation after cardiac death protocols: a national survey. Transplantation. 2011;91(4):386–9.
3. Rodriguez-Arias D, Smith M, Lazar N. Donation after circulatory death: burying the dead donor rule. Am J Bioeth. 2011;11(8):36–43.
4. Dhanani S, Hornby L, Ward R, Shemie S. Variability in the determination of death after cardiac arrest: a review of guidelines and statements. J Intensive Care Med. 2012;27 (4):238–52.
5. Morrissey PE. The case for kidney donation before end-of-life care. Am J Bioeth. 2012;12 (6):1–8.
6. Morrison W. Organ donation prior to death: Balancing benefits and harms. Am J Bioeth. 2012;12(6):14–5.
7. Truog RD, Miller FG, Halpern SD. The dead donor rule and the future of organ donation. N Engl J Med. 2013;369(14):1287–9.
8. Bernat JL, Capron AM, Bleck TP, Blosser S, Bratton SL, Childress JF, et al. The circulatory-respiratory determination of death in organ donation. Crit Care Med. 2010;38 (3):963–70.
9. Manara AR, Murphy PG, O'Callaghan G. Donation after circulatory death. Br J Anaesth. 2012;108(S1):108–21.
10. Gardiner D, Shemie S, Manara A, Opdam H. International perspective on the diagnosis of death. Br J Anaesth. 2012;108(S1):i14–28.
11. DeVita MA, Snyder JV. Development of the University of Pittsburgh Medical Center policy for the care of terminally ill patients who may become organ donors after death following the removal of life support. In: Arnold RM, Younger SJ, Schapiro R, Spicer CM, editors. Procuring organs for transplant: The debate over non-heart-beating cadaver protocols. Baltimore, MD: The Johns Hopkins University Press; 1993. p. 55–68.
12. Potts JT, Herdman R. Non-heart beating organ transplantation: medical and ethical issues in procurement. Washington, DC: Institute of Medicine; 1997.
13. Institute of Medicine. Organ donation: opportunities for action. Washington, D.C.: The National Academies Press; 2006.
14. American College of Critical Care Medicine/Society of Critical. Care Medicine. Recommendations for nonheartbeating organ donation. Crit Care Med. 2001;1(29):1826–31.
15. American Society of Transplant Surgeons. ASTS recommended practice guidelines for controlled donation after cardiac death organ procurement and transplantation. Am J Transplant. 2009;9:2004–11.
16. American Academy of Pediatrics. Policy statement: ethical controversies in organ donation after circulatory death. Pediatrics. 2013;131(5).
17. Gries CJ, White DB, Truog RD, DuBois J, Cosio CC, Dhanani S, et al. An Official American Thoracic Society/International Society for Heart and Lung Transplantation/Society of Critical Care Medicine/Association of Organ and Procurement Organizations/United Network of Organ Sharing Statement: Ethical and policy considerations in organ donation after circulatory determination of death. Am J Respir Crit Care Med. 2013;188(1):103–9.
18. Shemie SD, Hornby L, Baker A, Teitelbaum J, Torrance S, Young K, et al. International guideline development for the determination of death. Intensive Care Med. 2014;40:788–97.
19. Sheth KN, Nutter T, Stein D, Scalea TM, Bernat JL. Autoresuscitation after asystole in patients being considered for organ donation. Crit Care Med. 2012;40(1):158–61.
20. Harrison CH, Laussen PC. Controversy and consensus on pediatric donation after cardiac death: ethical issues and institutional process. Transplant Proc. 2008;40:1044–7.

21. Sarnaik AA, Clark JA, Meert KL, Sarnaik AP. Views of pediatric intensive care physicians on the ethics of organ donation after cardiac death. Crit Care Med. 2013;41:1733–44.
22. Curley MAQ, Harrison CH, Craig N, Lillehei CW, Micheli A, Laussen PC. Pediatric staff perspectives on organ donation after cardiac death in children. Pediatr Crit Care Med. 2007;8 (3):212–9.
23. Carcillo JA, Orr R, Bell M, Joffe A, Maffei FA, Sullivan J, et al. A call for full public disclosure and moratorium on donation after cardiac death in children. Pediatr Crit Care Med. 2010;11(5):641–3.
24. Joffe A, Carcillo J, Anton N, deCaen A, Han Y, Bell M, et al. Donation after cardiocirculatory death: a call for a moratorium pending full public disclosure and fully informed consent. Philos Ethics Humanit Med. 2011;6(17).
25. Nakagawa T. Letter to the editor. Pediatr Crit Care Med. 2011;12(3):375–6.
26. Mathur M, Abd-Allah S, Orr RD. The authors reply. Pediatr Crit Care Med. 2010;11(5):644–5.
27. Joffe A, Anton NR, deCaen AR. Survey of pediatricians' opinions on donation after cardiac death: are the donors dead? Pediatrics. 2008;122(5):e967–74.
28. Mathur M, Taylor S, Tiras K, Wilson M, Abd-Allah S. Pediatric critical care nurses' perceptions, knowledge, and attitudes regarding organ donation after cardiac death. Pediatr Crit Care Med. 2008;9(3):261–9.
29. Rhee JY, Ruthazer R, O'Connor K, Delmonico FL, Luskin RS, Freeman RB. The impact of variation in donation after cardiac death policies among donor hospitals: a regional analysis. Am J Transplant. 2011;11:1719–26.
30. OPTN/UNOS Organ Procurement Organization (OPO) Committee. Report of the OPTN/UNOS Organ Procurement Organization (OPO) Committee to the Board of Directors. Organ Procurement and Transplantation Network/United Network for Organ Sharing, 2006 March 23, 2006 Report No.
31. Children's Hospital Boston. Report of the task force on donation after cardiac death. Boston, MA: Children's Hospital Boston; 2006.
32. Durall AL, Laussen PC, Randolph AG. Potential for donation after cardiac death in a children's hospital. Pediatrics. 2007;119(1):e219–24.
33. Labrecque M, Parad R, Gupta M, Hansen A. Donation after cardiac death: the potential contribution of an infant organ donor population. J Pediatr. 2011;158(1):31–6.
34. The Center for Conflict Resolution. Building united judgment: a handbook for consensus decision making. Rutledge, MO: The Fellowship for Intentional Community; 1981.
35. Curley MAQ, Urion DK, Tamara V, Harrison CH, Laussen PC. The authors reply. Pediatr Crit Care Med. 2007;8(5):507–9.
36. Meyer EC, McCrary AM, Truog R. Pediatric organ donation after circulatory death: an innovative educational initiative. The Simulation of Dialysis Access. 2012.
37. Siminoff LA, Chillag K. The fallacy of the 'gift of life'. Hastings Cent Rep. 1999;29(6):34–41.
38. Randhawa G, Schicktanz S, editors. Public engagement in organ donation and transplantation (from the ELPAT Public Issues Working Group). Lengerich: Pabst Science Publishers; 2013.
39. Bastami S, Matthes O, Krones T, Biller-Andorno N. Systematic review of attitudes toward donation after cardiac death among health care providers and the general public. Crit Care Med. 2013;41(3):1–9.
40. Cowl AS, Cummings B, Yager PH, Miller B, Noviski N. Organ donation after cardiac death in children: acceptance of a protocol by multidisciplinary staff. Am J Crit Care. 2012;21(5):322–7.
41. Longnus SL, Mathys V, Dornbierer M, Dick F, Carrel TP, Taevaearai HT. Heart transplantation with donation after circulatory determination of death. Nat Rev Cardiol. 2014;11:354–63.
42. Rodriguez-Arias D, Veliz C. The death debates: a call for public deliberation. Hastings Cent Rep. 2013;43(6):34–5.
43. Zeiler K, Furberg E, Tufveson G, Welin S. The ethics of non-heart-beating donation: how new technology can change the ethical landscape. J Med Ethics. 2008;34:526–9.

Part III
Emerging Challenges

Chapter 9
Ethical Issues in Pediatric Small Bowel Transplantation

Annie Fecteau

Abstract Small bowel transplantation and short bowel syndrome are fields where many ethical issues come to light and leave clinicians with some moral distress. The ethical issues present themselves for most patients in the acute setting of a gastrointestinal catastrophe such as necrotizing enterocolitis totalis or mid-gut volvulus when urgent decisions with possible life-long implications need to be reached quickly. To better understand these ethical issues one has to understand the history of small bowel transplantation, the indications for the procedure, and have a grasp of the recent outcomes. This chapter will succinctly review those medical points before discussing the ethical issues pertaining to pediatric small bowel transplantation. The ethics of emerging technology, best interest principle, accessibility to care, and the societal cost of small bowel transplantation will be explored.

Keywords Intestinal transplantation · Emerging technology · Societal cost · Best interest · Accessibility to care · Pediatric

9.1 The History and Outcomes of Small Bowel Transplantation

Small bowel transplantation was first studied in dogs in the 1960 by Lillehei et al. [1] and Starzl and Kaupp [2]. The first technical success with en bloc liver-intestine transplantation was described by Starzl et al. [3] and Williams et al. [4] in 1989, but both articles only recorded a short post-transplant survival. The same year, Grant et al. [5] reported the first long-term survivor of combined liver-small bowel transplantation. The first successful isolated small bowel transplant was then reported by three groups in the ensuing two years [6–8]. The introduction of a new immunosuppressive medication, tacrolimus, in 1990, improved actuarial graft and

A. Fecteau (✉)
Division of General and Thoracic Surgery, The Hospital for Sick Children, University of Toronto, 555 University Ave, Toronto, ON, Canada
e-mail: Annie.fecteau@sickkids.ca

R.A. Greenberg et al. (eds.), *Ethical Issues in Pediatric Organ Transplantation*, International Library of Ethics, Law, and the New Medicine 66,
DOI 10.1007/978-3-319-29185-7_9

patients' survival rates following all types of intestinal transplantation [9]. The 1997 Intestine Transplant Registry report noted an overall survival rate of 50 % in 260 transplant recipients, two-thirds of whom where children or teenagers. The Intestine Transplant Registry report also emphasized the improved survival of intestinal transplantation patients when the transplant was performed at centers with experience of at least 10 previous intestinal transplants [10].

The use of rabbit anti-thymocyte globulin as induction therapy was initiated in 2002, raising the five-year survival rate to 76 % [11]. The most recent Intestine Transplant Registry report stated that for patients transplanted at centers of excellence the actuarial overall patient survival rate has now reached 93 % at one year, 70 % at five years and 50 % at 10 years (12). There has been no major advances in intestinal graft survival rates for the past five years: short-term outcomes are excellent but longer-term outcomes remain sub-optimal [12]. Rejection, infection, lymphoma, and increasingly chronic rejection, remain prime determinants of long-term survival.

Enteral autonomy and freedom for parental nutrition is achieved in 92 % of those who survive beyond five years [13]. Positive growth has been observed in most pediatric survivors, particularly those receiving steroid-free immunosuppression, but data has shown limited catch-up growth [14]. Neuro-development outcomes post intestinal and multivisceral transplant have been studied only by a few groups. A report from Miami showed that while some children were able to achieve normal developmental outcomes, most children suffered from significant motor delays and several patients experienced severe cognitive delays [15]. When compared to infant recipients of liver transplantation, intestinal transplant patients fared worse in their motor and mental development [16]. A group of investigators from Pittsburgh reported that 37 % of pediatric patients suffered neuropsychiatric impairment, including autism in 32 % and developmental delay in 27 % of patients. 24 % of pediatric patients who had reached adulthood experienced neuropsychiatric disorders consisting of significant hearing loss, developmental delay, major depression, substance abuse, and other axis 1 major mental disorders. Nonetheless, that group also described that 66 % of the survivors who were children at the time of transplantation completed high school or college, with the remaining attending special skills classes or high school in an Individualized Education Program at the time of the report [13].

There is very little data available on the quality of life after pediatric intestinal transplantation. After intestinal transplantation, patients and their families face the new postoperative challenges including the side effects of immunosuppressive drugs, non-adherence and rejection of the transplanted organ, and the psychosocial stresses and financial burden of continued medical care. For approximately 6–12 months post-transplantation, care routines of pediatric small bowel transplant patients may include up to 15 daily medications, tube feedings, intravenous fluids, and maintenance of gastrostomy tube, jejunostomy tube, ostomy and central venous catheter, as well as usual child-care needs [17]. A mean of seven daily medications is still required at more than three years post-transplantation, and more than 17 % of patients continue to require some enteral feeding due to oral aversion [17]. Adult intestinal transplant recipients had better quality of life, scoring significantly better on a Quality-of-Life Inventory, including twelve domains compared to home parenteral nutrition patients [13]. Psychiatric and psychosocial problems affecting quality of life following intestinal transplantation are a function of severity of disease, duration of preoperative parenteral nutrition, length of waiting period and prolonged postoperative course, and vary inversely with available social support [18].

The number of intestinal transplants performed has been steadily decreasing since 2007, when it peaked at 198 transplanted patients in the United States. In 2013, 109 intestinal transplants were performed. The proportion of pediatric patients receiving intestinal transplantation compared to adults is also falling, with the number of adults now equating the number of children [19]. This decrease in transplants is most likely reflective, at least in part, of the emergence and increasing expertise of intestinal rehabilitation teams throughout the world, who are able, for short gut patients, to achieve emancipation from parenteral nutrition and growth with intense medical and surgical therapies before the need for intestinal and liver transplantation arises.

9.2 Indications and Contraindications

Short bowel syndrome can be defined as malabsorption due to insufficient bowel length, after small bowel resection from congenital or acquired lesions. Inadequate bowel length can lead to intestinal failure, necessitating parenteral nutrition. The North American Home Parenteral and Enteral Nutrition Patient Registry indicated one- and four-year survival rates of 94 and 80 %, respectively, for all patients receiving parenteral nutrition at home on account of short bowel syndrome [20]. Most, but not all, infants and children with neonatal-onset short bowel syndrome, resulting mainly from congenital intestinal anomalies, volvulus and necrotizing enterocolitis, survive a period of parenteral nutrition of variable duration. The introduction of fish oil emulsions [21] and intestinal lengthening techniques such as the serial transverse enteroplasty procedure (STEP) procedures have increased the long-term survival for many of these patients [22]. The majority, about 80 %, wean from parenteral nutrition completely during childhood, usually between one and two years of age. A distinct minority, less than 10 %, die from complications, most notably from sepsis associated with central venous catheterization and from liver failure. Those children represent the target population for intestinal transplantation. The American Society of Transplantation has recommended referring children for intestinal transplantation when the intestinal failure is accompanied by impending life-threatening complications such as liver failure, recurring sepsis, and impending loss of central venous access [23]. They also recommend intestinal transplantation in conditions that virtually always result in early death despite optimal parenteral nutrition, such as extremely short bowel syndrome, congenital intractable epithelial disorders, or when the intestinal failure is associated with high morbidity and poor life quality.

Quality of life is a controversial indication for transplantation, particularly in children. When transplant outcomes are excellent, such as in liver transplantation, quality of life as an indication for transplant is accepted in most transplantation centers. In small bowel transplantation, where outcomes have not yet reached excellence, using quality of life is more controversial. It then become difficult for the parents and the care givers to decide what is in the best interest of the child as the transplant may shorten the child's life.

In its position statement the American Transplant Society stated contraindications to intestinal including profound neurological disabilities, life-threatening and other non-correctable illness not directly related to the digestive system, severe congenital or acquired immunological deficiencies or multisystem autoimmune diseases, non-resectable malignancies, and insufficient vascular patency to guarantee central venous access for up to six months following intestinal transplantation [23]. Neurological disabilities as a contraindication for transplant are being revaluated in a lot of transplantation centers. A chapter in this book is exploring neurodevelopmental status as a criterion for solid organ eligibility.

Goulet and Ruemmele [24] stress in their writings the concept that the patient must have the potential to derive an obvious benefit from the transplant procedure. The benefit must remain paramount in view of the high morbidity and mortality rates after intestinal transplantation, especially in cases of poor recipient conditions at the time of transplantation and the reduced donor pool size. Given that the outcomes and quality of life after intestinal transplant are not optimal, it is still challenging to assess when intestinal transplantation is in the best interest of the child. Families and the treating team need to decide if intestinal transplantation should be pursued using the ethical principles of beneficence, non-maleficence and best interest of the child.

9.3 Ethical Issues

9.3.1 Ethical Issues in Emerging Technology

Debate exists regarding the optimal management of patients with short bowel syndrome and there is a wide variation in practices between, and even within institutions, from comfort care to more aggressive surgical measures such as intestinal lengthening and intestinal transplant [25]. Short bowel is at the state of a paradigm shift in accepted practice, like the treatment of hypoplastic left heart, extracoporeal membrane oxygenation (ECMO), and resuscitation of extremely low birth weight were in the recent past. There is moral uncertainty about what is the best treatment option for the children suffering from short bowel syndrome.

Despite having a greater than 20-year history, small bowel transplantation is still considered an emerging technology and has not reached a widely accepted status of standard of care. This may be due in part to the limited number of centers with clinically active intestinal transplant programs. In the Scientific Registry of Transplant

Recipients (SRTR) and the Organ Procurement and Transplant Network (OPTN) 2012 annual report only 19 centers were actively performing intestinal transplants in the United States [19]. Two centers offer intestinal transplantation in Canada, Toronto and Edmonton. The 2013 Bi-Annual report of the Intestine Transplant Registry captured 38 pediatric centers worldwide [12].

There have also been a limited number of patients transplanted: since its inception in 1985, the Intestine Transplant Registry has acquired data on 1611 pediatric transplant patients with an estimated capture of 95 % [12]. Thus, in most neonatal intensive care units throughout the world, where potential candidates are first identified these programs are not available. Neonatologists, surgeons and gastroenterologists caring for the neonates with significant loss of bowel may be less attuned to the latest outcomes in the intestinal rehabilitation and small bowel transplantation fields and may not discuss these alternatives with the patients' families. Unfamiliarity with the pre-transplant and post-transplant course and outcome may have a negative influence on the referral of children with short gut syndrome from non-transplant centers to centers offering intestinal transplantation. In these non-transplant centers, there might be more discussions that the procedure might result more in harm than a benefit for the child with short gut syndrome.

Over the last 10 years, the emergence of intestinal rehabilitation teams and achievement of parenteral independence even in very short bowel patients has further shifted this paradigm. Though the process might be lengthy, the diagnosis of short gut for many patients no longer equates to life-long total parenteral nutrition. There is now heightened hope for avoiding transplantation altogether for many of these short bowel infants. The required length of gut to achieve full or partial emancipation from parenteral nutrition has shortened in the last decade after the introduction of surgical lengthening procedures such as the serial transverse enteroplasty procedure [26, 27]. Studies have suggested that length of bowel is no longer necessarily predictive of complications or death [28, 29]. Unfortunately, there is still the inability to predict which infants are likely to have a good outcome. Clinically, pediatric surgeons and pediatric gastroenterologists struggle to predict how any given infant will do on total parenteral nutrition at the outset of treatment and who will go on to require small bowel transplantation. It is impossible to predict if and when the complications of liver failure, recurrent sepsis and loss of catheter sites would make small bowel transplantation necessary for the infant's survival. All this conjecture has made the intraoperative decisions for surgeons dealing with necrotizing enterocolitis totalis and volvulus even more challenging. As the length of bowel necessary for survival and enteral autonomy is even less defined than a decade ago, the decision to "open and close" and opt for comfort care has become more ethically challengeable. This moving target for gut autonomy creates an ethical dilemma in the acute intraoperative decision-making and has developed into a renewed source of moral distress for pediatric surgeons.

There is incertitude whether treatment outcomes have reached a level at which physicians should strongly recommend to guardians of infants suffering from short gut syndrome the alternative of intestinal rehabilitation and possibly transplantation. In 1999, Cooper et al. studied the managing disagreements between clinicians in the management of short bowel patients. After a survey of directors of neonatology and gastroenterology in the United States, the data revealed substantial disagreement in the treatment options discussed and recommended by neonatologists and gastroenterologists [30]. The authors raised the issue of the required amount of data a clinician needs about new therapeutic interventions before beginning to discuss and recommend them to patients. There seems to be no standard emerging in either the legal or ethical literature, for discussing and recommending alternatives not yet adequately supported by data [30]. At what point do therapies become sufficiently plausible to have them discussed as an alternative with the parents? The authors also raised the issue of potential for much parental confusion, at a difficult time, when counseling physicians at the same institution discuss and recommend different and often opposite options. Cooper recommended that colleagues as individuals, and subspecialty services as groups, share their views with each other before individual discussions with families, with the goal of ensuring that there is a clear understanding of differing beliefs. If those discussions reveal true disagreements, they can be presented reasonably to families in a collegial fashion [30]. Further, discussions in case debriefings exploring the ethical issues related to each patient may be a valuable way for physicians to address explicitly their moral angst, uncertainty and distress with the case.

Surprisingly, in 2013, 14 years after Cooper's paper, Cummings et al. [31] still reported wide counselling variation among physicians regarding intestinal transplant for short bowel syndrome. Neonatologists and pediatric surgeons were queried on their knowledge, attitudes and counselling practices regarding intestinal transplant for infants with short bowel syndrome. Clinicians were also asked whether they viewed intestinal transplant as ethically impermissible, optional or obligatory. A majority of clinicians felt that parents of a neonate with severe short bowel syndrome should be offered the option of TPN with the goal of reaching intestinal transplantation. That is, the procedure was believed to be 'ethically permissible', as it was not deemed wrong. Furthermore, 90 % of respondents agreed that family preference should influence this decision. However, only 33 % reported that they 'often/always' gave parents that option, and fewer than 10 % recommended it. Thus with regards to parent counselling, there was a sharp disconnect between what physicians stated ought to be done and what they reported doing in practice. Pediatric surgeons were more likely to offer intestinal transplantation and intestinal rehabilitation including bowel lengthening procedures than neonatologists.

Small bowel transplantation is a permissible treatment as it offers a reasonable chance of survival in children who otherwise would not survive [32]. Because of the scarcity of data for long-term survival and quality of life post-transplantation in this population, small bowel transplantation is not felt to be obligatory. The burden to the patient and his/her family of pursuing this alternative is also significant. Responding physicians identified quality of life and social and financial burden for patients and their families as significant factors in decision-making and counselling. Moral uncertainty exists because it is unclear if the burden of such a long journey, both getting to the transplant procedure and post-transplant course, is worth the benefit of survival [31]. Moral uncertainty is also compounded by the suboptimal long-term outcomes and the comparable short term outcomes of TPN and transplantation. Depending on their ethical perspective, treating physicians and families of infants with intestinal failure will weigh alternative therapies and options based on what matters most to them: survival alone, maximising individual happiness, maximising comfort, etc.

9.3.2 Informed Consent for Transplantation and the Best Interest Principle

Clinicians and parents together must make difficult decisions guided by their understanding of the child's best interest after open discussions with regards to the medical status, prognosis and treatment options for the patient [33]. What is in the best interest of the child is subjective and might be very different for everyone involved in the decision making. The medical best interest of the child is difficult to determine because of lack of solid data for treatment alternatives. There is more to consider in the best interest of the child than only medical outcomes: quality of life, suffering, support from family, etc.

Authors have compared the choice of total parenteral nutrition and small bowel transplantation in children to the use of infants' dialysis and kidney transplantation to help center the discussion on the best interest of the child [34]. Dialysis and renal transplantation have been offered for decades and there is more familiarity and comfort with the ethical decision process involved. There are many similarities between the two therapies: both dialysis and total parenteral nutrition can be life-saving, be used as a temporary or lifelong measure, or as a bridge to transplant, and both are used in the treatment of infants with relatively high mortality rates. Both dialysis and total parenteral nutrition are physically invasive, requiring significant time spent attached to medical equipment. They are both associated with prolonged and frequent hospitalizations, diversion of family time and energy to technical care-giving and time spent away from work and other children, and psychological stress and monetary expense. On the other hand, there are major differences between dialysis/renal transplantation and small bowel transplantation. Firstly, there are proposed ethical guidelines for the use of dialysis, though not explicit in the care of pediatric patients, where similar guidelines for the use of total parenteral nutrition in infants with intestinal failure have not yet been formulated. Secondly, there is an option for reverting to dialysis for a failed kidney transplant, compared to a failed small bowel transplant, where returning to long-term total parenteral nutrition is usually no longer an option as intravenous access is exhausted [34]. The central challenge in both continuum of dialysis and kidney transplantation and of total parenteral nutrition and small bowel transplantation is to determine what is in the best interest of the child in view of that child's comorbidities and expected outcomes. The decision to proceed with comfort care must be made when it is in the best interest of the child to forego aggressive therapy.

As small bowel transplantation is still an emerging technology and not standard of care, such option should be offered to families but not should not be imposed. When the prognosis is unclear, informed parents should dictate treatment directions [33]. The American Academy of Pediatrics in their Guidelines on Forgoing Life-Sustaining Medical Treatment, supports the right of parents to refuse life-sustaining treatment in instances in which non-treatment is in the best interest of the child [35]. The guidelines clearly include organ transplantation, nutrition and hydration provided intravenously or by tube in their definition of life-sustaining interventions that may be withheld or withdrawn.

Glover et al. [34] also suggested that reliance on a narrow medical best interest principle fails to take into account other moral features that are the necessary part of ethical decision-making in pediatrics, including respect for family decision-making and the just distribution of health care resources. An expanded best interest standard is an attempt to balance the benefits and burden of a health care intervention according to the values of the parents, pediatric physicians, surgeons, members of the team, and the larger society.

Families and the treating team should keep the burden of continued treatment for the patient in focus. Decisions for continued or aggressive treatment taken at the time of the acute events are often made while parents are emotionally overwhelmed and need to absorb a great deal of medical information quickly. Re-evaluation of the decision to pursue intestinal transplantation should be done on a regular basis based on the status of the patient, the burden to patient and the family. The comparative weight of benefits, risks and harms changes continuously over time. Redirection of the care plan to comfort care should be entertained when the prognosis changes and the patient's benefit from continued aggressive care is unlikely. Continued involvement of the family in discussions about the direction of care is paramount. Consent to a course of treatment is fluid and not static: it should be based on risks, benefits, burden etc. If the balance of these criteria changes it is permissible for families to no longer provide consent to the plan of treatment.

There is no ethical or legal difference between not starting (withholding) and stopping (withdrawing) therapy. Authors have suggested that if a treating physician had enough justification not to start a therapy in the first place, then there is more than enough justification to stop it after a trial of therapy has clarified the limited benefits and substantial burdens of that same therapy [34].

9.3.3 Accessibility to Intestinal Transplant

With few centers performing intestinal transplant around the world and difference in referral patterns, accessibility to care is unequal for all patients with short gut syndrome in need of a transplant. This disparity raises issues of justice.

Optimally patients should have the right to equal and just treatment. This would be in accordance to article #24 of the United Nation Convention for the right of the child which states that "states parties recognize the right of the child to enjoyment of the highest attainable standard of health and to facilities for the treatment of illness and rehabilitation of health. States Parties shall strive to ensure that no child is deprived of his or her right of access to such healthcare services" [36]. This would also be in accordance to the Canadian Health Act which is based on the principle of universality, accessibility and portability [37]. The rights of the newborn include the right to life, including medical interventions that have a reasonable potential to preserve life. On the other hand, newborns also have the right to mercy, specifically the right to be free from pain and suffering if these interventions are very unlikely to yield benefit to the child. Disparities and dissonances in physicians' recommendations and referral patterns may represent an injustice for patients and their families. As previously discussed, there is a limited number of centers around the world performing intestinal transplantation and families of children not born at these center may not be given all the information about intestinal transplantation as physicians tend to concentrate on treatment options available at their center. Distributive justice would argue that similar cases should be treated similarly and different cases differently. Accordingly geographical location or physician's point of view should not cause disparity in care.

Right to treatment in some settings may include the option of transfer, if feasible, to a center of excellence in intestinal rehabilitation and offering small bowel transplantation. It seems unjust to offer a specific treatment to a patient at one center where it is provided, yet not offer the same treatment to a similar patient at another center simply because the treatment is unavailable there [32]. Families at non-transplant centers have the right to accurate information regarding prognosis and all available treatment options, including those at other centers, as well as the right to guidance and support throughout this decision making process [38]. This entails that physicians at non-transplant centers stay abreast of the advances in bowel rehabilitation and small bowel transplantation. Nowadays, parents are able to independently search the internet for more information on the condition affecting their child and the treatment options available. Responsible physicians should be able to guide the parents in their search and help them make sense of the information gathered. Guardians depend upon relevant information on all available and reasonable therapy alternatives to make autonomous and informed decisions for their children. For short bowel syndrome patients, this will include comfort care, parenteral nutrition and intestinal adaptation therapies, as well as newer technologies such as intestinal lengthening procedures and intestinal transplantation [32].

Transfer to a distant center has a lot of implications for families: prolonged admission in an unfamiliar city, a large financial burden, loss of income, loss of family support and often disruption of the family unit if uprooting the entire family is not possible. Some concerned parents, after thoughtful evaluation of the outcomes information, the cost to the patient in terms of suffering and the cost to the family unit, may decide that such an aggressive treatment is not in the best interest of their child and their family.

Despite the fact that the best interest of the child should remain paramount in the decision-making, many parents naturally take into account the interests of the entire family when making decisions for their children [32].

Justice in care would entail the medical system to support families deciding to pursue small bowel transplantation not available in their center. As outcomes of small bowel transplantation are related to center volume, informed parents may look at pursuing care for their infant in centers not covered by their state Medicaid or insurance, leaving them to find additional funds from private, charitable or public sources. The alternative for families is to choose care at centers covered by their insurance with less experience in small bowel transplantation and become part of the learning curve for these new programs, with possibly less than optimal outcomes.

Accessibility to small bowel transplant also implies getting a suitable donor organ in a timely fashion. Because of their small size, infants candidate for bowel transplantation have a much lower chance of finding a suitable match from the potential donor pool, as the donor and recipient need to be a close size match. The number of infant donors available each year is limited. Children requiring both liver and small bowel transplantation are also in competition for organs with other recipients waiting for liver grafts only. The liver transplant waiting list is substantially larger than the combined liver-small bowel list. Appropriate timing of referral to transplant center and listing is thus of critical importance: patients have to be listed early enough in their disease process to ensure that they will survive long enough to identify a suitable donor. Little data is available for survival while awaiting intestinal transplantation, after the decision is made to list the patient. This hidden mortality definitely leads to an overestimation of survival of intestinal transplantation. Some of the available data shows that shortage of cadaveric intestinal grafts continues to result in appreciable morbidity and mortality for patients awaiting transplantation [39]. The UNOS report indicates that mortality on the intestinal transplant waiting list is higher than on other waiting lists [40]. Emerging data from the recent SRTR and OPTN report that pre-transplant mortality rate for intestinal transplant candidates has decreased dramatically over time, from 51 per 100 wait-list years in 1998–99 to 6.7 per 100 wait-list years for patients listed in 2012. However, pre-transplant mortality is notably higher for intestine-liver transplant candidates than for intestinal transplant candidates (respectively 14.2 vs. 1.5 deaths per 100 wait-list years in 2012). In 2012, wait times were less than 1 year for 37 % of the patients listed for small bowel transplantation, between 1 and 2 years in 21 % of the patients and more than 2 years for 42 % of the patients [19]. This prolonged wait contributes to emotional and financial cost, and burden of care to the patients and their families. Long waiting periods and wait-list mortality weigh heavily when evaluating the burden of treatment options and the patient's best interest. Families might be reluctant to embark in a course of treatment if there is little hope of the child getting an organ in a timely fashion.

9.3.4 *Living Donor for Intestinal Transplants*

Living-related donation is a proven way to reduce mortality on the deceased donor waiting list, allowing for timely transplantation and prevention of progression of the complications caused by total parenteral nutrition. Living donation is the sole source of organs in many countries. Living-related small bowel transplantation is possible but only a few cases have been reported [41, 42]. If the recipient requires both a liver and small bowel transplantation living donation is still possible. Donation can be sequential, hence preserving the possibility of removal of one of the grafts if it fails without affecting the other component. This comes at the price of multiples surgeries for the donor, increasing the rehabilitation time and burden of donation. There is no available data regarding complications and long-term outcome in living intestinal transplant donors, which makes the definition of the risk-benefit ratio for potential donors impossible to estimate [43]. Certainly, a complete discussion of the ethical principles of living donation is beyond the scope of this chapter. A more complete discussion can be found in the report from The Vancouver Forum, a consensus group who enunciated the principles of living donation: the responsibilities of the transplant team performing live donation, the prime importance of informed consent and of donor autonomy and selection [44].

9.3.5 *Resource Allocation and Transplantation*

When considering new technologies, part of the ethical reflection should always entail the cost to the health care system. For publicly funded health care systems, society needs to decide on the allocation of scarce resources. Accordingly the allocation of these resources becomes an ethical dilemma as there are many competing demands. Moreover, for treatment with high cost, sub optimal outcome and poor ability to prognosticate, like intestinal transplantation, these issues of resource allocation are challenging. Improvements in small bowel transplantation have resulted in decreased financial costs for the patients, insurers and the health care system. Between 1990 and 1994, the average cost of an intestinal transplantation was $203,111 for an isolated intestinal transplant, $252,453 for a combined liver-small bowel transplant, and $284,452 for a multivisceral transplant. By 1999, the average costs had decreased appreciably and were $132,285 for an isolated intestinal transplant, $214,716 for a combined liver-small bowel transplant, and $219,098 for a multivisceral transplant [45]. In 2010, Medicare costs for all intestinal transplant patients in the United States totalled at $13,991,380 for 217 patients. The cost per patient per year was evaluated at $73,479 [19]. These costs only encompass the cost of the transplant care and do not include the cost of the preceding usually lengthy neonatal intensive care admission, the medical and surgical care during the intestinal rehabilitation, and the cost of parenteral nutrition.

There is no published estimate of the total cost of care in pediatric patients coming to intestinal transplantation.

It is widely accepted that medical and technological innovations escalate the cost of health care while the available health care dollars are limited. There is no consensus on how resources should be allocated or priorities set. However, if these innovations are the patient's only hope for survival, limiting their use becomes problematic [46]. Various approaches to allocating health care resources included: (1) Fair Innings—Some ethicists argue that preference for limited health care dollars should be given to younger patients to compensate for years they did not live [47]. (2) Additional Years of Life—Others would advocate that resources should be used to maximize the net additional years of life, regardless of the patient's age. (3) Distributive justice—Additionally, some authors feel that there should be a dollar limit on the amount of money that society should spend to save a single patient. (4) Maximization of benefits—Others feel we should limit expenditures relative only to the benefit to be gained [48]. (5) Maximization of productive years—The author would argue that we should maximize the number of productive years gained to help replenish used resources and keep the health care costs sustainable.

In the case of intestinal transplantation, both the length of survival and the neurocognitive and functional outcome of intestinal transplant survivors should then be taken into consideration when evaluating the societal cost and return in benefit of this emerging procedure and deciding to assigning limited health care dollars. Should ex-premature patients with significant neurological injuries and chronic respiratory problems be considered for intestinal transplantation on the same basis as term infants, without any neurological impairment or other comorbidities, who suffered massive bowel loss from mid gut volvulus? Can we, trying to be good stewards of limited health care resources and a limited donor pool, withhold this therapy for some patients? The American Society of Transplant did address this issue in part in its position statement by noting profound neurological disabilities as a contraindication to intestinal transplantation [23]. These questions remained mostly unanswered and will need to be the focus of continued medical and societal debates. Chapter 12 by Wightman et al. further explores neurodevelopmental status as a criterion for solid organ transplant eligibility.

9.4 Conclusion

This chapter has tried to illustrate the ethical issues in small bowel transplantation in the context of the advances in the field and recent outcomes. Small bowel transplantation is not widely accessible due to the limited number of centers of excellence performing the procedure. This in part and the lack of long-term outcome and quality of life data have kept intestinal transplantation in a state of emerging technology. There is no consensus on the duty to refer to a transplant center, but the availability of this treatment alternative should be discussed with families to ensure informed decision-making. Recent advances in the management of short bowel syndrome have made decisions at the onset more challenging for the treating team and families. The principle of best interest of the child should guide care, weighing the chance and quality of survival and burden of care, while keeping the decision in the context of the family. The societal cost of caring for these extremely complex patients needs further exploration.

References

1. Lillehei RC, Goott B, Miller FA. Homografts of the small bowel. Surg Forum. 1960;10:197–201.
2. Starzl TE, Kaupp HA Jr. Mass homotransplantations of abdominal organs in dogs. Surg Forum. 1960;11:28–30.
3. Starzl TE, Rowe MI, Todo S, et al. Transplantation of multiple abdominal viscera. JAMA. 1989;261:1449–57.
4. Williams JW, Sankary HN, Foster PF, et al. Splanchnic transplantation. An approach to the infant dependent on parenteral nutrition who develops irreversible liver disease. JAMA. 1989;261:1458–62.
5. Grant D, Wall W, Mimeault R, et al. Successful small bowel/liver transplantation. Lancet. 1990;335:181–4.
6. Deltz E, Schroeder P, Gedhardt H, et al. Successful clinical small bowel transplantation: report of a case. Clin Transplant. 1989;3:89–91.
7. Goulet O, Revillon Y, Jan D, et al. Small-bowel transplantation in children. Transplant Proc. 1990;22:2499–500.
8. Starzl TE, Todo S, Tsakis A, et al. The many faces of multivisceral transplantation. Surg Gynecol Obstet. 1991;172:335–44.
9. Abu-Elmagd K, Reyes J, Todo S, et al. Clinical intestine al transplantation: new perspectives and immunological considerations. J Am Coll Surg. 1998;186:512–27.
10. Grant D. Intestinal transplantation: 1997 report of the international registry. Transplantation. 1999;67:1061–4.
11. Nayyar N, Mazariegos G, Ranganthan S, et al. Pediatric small bowel transplantation. Sem Ped Surg. 2010;19:68–77.
12. Intestinal transplant registry bi-annual report 2013 available to members at www.tts.org/ita/resources.
13. Abu-Elmagd KM, Kosmach-Park B, Costa G, et al. Long-term survival, nutritional autonomy, and quality of life after intestinal and multivisceral transplantation. Annals of Surg. 2012;256:494–508.
14. Nucci AM, Strohm S, Squirer RH, et al. Growth pre- and post-implementation of a steroid-free induction protocol in a large pediatric intestinal transplant population. J Pediatr Gastroentrol Nutr. 2011;52:601–6.
15. Thevenin DM, Mittal N, Kato T, et al. Neuro-developmental outcomes of infant intestinal transplant recipients. Transpl Proc. 2004;36:319.
16. Thevenin DM, Baker A, Kato T, et al. Neurodevelopmental outcomes for children transplanted under the age of 3 years. Transpl Proc. 2006;38:1692–3.
17. Kosmach B. Care routines following pediatric intestine transplantation. In: International symposium on intestinal transplantation. UK: Cambridge; 1997.
18. Stenn PG, Lammens P, Houle L, et al. Psychiatric psychosocial and ethical aspects of small bowel transplantation. Transpl Proc. 1992;24:1251–2.

19. Intestine chapter SRTR & OPTN annual data report; 2012. Available on http://optn.transplant.hrsa.gov.
20. Howard L, Malone M. Current status of home parenteral nutrition in the United States. Transplant Proc. 1996;28:2691–5.
21. Diamond IR, Sterescu A, Pencharz PB, et al. Changing the paradigm: omegaven for the treatment of liver failure in pediatric short bowel syndrome. J Pediatr Gastroenterol Nutr. 2009;48:209–15.
22. Oliveira C, De Silva N, Wales PW. Five-year outcomes after serial transverse enteroplasty in children with short bowel syndrome. J Pediatr Surg. 2012;47:931–7.
23. Kaufman SS, Atkinson JB, Bianchi A, et al. Indications for pediatric intestinal: a position paper of the American society of transplantation. Ped Transplant. 2001;5:80–7.
24. Goulet O, Ruemmele F. Causes and management of intestinal failure in children. Gastoentero. 2006;130:S16–28.
25. Cummings CL, Mercurio MR. Ethics of emerging technologies and their transition to acceptable practice: intestinal transplant for short bowel syndrome. J Perinatol. 2012;32: 752–6.
26. Mercer DF, Hobson BD, Gerhardt BK, et al. Serial transverse enteroplasty allows children with short bowel to wean from parenteral nutrition. J Pediatr. 2014;164:93–8.
27. Wales PW, Jancelewicz T, Romao RL, et al. Delayed primary serial transverse enteroplasty as a novel management strategy for infants with congenital ultra-short bowel syndrome. J Pediatr Surg. 2013;48(5):993–9.
28. Gambarara M, Ferretti F, Bagolan P, et al. Ultra-short-bowel syndrome is not an absolute indication to small-bowel transplantation in childhood. Eur J of Pediatr Surg. 1999;9:267–70.
29. Glover JJ, Caniano D, Balint J. Ethical challenges in the care of infants with intestinal failure and lifelong total parenteral nutrition. Seminars in Ped Surg. 2001;10:230–6.
30. Managing disagreements in the management of short bowel and hypoplastic left heart syndrome. Pediatr 1999;104:e48–e55.
31. Cummings CL, Diefenbach KA, Mercurio MR. Counselling variation among physicians regarding intestinal transplant for short bowel syndrome. J Med Ethics 2013;0:1–6.
32. Cummings CL, Mercurio MR. Ethics of emerging technologies and their transition to acceptable practice: intestinal transplant for short bowel syndrome. J Perinatol. 2012;32: 752–6.
33. American Academy of Pediatrics Committee on Fetus Newborn. Policy statement. Noninitiation or withdrawal of intensive care for high-risk newborns. Pediatrics. 2007;119 (2):401–3.
34. Glover JJ, Caniano D, Balint J. Ethical challenges in the care of infants with intestinal failure and lifelong total parenteral nutrition. Seminars in Ped Surg. 2001;10:230–6.
35. Committee on Bioethics. American academy of pediatrics. Guidelines on forgoing life sustaining medical treatment. Pediatrics. 1994;93:532–6.
36. http://www.unicef.org/crc/files/Rights_overview.pd.
37. http://www.parl.gc.ca/content/sen/committee/372/soci/rep/repoct02vol6part7-e.htm.
38. Cummings CL, Mercurio MR. Ethics for the pediatrician: autonomy, beneficence and rights. Pediatr Rev. 2010;31:252–5.
39. DeLegge M, Alsolaiman MM, Barbour E, et al. Short bowel syndrome: parenteral nutrition versus intestinal transplantation. Where are we today? Dig Dis Sci. 2007;52:876–92.
40. Fryer J, Pellar S, Ormond D, et al. Mortality in candidates waiting for combined liver-intestine transplants exceeds that for other candidates waiting for liver transplants. Liver Transplant. 2003;9:748–53.
41. Testa G, Holterman M, Abacarian H, et al. Simultaneous or sequential combined living donor-intestine transplantation in children. Transplantation. 2008;85:713–7.
42. Gangemi A, Tzvetanov IG, Beatty E, et al. Lessons learned in pediatric small bowel and liver transplantation from living-related donors. Transplantation. 2009;87:1027–30.
43. Panocchia N, Bossola M, Silvestri P. et al. Ethical evaluation of risks related to living donor transplantation programs. Transpl Proc 2013;45:2601–3.

44. Pruett TL, Tibell A, Alabdulkareem A, et al. The ethics statement of the Vancouver forum on the live lung, liver, pancreas and intestine donor. Transplantation. 2006;81:1386–7.
45. Abu-Elmagd K, Reyes J, Fung J, et al. Evolution of clinical intestinal transplantation: improved outcome and cost effectiveness. Transpl Proc. 1999;31:582–4.
46. Cooper TR, Caplan WD, Garcia-Prats JA, et al. The Interrelationship of ethical issues in the transition of old paradigms to new technologies. J Clin Ethics. 1996;7:243–50.
47. Kamm F. Morality, Mortality volume 1: Death and whom to save from it. New York: Oxford University Press 1998. 353p. ISBN-13: 9780195119114.
48. Menzel PT. Strong medicine: the ethical rationing of health care. New York: Oxford University Press, 1990. 256p. ISBN-13: 9780195057102.

Chapter 10
Ethical Issues in Pediatric Vascularized Composite Allotransplantation

Randi Zlotnik Shaul, Gregory H. Borschel, Jennifer Flynn, Mark D. Hanson, Linda Wright and Ronald M. Zuker

Abstract Vascularized Composite Allotransplantation (VCA) enables improved function and quality of life for those who have lost anatomical parts such as a limb and the relief of profound suffering for those with severe facial disfiguration. VCA differs from vital organ transplantation in two important ways: (1) these transplants are not necessary for recipient survival, impacting risk-benefit ratios; and (2) face and limb transplants are visible and touchable, leading to issues of identity changes, social interaction and psychological adjustments. This chapter will consider the ethical implications of VCA in pediatrics. This chapter will be divided into the following six sections: (1) introduction; (2) the state of the science and experience with VCA; (3) risk-benefit assessment; (4) informed consent (i) donors and donor surrogate decision makers (ii) recipients and recipient surrogate decision makers;

R.Z. Shaul (✉)
Department of Bioethics, The Hospital for Sick Children, Toronto, Canada
e-mail: randi.zlotnik-shaul@sickkids.ca

R.Z. Shaul
Department of Paediatrics, University of Toronto, Toronto, Canada

G.H. Borschel
Division of Plastic and Reconstructive Surgery, The Hospital for Sick Children, Toronto, Canada
e-mail: gregory.borschel@sickkids.ca

G.H. Borschel
Department of Surgery, University of Toronto, Toronto, Canada

J. Flynn
Faculty of Medicine, Memorial University, St. John's, Canada
e-mail: jflynn@mun.ca

M.D. Hanson
Department of Psychiatry, The Hospital for Sick Children, Toronto, Canada
e-mail: mark.hanson@sickkids.ca

M.D. Hanson
Faculty of Medicine, University of Toronto, Toronto, Canada

R.A. Greenberg et al. (eds.), *Ethical Issues in Pediatric Organ Transplantation*, International Library of Ethics, Law, and the New Medicine 66,
DOI 10.1007/978-3-319-29185-7_10

(5) establishing a pediatric program (i) recipient selection (ii) allocation; (iii) accountability for surgical innovation and (6) conclusion.

Keywords Vascularized composite allotransplantation · VCA · Face transplant · Limb transplant · Pediatric transplant ethics

10.1 Introduction

Vascularized Composite Allotransplantation (VCA) enables improved function and quality of life for those who have lost anatomical parts, such as a limb, and the potential relief of profound suffering for those with severe facial disfiguration. VCA differs from vital organ transplantation in two important and ethically significant ways: (1) these transplants are not necessary for recipient survival, impacting risk-benefit ratios; and (2) face and limb transplants are visible and touchable, leading to issues of identity changes, social interaction and psychological adjustments.

This chapter examines ethical issues in pediatric VCA, especially the upper extremity and face transplants as these will likely be the most common. It focuses on the state of the science and international experience with VCA; the balancing of risks and benefits; issues of informed consent; and establishing pediatric VCA programs which includes donor and recipient selection, resource allocation and accountability pathways for surgical innovation. At the time of writing, there have not been any VCAs requiring immunosuppression in children, so much of what we discuss is derived from experience in the pediatric surgical area, experience with solid organ transplantation and what we foresee as the ethical issues going forward. Most of the VCA experience involves hand and face transplants, but these experiences probably have crossover with other forms of VCA such as abdominal wall transplantation.

L. Wright
Department of Surgery, University of Toronto, Toronto, Canada
e-mail: Linda.Wright@utoronto.ca

R.M. Zuker
The Division of Plastic and Reconstructive Surgery, The Hospital for Sick Children, Toronto, Canada
e-mail: ronald.zuker@sickkids.ca

R.M. Zuker
Department of Surgery, University of Toronto, Toronto, Canada

10.2 State of the Science and Experience with VCA

10.2.1 General Overview

VCA is a true multidisciplinary intervention. It does not simply involve the technical aspects of transplanting a segment of tissue in a recipient. Preparation for this procedure requires physical evaluation from a regional and general standpoint, psychosocial evaluation, rehabilitative potential and a full disclosure of and attention to the ethical issues involved. The surgical procedure itself involves multiple teams of specialists but relies upon the manipulation of the immunological system to prevent rejection of the transplant. Medical transplant specialists focus on this aspect of care while the surgical team looks after the postoperative care and rehabilitation experts work to ensure that the transplant will reach its maximum function. In addition to this, the immunological manipulation renders the patient not only susceptible to long-term complications such as diabetes, renal failure, skin cancers and lymphomas, but also renders the patient susceptible to a number of infective agents ranging from viruses to bacteria, to parasites and fungi. All of these issues associated with VCA underscore the need for appropriate ethical considerations, as potential harm can result at almost every stage.

Similar to solid organ transplantation, at the present time, several advances are being made in VCA. Firstly, from an immunologic standpoint, new regimens are being formulated and studied that will be less toxic to the recipient and will also reduce the long-term complication rate [1]. The goal is to achieve chimerism where the recipient and the transplanted tissue live together in immunological harmony without the need for immunosuppressant drugs. There has been some progress in this respect with promising results seen in the very rare kidney and liver transplant recipients who are now off their immunosuppressant medications [2]. Even a reduction in dosage of immunosuppressant regimens will make a huge difference to complication rates, both infective and long-term. An enormous amount of research is being conducted worldwide to address this issue.

Secondly, new immunosuppressive drugs are being developed and will be much more effective [3]. These new medications are currently being introduced when the standard medications become ineffective or incite complications that are often severe and life-threatening themselves.

Thirdly, there are new immunological tolerance induction techniques being evaluated and assessed both for solid organ transplants and for VCA. This process, known as immunomodulation [4, 5], has met with some success and may well be expanded in the future. It is a technique whereby donor bone marrow is also used to modulate the immune system of the recipient, again attempting to achieve chimerism.

Lastly, a great deal is being learned about the infective complications associated with the immunosuppression necessary for VCA. Transplants that are exposed to air, such as the extremity, or exposed to mucosal surfaces, such as the face, are subject to different, more diverse susceptibility to viral, bacterial and fungal agents. As experience is gathered in these transplant techniques, more specific antimicrobial therapy is being developed [6].

10.2.2 Technical Advances in VCA

Advances in the science of VCA are also being made on the technical front. A greater focus on functional outcomes has led to adjustments in surgical techniques that should improve the functional capabilities of the extremity, face, and abdominal wall. For example, in the upper extremity, a mid-forearm amputation often has poor residual forearm musculature. Thus, it would seem reasonable to transplant the entire muscle tendon unit to achieve an optimal functional outcome. This has been carried out in recent transplants and has been found to be most effective [7]. Similarly, in face transplants, the facial nerve and facial musculature component of the transplant is often critical to success. Again, the entire muscle unit with its specific nerve can be transplanted with reinnervation via nerve repair rather than attempting to achieve function with muscle repair. This has proven to be technically feasible and, in some of the more recent facial transplants, has proven to be very effective [8]. It seems that the sensory component is not as critical as the sensation tends to return, even when no sensory nerve repairs are carried out.

The technical considerations underscore the importance of practice in the cadaver laboratory. All major centers now perform such dissections with varying degrees of sophisticated imaging and simulation. Particularly with technically difficult body regions, repeated cadaver simulations have proven to be very helpful in reducing technical errors and in reducing judgment decisions on the operating table under stressful conditions [9].

10.2.3 History of Vascularized Composite Allotransplantation

The modern era of transplantation started in 1954 when Plastic Surgeon Dr. Joseph Murray successfully transplanted a kidney from one identical twin to another—the Herrick boys. It was followed up in 1959 with a kidney transplant to a fraternal twin after he received radiation to reduce the immunogenic load. Thus no immunosuppression was needed. However, in 1962 the true era of transplant surgery began when Dr. Murray transplanted a cadaver kidney, maintaining its viability and function with early immunosuppressant drugs. He was awarded the Nobel Prize in 1990 for his work [10]. More recently, the concept of transplantation was then married to the new subspecialty of reconstructive microsurgery. The latter involves magnification of the critical elements that need to be repaired such as blood vessels and nerves with the aid usually of a microscope, micro instruments specifically developed for repair of vessels and nerves, and innovative strategies to address previously untreatable or poorly managed conditions. Recently this marriage of transplantation and reconstructive microsurgery has been termed "Reconstructive Transplantation or Vascularized Composite Allotransplantation." It offers many advantages over traditional reconstructive techniques in that "like tissue" can be used to replace that which is deficient. The precise deficit is replaced with the same body part i.e. a deficient hand is replaced with a transplanted hand. The tissues are revascularized in the recipient in their same innate location by arterial and venous repairs. Form is restored by the transplant and function can be restored by careful positioning of these structures and careful nerve coaptation. The actual part is restored and its movement reconstructed.

However, new issues have been identified in this life-enhancing but not life-saving process. Firstly, there are the immunological issues of all transplant recipients requiring lifelong immunosuppression with the infective complications and long-term complications that are associated with this treatment. Secondly, a number of ethical issues arise in patient selection, focusing on the principles of utility, autonomy, best interests and justice that are addressed elsewhere in this chapter. Thirdly, there are a number of psychosocial issues that present themselves as these transplants can be seen, can be touched, and move to varying degrees. Issues of identity, sense of self and social interactions arise in this form of transplantation and need to be studied, evaluated and directly addressed.

The first attempt at an upper extremity transplant in recent times was in 1963 in Ecuador. This transplant was carried out without immunomodulation and failed, almost killing the patient. The first upper-extremity transplant that was successful—at least in the short-term—was carried out in 1998 in France [10]. At the present time, there have been 75 hand and upper-extremity transplants carried out worldwide [10]. There has been a progression from simple transplants in adults at the wrist level to more complex transplants at the mid-forearm level and eventually complicated transplants at the proximal arm level, and even bilateral transplants. There have been meaningful outcome analyses which provide us with clear data as to how we can move forward in this field. These will be discussed later in this chapter.

The first face transplant was carried out in 2005 for an adult female in France. There have now been 26 cases worldwide (all adult) with varying degrees of success [10]. Three patients have died. Again, progression has been from a smaller segment such as a partial face transplant with minimal musculature, to a full face transplant with the entire compliment of skin, facial musculature and even large segments of the facial bones, including the tongue.

Other areas of VCA involve anatomical structures that are difficult to adequately reconstruct with conventional techniques. Transplantation of the larynx has been met with considerable success since its initiation in 1998 [11]. Similarly, transplantation of the abdominal wall has proven to be helpful in difficult clinical situations. Uterine transplants have been carried out to facilitate child bearing in certain situations. This has met with some success and opens the door to a new form of fertility restoration. Advances are being made in tracheal transplantation whereby combined techniques utilizing VCA in combination with tissue engineering may provide optimum reconstruction. VCA has also been used for lower extremity transplants, a penile transplant and a tongue transplant. Unfortunately, these have not been met with success [10]. Again it underscores the importance of ethical considerations that may not justify pursuing new surgical procedures that are technically feasible but may not be practical or effective.

The bottom line for patients, practitioners and funding agencies (be they governments, insurance companies or private individuals) is outcomes. In upper extremity transplants, a significant amount of information has been put into and reported by the International Registry for Hand and Composite Tissue Allotransplantation [12]. A great deal has been learned about short- and long-term function, longevity of the allograft, and particularly the complications of both the surgery and long-term immunosuppressants. In upper extremity transplantation, 10-year outcomes have been reported. It has been shown that in the unilateral situation, and even more so in the bilateral situation, upper extremity disability goes down over a 10-year period, as measured by the standardized outcome measure of disabilities of the arm, shoulder and hand—the DASH score [13] and functional transplantation score improves over a 10-year period, as measured by the Hand Transplantation Score System (HTSS) [12]. The early complication rates relating to adequate circulation, early infections are quite low, whereas the late complication rates are significant. These late complications involve metabolic issues that occur in the majority of patients, opportunistic infections that occur in over 80 % of patients and at least one episode of acute rejection requiring immunomodulation, which also occurred in over 80 % of patients reported [12].

The experience in upper extremity transplantation differs between different centers. The US and European experience has been quite positive and involves appropriate patient selection, postoperative rehabilitation, and compliance in taking the immunosuppressant drugs [14]. However, in other areas of the world, a very different pattern of graft failure and amputation has been reported where these parameters have not been followed [14]. Thus, the upper extremity transplantation data reinforces the concepts of appropriate patient selection anticipating that a complex rehabilitation program and stable psychosocial environment will be required, and the immunosuppressant drug regimen must be adhered to very carefully. The first upper extremity transplant, carried out in France, was an initial vascular success in 1998, but proved to be a long-term failure requiring amputation at the patient's request in 2001. After a year and a half, the patient was dissatisfied with the function attained and the need for ongoing immunosuppression. Similar amputations have been required because of psychosocial issues where the transplant could not be maintained and immunological rejection when the immunosuppressant regimen was not appropriately followed [14].

10.2.4 Face Transplantation

With respect to face transplants, some of the outcomes have been incredibly successful. Again, the importance of patient selection has been identified and the concept of focusing on improved function has proven very beneficial [15]. The physical transformation of severely deformed individuals who were felt largely housebound with severe psychological issues has been quite dramatic. Failures where the transplant and the patient have died, however, have occurred from infective complications and from a discontinuance of immunosuppressant drugs [16, 17]. Recent evaluations of face transplants where tissue has survived, however, have shown enormous benefit from a psychosocial standpoint, allowing patients to re-enter society and even obtain gainful employment thanks to their facial rehabilitation [18]. Dramatic post-surgical outcomes have been demonstrated [16, 18], thus confirming the value of such interventions.

Moving forward, outcomes will continue to be evaluated. The evidence [19] now clearly points in a very positive direction, both for upper extremities and for face transplants in European and American centers. Further advances are anticipated in immunosuppressive regimens, new immunosuppressive agents, progress on the road to achieving true chimerism, and improved knowledge and treatment of the infective complications of transplantation. New technical advances are being made that will allow for even better functional outcomes in the future. The VCA field is steadily progressing. That said, the ethical issues need to be discussed openly so that patients, practitioners and the public have a clear understanding of what can be done, how it will be done, and realistic expectations regarding expected outcomes and risks.

10.3 Risk-Benefit Assessment

VCA provides the potential for recipients' benefiting from improved functional and cosmetic outcomes. Yet VCA is not a life-saving procedure and presents risks emanating from the need for life-long immunosuppression [10]. Pediatric VCA experience is limited. Two cases of limb transplantation have been reported, neither of which required immunosuppression because they were twin-twin isografts [20, 21]. However, adult VCA experience is more extensive. VCA of face and hands have been performed with a total of 25 partial or full-face transplants and more than 70 hand transplants being reported to date [10]. The mortality rate associated with face transplantation of close to 10 % is considered high [10] and two reports of cancer in patients with face transplants are reported [22]. Face transplantation is designated an experimental procedure [23]. Guidelines for Face and Hand Transplant Programs are published and emphasize institutional commitment to multidisciplinary team assessment and care [23, 24]. This institutional commitment will enable development of the team expertise necessary to not only elucidate the risks and benefits associated with hand and face transplantation but the essential analysis of the balance of benefits versus risks for each individual potential recipient.

As noted, two pediatric reports of twin-twin isograft limb transplantation exist; hand transplant between identical twins [20] and leg transplant across conjoined twins [21] immunosuppression was not required in these contexts. In the hand transplantation case, with a set of one-month-old female twins, patient and graft survival are reported without acute or chronic rejection [20]. In the leg transplantation case, with the surviving three-month-old conjoined twin, the limb is reported to be fully sensate, with muscle function returning and limb re-innervation under cortical control—all positive benefits of this limb transplantation [21]. Limb transplantation by means of VCA therefore presents potential case-specific benefits as noted above over prostheses when other surgical options are not present [21]. At the present time, VCA should only be offered to amputees who have been given a full trial with a prosthesis but whose prosthesis failed to provide adequate function. Patient selection is again emphasized as crucial to overall and lasting success. Also, as with adult populations, face transplantation is noted to be experimental within pediatric populations [25]. Potential case-specific benefits and risks associated with VCA warrant investigation within pediatric patient populations. Furthermore, benefits and risks distinct from those of adult patient populations may be associated with VCA within pediatric patient populations.

Investigation of VCA for children with Congenital Craniofacial Defects is warranted given the potential adverse impact upon the quality of life of affected children and their families [26, 27]. However, individualized case selection in this context is necessary in light of pediatric case reports that reflect children's adaption to their lives with congenital anomalies and the ensuing "dilemma of getting better" subsequent to non-VCA corrective surgery for their conditions [28]. Developmental adaptation and incorporation of a congenital condition into a child or adolescent's sense of self is to be anticipated and should be incorporated into preoperative psychiatric/psychological VCA benefits/risk assessment and subsequent care. Developmental adaptations in the cases of congenital deformities may be more relevant for VCA within pediatric patient populations as it is noted that adult face transplantations are performed most commonly in acquired traumatic injuries, with fewer being performed for congenital conditions such as neurofibromatosis [10].

Another pediatric patient population warranting investigation as regards VCA's potential benefits are patients with major burn injuries [29]. Major burn injuries in pediatric populations are not only associated with psychiatric morbidity; including anxiety, depression and PTSD, but commonly patients with burn injuries and associated psychiatric conditions do not avail themselves of psychiatric care [29]. Pediatric face transplantation for these patients may reduce the risk of depressive and anxiety disorders via reduced surgeries, hospitalizations and disfigurement. The population of pediatric patients with major burn injuries affecting both face and hand may be considered for concomitant face and hand transplantation [29]. Adult face and hand transplantation reports identify risks for consideration within pediatric face transplantation for burn injury. One is the hypothetical risk of hyper-acute rejection of VCA transplants for patients with burns receiving cadaveric skin allografts during acute burn care prior to VCA transplantation [30]. Second is the risk of resurgence of a resistant infection from the acute burn phase under the subsequent VCA condition of immunosuppression; this is the suspected cause of death in an adult patient with concomitant face and hand transplantation [30]. In the burn injury context, the potential benefits of VCA include reduced surgeries, hospitalization, disfigurement and risk of psychiatric morbidity. Yet, benefits must be balanced against the known general VCA risks associated with life long immunosuppression, plus the potential VCA burn-specific risks of hyper-acute rejection and resurgent infections.

In general, assessment for pediatric VCA, irrespective of whether it be secondary to an acquired or a congenital condition, requires appreciation of each recipient's developmental context as it will be relevant to a recipient's capacity to participate in post-operative immunosuppression and rehabilitation protocols; both of these VCA care components are important to procedural success and associated benefits [9, 10]. Developmental considerations regarding a child's concept of time, understanding (including fantasies) of surgery and surgical recovery and an estimate of the child's capacity to participate in post-operative care in the face of pain and discomfort should be a required institutional component of individualized VCA pediatric assessments. This developmental assessment constitutes a unique requirement in analyzing the balance across benefits versus risks for each potential pediatric recipient. Such a developmental assessment will be challenging, and perhaps unknowable for young children, but warrants a specific assessment that should include parental, or surrogate decision maker input.

For pediatric face transplantation, issues of medical uncertainty and graft longevity present unique risks distinct from adult face transplantation, requiring assessment in pediatric risk/benefit analysis [25]. In terms of graft longevity there are questions that cannot yet be answered. Will the graft burn out in time as recently occurred in an upper extremity transplant and have to be removed. Will it grow approportionately? These risks should be assessed within a developmental context. For instance, a young child with a first face transplant may need a second face transplant during adolescence when issues of identity are heightened. In contrast, an adolescent with a first face transplant during this developmental stage may need a second face transplant as a young adult when issues of identity are more consolidated. In terms of medical uncertainty, two unique issues are identified for pediatric patients with face transplants [10]: fewer surgical options exist for pediatric patients than adult patients in the event of graft failure in the face transplant context, and [11] the identification of bone growth centers in pediatric face transplants including bone will be complex and not easy to predict.

Continued exploration of pediatric VCA is warranted to not only clarify associated benefits and risks, but also for the essential step of risk/benefit analysis for each potential pediatric recipient. To date, pediatric experience with VCA is with limb transplantation within clinical contexts not requiring immunosuppression treatment. Case specific graft success and patient survival are documented with lower-limb transplantation versus prostheses. In addition, case-specific benefits of limb transplantation are reported in terms of return of limb sensation, muscle function and limb re-innervation is under cortical control. Pediatric VCA experience is not reported in pediatric cases requiring immunosuppression for a VCA procedure alone or in pediatric cases requiring immunosuppression in the context of a required life-saving solid organ transplantation procedure plus a concomitant VCA procedure. Advancing the exploration of pediatric VCA will therefore require a recalibration of the risk/benefit analysis experienced to date with twin-twin isograft limb transplantation as advanced VCA procedures will require incorporating the real risks associated with immunosuppression. As guidelines are necessary and outlined for both hand and face transplantation programs for adult populations, [23, 24] advancement of pediatric VCA will require the commitment of similarly wide-ranging resources within Pediatric Academic Health Sciences Centers to best address the unique benefits, risks and risk/benefit analysis associated with complex VCA care protocols for children, teenagers and their families.

10.4 Informed Consent

10.4.1 Donors and Donor Surrogate Decision Makers

Organ donation within pediatrics raises many difficult ethical issues from the point of view of informed consent. For a variety of reasons, including the fact that the transplants discussed in this chapter involve organs that are not internal but visible, VCA raises an additional set of ethical issues to the standard ones. (Such standard issues include the complexity of assessing whether one consents in a truly uncoerced way, or whether one truly understands the attendant risks to the extent one claims to). The externality of these organs raises certain issues vis à vis informed consent. Informed consent as it pertains to pediatric facial transplantation and donation will be considered first. Generally speaking, the overarching and most fundamental ethical issue surrounding informed consent in any context is the extent to which those offering informed consent are doing so in a fully informed, and voluntary, way. One must note here that within the context of pediatric facial transplantation, donors will be deceased, and donor surrogate decision makers will be the recently deceased child's parent(s), or another surrogate decision maker.

One issue to consider is whether the donor surrogate decision maker (in the current context, the person making a donation decision on behalf of a deceased child) is aware of the medical uncertainty involved in facial transplantation, which in the case of pediatric facial transplantation is particularly pronounced [25]. The uncertainty of facial transplant outcomes is relevant to the ethics of informed consent in that the donor surrogate decision makers (again, this would in most cases be the deceased child's family) might well want assurance that the donation will be put to good use. One can expect the desire for this assurance given the emotional dimension of donor families considering the donation of their recently deceased child's *face*. In order for consent to be informed and truly voluntary, it is crucial that all efforts are made to convey the pronounced uncertainty involved in the transplant. Grief and emotional strain of donor decision-makers can detract from truly voluntary decisions, and it is important to recognize such factors given the recent death of a child, and the fact that a decision is being made about an organ with such close connections to that child's identity.

A sense of the longevity of the facial graft would be crucial to informed consent on this front: donor surrogate decision makers could be surprised to know that transplanted external organs might have a limited lifespan. The chance of subsequent re-transplantation is all the more pronounced with child recipients, given the young age and increased years of exposure to immunosuppressive regimes. What is particularly worth noting here is that given the young age of the patient, a subsequent transplantation would occur at a new psycho-social stage of development, posing an additional set of post-transplant emotional adjustment challenges. This information could have a significant emotional impact on the thinking of donor surrogate decision makers, who may well have expected the donation to last a lifetime with the recipient.

Finally, it has been anticipated by some in the literature [17] that donor families might be motivated to donate their child's face in the hopes of seeing their child's face again [31]. Fully informed consent for donor surrogate decision makers within this context would involve them consenting to donation without being driven by such goals, mainly because facial resemblance depends upon resemblance in bone structure and it is likely that only a segment of the face will need to be transplanted and thus not be easily recognizable. Being driven to see one's child's face again would indicate a lack of understanding of the procedure at stake, and severely compromise genuine informed consent. There is an issue related to this donation rationale that we will just note here. This rationale for donation—"I want to see my child's face again"—might well set the stage for unwelcome contact of recipients by donor surrogate decision makers. (Though donation from deceased donors is anonymous at the time of donation in many jurisdictions, practices vary with regard to breaking anonymity post donation.)

When it comes to limb donation for transplantation within pediatrics, there are a few relevant differences from facial transplantation. First, there is more experience with such procedures, and so the transplant team can have a more specific conversation about outcomes with donor surrogate decision makers. Perhaps the (possible) drive to see their child's limb on another child (parents are referred to because they are the likely substitute donor substitute decision-makers) may differ from the drive to see their child's face again as discussed above—this has to do with our emotional connections to faces, and how those connections relate to the personal identity of the deceased donor—though potential donor surrogate decision-makers should be encouraged to reflect upon whether the desire to see their child's limb on a recipient is motivating the donation. It would seem that in some ways this consideration is differently relevant for a donated limb as it could retain its pre-donation appearance to a greater extent than would a donated face. Perhaps in this case the expectation is more accurate and less indicative of a lack of understanding of the procedure to which one is asked to offer consent. This matter would ideally be discussed during informed consent conversations. The issue noted in the above discussion of facial transplants—the issue of unwelcome contact of recipients by donor surrogate decision makers—applies here too, in that donor surrogate decision makers may, for example, wish to hold their child's hand again.

10.4.2 Recipients and Recipient Surrogate Decision Makers

Here again, the ethical issues around informed consent center around recipients or recipient surrogate decision makers being both adequately informed about the procedure (the team's responsibility), and demonstrating an understanding of the likely risks (the purview of the surrogate decision makers). Within the pediatric context, most if not all potential recipients will require a surrogate decision maker, given the complexity of facial transplantation.

Starting with facial transplantation, a quite basic challenge is the newness of the procedure; simply because the procedure has not been performed in the pediatric context, the informed consent process at this point cannot involve reports on outcomes of previously performed procedures. Medical uncertainty is relevant as it was above in the discussion of facial donation, as is the more specific uncertainty regarding the longevity of the graft.

The possibility of future resentment on the part of the recipient toward the recipient surrogate decision makers should be discussed as part of the informed consent process. At any time following the transplant, the patient may resent that the choice resulted in him or her having to take certain medications for a lifetime, and/or the patient may resent having this "new face" [25]. There could also be resentment from the patient toward the transplant team for performing the transplant. Of course, the recipient's attitude might just as well be gratitude as much as it could be resentment, yet this chance should likely still be discussed as part of the informed consent process.

Attention will need to be given to assent requirements in the jurisdiction where the VCA is taking place. Opportunities for a child to assent or dissent should VCA be offered in the context of a research study may be different from the child's opportunity to assent or dissent should VCA be offered within a purely clinical context. Those with the cognitive ability to assent [22] are generally given greater opportunity to refuse participation in a research study than they are given to refuse a treatment that their guardians and health care providers consider to be in their best interests. The widely held, and documented, connection between the face and both personal identity formation and personal identity itself is relevant to informed consent [31–35]. That the face is an external organ is highly relevant here; recipients (to the extent that this is developmentally feasible) and recipient surrogate decision makers should discuss with the transplant team the possible implications of this connection to personal identity, which could include post-transplant social integration challenges, considering the developing personal identity of the child along with the face's psycho-social significance. (Whether such challenges would be any more difficult than those encountered by not opting for the transplant is a good question, which will not be addressed here.) Relatedly, a discussion might be had with potential recipient surrogate decision makers about the relevance of the speed at which children and adolescents pass through developmental stages, and how such periods of development might relate to the assessment of recipient suitability. These periods of development are ethically relevant because they can create uncertainty as to whether a candidate deemed suitable at one stage of development will continue to be suitable across periods of development. Though transplant teams ought to aim for as current a patient assessment as possible, the relevant challenge issuing from the uncertainty regarding candidate suitability resulting from this developmental speed could be discussed with potential recipient surrogate decision makers.

Recipient surrogate decision makers should also be made aware of the importance of family support to patients' post-surgery recovery [36–40]. Informed consent will require a certain knowledge and understanding of the relevance of patient pre-trauma psychological health to the success of the post-transplant recovery: it is generally held that children and adolescents who are psychologically healthy and resilient before transplantation will recover from the transplant more readily than those who are not as psychologically healthy before transplantation. Assessing recipient surrogate decision makers' insight into their own child's pre-trauma psychological state, as well as into their own ability to support the child post-transplant, could be important for the transplant team to discuss.

When it comes to the voluntary aspect of consent, one must take into account the scarcity of donations. Scarcity is connected to how closely matched donors and recipients will need to be; they will need to be matched for approximate age because of tissue growth rates, and also for complexion and sex. Such scarcity may on some level pressure recipient surrogate decision makers to accept the donation and move ahead with the transplant, as there could be the fear that this is one's only chance to receive a donation. Such pressure could be internal to oneself, or come from external sources such as one's family or one's health care team. Even when a perfect match is found, and one could be hesitant to pass up a perfect match donation, no recipient substitute decision-maker, or young patient him or herself, should be pressured into accepting a transplant on the grounds that a perfect match has been found. A recognition and an appreciation of the emotional stress faced by recipient surrogate decision makers (the child will be severely injured or seriously ill) would help encourage clarity of thought on the part of the recipient substitute decision-makers, which would in turn enhance the likelihood of a voluntary decision. One should note that sometimes transplant teams will not allow a decision-maker to make a decision if there is not adequate first-hand recognition of emotional stress.

Moving on to limb transplantation from the recipient or recipient surrogate decision maker point of view, the most relevant difference relates again to a limb's seemingly less close connection to personal identity as compared to the face. Informed consent would enter less around the connection of post-transplant recovery to challenges surrounding personal identity and psycho-social adjustment. One ought not underestimate, though, the psychological impact of a child receiving a new limb. Such impact and the related potential challenges would require discussion between the transplant team, the potential recipient substitute decision-makers, and the potential patient (if his or her medical condition allows for decision making capacity), in the context of informed consent. As well, the transplant team ought to discuss with the potential child recipient where the donated limb comes from. This could help prepare the child for possible questions or teasing from his or her peers, in the event the transplant is carried out. It could also help the child recipient with his or her own understanding of, and potential fear around, the origins of the donated limb. Another relevant difference between limb and facial transplantation when it comes to informed consent is that prosthetics are available for limbs, but not for faces and while a limb graft can be removed if it fails, this is not an option for large face transplants. This should be discussed with potential recipients and recipient surrogate decision makers, because it relates to the nature of their options and thereby to being fully informed.

10.5 Establishing a Pediatric VCA Program

10.5.1 Donor and Recipient Selection

Experience in VCA has shown that patient selection is critical if success is to be realized [18]. Firstly, these patients must be in good general health. These operations are long and tedious and can involve significant metabolic shifts due to their magnitude.

Secondly, these patients should have a specific area in need of reconstruction that cannot be reconstructed by other means. Complex reconstructive techniques have improved dramatically in the last two decades. However, there are some areas, particularly of the face, that cannot be effectively reconstructed to date. Areas such as the central face and lips are particularly difficult to achieve an adequate result. In the upper extremity, even those patients who have had amputations need to have a trial of prosthetic use. Some patients do extremely well and do not require a transplant. Others, however, are not able to utilize the prostheses appropriately and may well be better served by upper extremity transplantation.

Thirdly, these patients require an intensive psychosocial evaluation. It is clear that successful transplantation involves a dedicated rehabilitation program and consistent use of immunosuppressant drugs. The psychosocial make-up of not only the patient but also his or her support system and home environment is critical. Thus, a full and detailed psychosocial assessment and rehabilitative assessment needs to be undertaken.

The specifics of the selection program will vary from unit to unit. However, it is important to have a screening process for eligibility on the basis of the patient's physical needs and physical state of health, the patient's behavioral history and psychological state, as well as the social systems that are in place for the patient and his or her family, in order to ensure success of the transplant.

For upper extremity transplants, appropriate candidates would be those with bilateral amputations at the level of the wrist or above, selected cases of unilateral transplants that are already on immunosuppressants, patients who have had multiple failed procedures and have not been successful in the utilization of prostheses. There should be significant functional impairment and the patient and or guardian must be willing and able to dedicate up to two years toward a postoperative rehabilitation program.

The donor for the upper extremity transplants must have documented brain death with hemodynamic stability. There should be minimal co-morbidities prior to death and the routine lab work should be normal. The donor should have suitable immunological typing for transplantation and be negative for a variety of viruses that could be transplanted to the recipient. From a physical standpoint, the donor extremity must have an acceptable bony configuration, and preference would be for the same sex and as close a match as possible for age, race, skin tone, hair color and physical structure. The motives of the donor family should also be carefully considered and unrealistic expectations of their child living on in someone else should be screened and thoroughly discussed with the donor family.

The specific clinical criteria for face transplants include the same principles. The recipient should have a significant facial deformity that has functional implications and cannot be reconstructed by conventional techniques. Very often, multiple attempts have been carried out without reconstructive success, and the patient remains with significant structural deficits and psychosocial isolation due to the fact that reconstructive efforts have not alleviated the functional deficits and the individual's appearance has not been adequately restored. The donor requirements are much the same as those listed for the upper extremity. An appropriate underlying bony configuration is necessary if bone is to be transplanted. It is preferable to have as close a physical match as possible. While for upper extremity transplants, preference would be for the same sex in the donor and recipient, a female donor for a male recipient would be acceptable. Regarding face transplants, however, it would be imperative to have the same sex. Regarding abdominal wall transplants, the sex would not really matter. There may be some variation in age that would be acceptable. Probably a range of two years above or below the recipient's age would be acceptable. Although it may be too early in the evolution of VCA to propose its use in infants, there may be sufficient benefit to warrant consideration of the application of VCA in children.

In all cases of VCA, a thorough evaluation of the physical, psychological and social needs of the recipient must be taken into consideration. This underscores the value of a multidisciplinary team working together with a common goal and common understanding of what constitutes a successful outcome. Only by working in concert can appropriate patient selection be effectively accomplished.

10.5.2 Allocation

As organs for transplantation are an absolutely scarce resource, the goal in distributing them is to save, extend or improve the quality of as many lives as possible, as fairly as can be achieved. In most jurisdictions, organs from deceased donors (DD) are donated anonymously, and allocated according to principles of distributive justice and set criteria applying to all potential recipients. Among vascularized composite allografts, transplantable limbs and faces are donated by deceased donors.

10.5.2.1 Allocation of VCA from Deceased Donors

Utility and justice are major ethical principles in addressing allocation from deceased donors [41]. Utilitarianism, most closely associated with John Stuart Mill, is a theory of ethics concerned with the goal of maximizing benefit and reducing suffering [42]. Applied to organ donation, benefit may be measured in outcomes such as life years post transplantation, the benefit experienced by each recipient, or the number of lives saved. Justice refers to the desire to be fair in allocating organs and may be viewed in several different ways. John Rawls' theory of justice considers giving more to the least well off in an attempt to make all as equal as possible [43]. In a health care setting, it is the sickest who are usually seen to be the least well off, as they lack, and therefore need, the greatest amount of good health. In the VCA context, the neediest are the patients for whom the magnitude of their loss impacts most on their daily function. This can include the ability to speak, eat, or use a limb(s). Among those listed for face transplant, the greatest need refers to those whose function is most reduced, such as the ability to eat, and those who suffer social stigma and difficulty with social integration due to the reaction of other people to their facial features. This includes a subjective component of how the patient perceives their condition and copes with it, which varies from person to person and may not be consistent with an objective evaluation of the situation by another person. Thus, health care professionals may evaluate these qualitative dimensions of a patient's function differently. In practice, the principle of justice would mean we transplant the sickest or neediest first. The utility principle does not correspond with this, as those most likely to do well may be the least sick who have more residual strength to aid recovery, or better predicted outcome. Allocation systems have struggled to reconcile these competing forces [41].

Generally, utility is reflected in the establishment of listing criteria, i.e. we list those whose diagnoses have been shown previously to respond well to transplantation and reach good outcomes. The Cleveland Clinic FACES Score provides an early tool to evaluate all candidates for a face transplant on relevant criteria [44]. Equity of access is met if all candidates have the opportunity to be considered for listing. Justice in allocation may be achieved if we treat like cases alike and give the most to the neediest.

10.5.2.2 Allocation Issues Particular to VCA

The visibility of limbs and faces add an extra dimension to their allocation, including the matching of skin tone and colour as well as gender specific attributes such as musculature, appearance and hirsutism. The more that these elements resemble those of the recipient, the less obvious the transplant may be to others and the easier it may be to achieve recipient acceptance. These dimensions contribute to social acceptance and the appearance of bodily integrity which contribute to the recipient's psychological acceptance of the limb or face [45]. They add new elements to the decision-making around choosing, ethically, the most suitable recipient for the VCA [46]. Functional ability is more of a deciding factor in decisions to allocate two arms or legs to one recipient whose need is likely to be considered greater than someone who lacks one limb. Such an allocation reduces inequity, but utility could be increased by giving one limb to each of two people if that improved each recipient's function significantly. These choices may be few when limb and face transplants are rare, as they are today. However, if these procedures become more numerous, and more than one potential recipient suits an available body part, more discussions will be needed on how to measure need and how to compare levels of suffering.

10.5.2.3 VCA Allocation Issues Particular to Children

The question of consent to a grafted body part for a child poses particular ethical issues. A child's wishes may be known, but will they endure as he/she matures and how will he/she feel later towards the acceptability of the transplanted body part? Patients are generally listed for transplantation after they have exhausted all other treatment options. A limb transplant would not be indicated until the patient has tried a prosthetic limb, as this will avoid the use of immunosuppressant therapy and surgery. A question that arises is whether a transplant should be postponed until the child reaches adolescence or early adulthood. This may enable the patient to make a more informed decision for herself, but it could involve unnecessary years of lesser function with a prosthesis or without any limb. An important listing criterion is the proven record of adhering to a medical treatment plan. Clearly, this is vital to ensure the success of the transplant. If a child's history suggests serious issues of non-adherence to treatment and/or coping difficulties, utility would dictate that the limb, face or stem cells be allocated elsewhere. However, denying a younger child these opportunities due more likely to the actions of the parents/guardians rather than the child raises questions of justice as this would disadvantage the child due to the actions of the parents/guardians.

10.5.3 Accountability for Surgical Innovation

In recent years there has been significant engagement in the surgical ethics literature regarding the need for accountability in innovation [47–52]. This section will discuss the importance of innovation pathways for introducing innovative surgeries and the need for provider accountability. Among others, Johnson and Rogers have identified four major areas of ethical concern in innovative surgery: harm to patients, informed consent, conflict of interest, and distribution of health care resources [53]. Each of these is significant for the introduction of VCA and is discussed in this chapter along with considerations regarding the need for and nature of the pre-operative and ongoing oversight or review.

Many have written about the ethics of surgical innovation [54], pointing out that surgical advances have often not been subjected to rigorous study design or regulatory scrutiny, and that nonetheless, procedures have improved and evolved. While it is clear that innovation depends on the inspiration of surgeons to problem solve and push the boundaries of what may be possible, best practices in seeking generalizable learnings in regard to innovative surgeries require a systematic, methodologically sound process of rigorous planning and evaluation [55].

Historically, surgical innovation has not been subject to systematic prospective review and was seen as unsuited to be the subject of formal research protocols aimed at providing high quality evidence to inform future use [56]. While there was a time when many feared that surgical innovation pathways, whether through research or other reviews, would stifle innovation, recent literature and data [57] suggests a recognition that review pathways and research are consistent with efforts to capture learnings from each procedure and to act in the best interest of the child [58].

Based on practice to date, for any institution that performs face or limb transplantation in pediatrics, the need to put in place appropriate accountability processes or research protocols seems widely accepted [58–61]. Such protocols would include preoperative evaluation, the surgical procedure itself, postoperative clinical care, and a detailed surgical and follow-up evaluation that address both the surgical result, as well as the psycho-social effects of the transplant.

Notwithstanding the broader experience with VCA in the adult context, the limited experience applying VCA in the pediatric context, should have it still be considered an experimental procedure. The boundaries between experimentation and innovation are rarely clear in the advancement of surgical care. In addition to adhering to the safeguards provided by an innovation pathway or research ethics requirements, a family-centered approach, in which a transparent discussion of expected risk and benefits addresses gaps in present knowledge, would attempt to empower patients and guardians to make an informed decision either for or against VCA. Because of the high degree of uncertainty related to many aspects of the overall outcome, pediatric VCA protocols should build in the careful study of outcomes that include both physiological measures and patient-reported outcomes (i.e., health-related quality of life) [25]. Given the innovative nature of the full surgical plan in pediatrics, some sort of pathway designated to review, approve and provide oversight over either research or innovative practices would be appropriate. While many of the technical components of VCA may essentially be standard practice for transplant and plastic surgeons, there is very limited experience performing VCA with children and therefore research in pediatric VCA should be conducted with institutional approval through an Institutional Review Board [25]. Given that the innovative nature of limb and facial transplantation goes beyond the complex technical reconstruction, preparing the innovative surgical protocols might require the involvement and expertise of specialists across a wide spectrum of practices, including anaplastology, anesthesiology, bioethics, burn surgery, craniomaxillofacial surgery, critical care, dentistry, facial plastic and reconstructive surgery, facial and limb rehabilitation, infectious disease, maxillofacial prosthodontistry, nursing, oral surgery, orthopedics, psychiatry solid-organ transplant, social work, and speech therapy.

Review of such surgical innovations by Institutional Review Boards will address issues including the justification, design and conduct of the research [62] the balance between potential risks and likely benefits, consent processes, data management, the distribution of resources, and the management of dualities and conflicts of interests [63]. Evaluating surgical innovations according to research ethics frameworks would not be limiting the innovative nature of the surgeries but would rather bring oversight and rigor to the innovative surgical research protocols and contribute to the trustworthiness of data and confidence of patients, their families and the general public. All parties involved must be able to have confidence that the surgery itself, and the innovative practices that are introduced within it, are not subject to distortions associated with personal, wider professional, industry or institutional interests [64].

An ideal pathway for introducing innovative surgeries adheres to both the ethical values of the institution and relevant legal standards in a way that can preserve the innovators' imagination and enthusiasm while ensuring scientific rigor and accountability [65].

Relevant to the innovation pathway that should be put in place for pediatric VCA, is a study that examined the experiences of the health care team members caring for patients undergoing facial transplant surgery [66]. This was the first study to explore the personal attitudes and experiences of the health care team members involved in facial transplantation and it recommended a focus on inter-professional practice and education [67]. The study participants identified several areas where they felt providers needed more knowledge. These areas included learning to become more comfortable with discussions about the new face with the patient and, in particular, about unique topics such as how one might want to deal with past photos, a blended hair color, and the patient's ability to be recognized in public. Although multidisciplinary protocols for roles and responsibilities for VCA program development are now found in the literature [68] this study pointed to the need for the research and publication of discipline-specific protocols [69]. VCA adds a new dimension of complexity beyond the technical and physiologic demands of the transplant itself. Limb and face transplantation introduce complex preparatory requirements for health care providers as well as complex issues for recipients regarding identity, ability to express emotion, and social integration [69]. All of these pieces will need to be addressed to move forward in an accountable manner with such innovative surgical procedures.

Favorable outcomes with VCA internationally have made this innovative surgical procedure a clinical possibility for severely disfigured pediatric patients. Further "research and scientific progress, especially in the field of transplant immunology, will be the key to continuous growth of this unique reconstruction option before it can be considered a standard of care" [22].

10.6 Conclusion

While many of the ethical issues relevant to VCA also apply to solid organ transplantation in pediatrics, VCA has several uniquely challenging dimensions. The fact that face and limb transplants are not necessary for a patient's survival and the fact that the transplants are visible and touchable, add a complex social and psychological dimension to evaluating readiness or ethical defensibility for such transplants. While new issues will no doubt be raised in the course of performing VCA, at a minimum, areas currently in need of consideration include issues related to informed consent, the balancing of risks and benefits, establishing pediatric VCA programs, donor and recipient selection, resource allocation and accountability pathways for surgical innovation.

References

1. Kawai T, Leventhal J, Madsen JC, Strober S, Turka LA, Wood KJ. Tolerance: one transplant for life. Transplantation. 2014;98(2):117–21.
2. Kerkar N, Danialifar T. Changing definitions of successful outcomes in pediatric liver transplantation. Curr Opin Organ Transplant. 2013;19(5):480–5.
3. Taylor GS. T cell-based therapies for EBV-associated malignancies. Expert Opin Biol Ther. 2004;4(1):11–21.
4. Geissler EK, Hutchinson JA. Cell therapy as a strategy to minimize maintenance immunosuppression in solid orgain transplant recipients. Curr Opin Organ Transplant. 2013;18(4):408–15.
5. Barth RN, Rodrigues ED, Mundinger GS, Nam AJ, Hui-Chou H, Jones LS, et al. Vascularized bone marrow-based immunosuppression inhibits rejection of vascularized composite allografts in nonhuman primates. Am J Transpl. 2011;11(7):1407–16.
6. Avery RK. Update on infections in composite tissue allotransplantation. Curr Opin Organ Transpl. 2013;18(6):659–64.
7. Shores JT, Higgins JP, Lee WP. Above-elbow (supracondylar) arm transplantation: clinical considerations and surgical technique. Tech Han Up Extrem Surg. 2013;17(4):221–7.
8. Smeets R, Rendenbach C, Birkelbach M, Al-Dam A, Grobe A, Hanken H, et al. Face transplantation: on the verge of becoming clinical routine? Bioemed Res Int. 2014;2014:907272.

9. Khalifian S, Brazio PS, Mohan R, Shaffer C, Brandacher G, Barth RN, et al. Facial transplantation: the first 9 years. Lancet. 2014;384(9960):2153–63.
10. Murphy BD, Zuker RM, Borschel GH. Vascularized composite allotransplantation: an update on medical and surgical progress and remaining challenges. J Plast Reconstr Aesthet Surg. 2013; 66(11):1449–552.
11. Jo HK, Park JW, Hwang JH, Kim KS, Lee SY, Shin JH. Risk acceptance and expectations of laryngeal allotransplantation. Arch Plast Surg. 2014;41(5):505–12.
12. Petruzzo P, Dubernard JM. The international registry on hand and composite tissue allotransplantation. Clin Transpl. 2011:247–53.
13. The DASH Outcome Measure. Disabilities of arm shoulder and hand. Available from: http://dash.iwh.on.ca.
14. Shores JT, Brandacher G, Lee WA. Hand and upper extremity transplantation: an update of outcomes in the worldwide experience. Plast Reconstr Surg. 2014.
15. Mohan R, Fisher M, Dorafshar A, Sosin M, Bojovic B, Gandhi D, et al. Principles of face transplant revision: beyond primary repair. Plast Reconstr Surg. 2014;134(6):1295–304.
16. Dahlborg EJ, Diaz-Siso JR, Bueno EM, Sisk GC, Pomahac B. The value of innovation: face and hand transplantation programs at Brigham and women's hospital. Plast Reconstr Surg. 2014;134(1):178e–9e.
17. Broyles JM[1], Alrakan M, Ensor CR, Khalifian S, Kotton CN, Avery RK, et al. Characterization, prophylaxis, and treatment of infectious complications in craniomaxillofacial and upper extremity allotransplantation: a multicenter perspective. Plast Reconstr Surg. 2014;133(4):543e-51e.
18. Byrne PJ, Gordon C[2], Lee WP. Face transplant teams and the collaboration imperative. JAMA Facial Plast Surg. 2014;16(2):83–4.
19. Khalifan S, Brazio PS, Mohan R, Shaffer C, Brandacher G, Barth RN, et al. Facial transplantation: the first 9 years. The Lancent. 2014: 384(9960):2153–63.
20. Gander B, Brown CS, Vasilic D, Furr A, Banis JC, Cunningham M, et al. Composite tissue allotransplantation of the hand and face: a new frontier in transplant and reconstructive surgery. Transpl Int. 2006;19(11):868–80.
21. Zuker RM, Redett R, Alman B, Coles JG, Timoney N, Ein SH. First successful lower-extremity transplantation: technique and functional result. J Reconst Microsurg. 2006;22(4):239–44.
22. Siemionow M, Ozturk C. Face transplantation: outcomes, concerns, controversies, and future directions. J Craniofac Surg. 2012;23(1):254–9.
23. Siemionow M, Gordon CR. Overview of guidelines for establishing a face transplant program: a work in progress. Am J Transplant. 2010;10(5):1290–6.
24. Gordon CR, Siemionow M. Requirements for the development of a hand transplantation program. Ann Plast Surg. 2009;63(3):262–73.
25. Flynn J, Zlotnik Shaul R, Hanson MD, Borschel GH, Zuker R. Pediatric facial transplantation: Ethical considerations. Can. J Plast Surg. 2014;22(2):67–9.
26. Washington KM, Zanoun RR, Cadogan KA, Afrooz PN, Losee JE. Composite tissue allotransplantation for the reconstruction of congenital craniofacial defects. Transpl Proc. 2009;41(2):523–7.
27. Weinstein JM, Chamlin SL. Quality of life in vascular anomalies. Lymphat Res Biol. 2005;3(4):256–9.
28. Terr LC, Abright AR, Brody M, Donner S, Eth S, Fine L, Fornari V, Jetmalani A, Livingston R, Powers JH, Robson K. When formulation outweighs diagnosis: 13 "moments" in psychotherapy. J Am Acad Child Adolesc Psychiatry. 2006;45(10):1252–63.
29. Hanson MD, Zuker RM, Zlotnik Shaul R. Pediatric facial burns: is facial transplantation the new reconstructive psychosurgery? Can J Plast Surg. 2008;16(4):205–10.
30. Leonard DA, Gordon CR, Sachs DH, Cetrulo CL Jr. Immunobiology of face transplantation. J Craniofac Surg. 2012;23(1):268–71.
31. Huxtable R, Woodley J. Gaining face or losing face? Framing the debate on face transplants. Bioethics. 2005;19(5–6):505–22 (PubMed PMID: 16425487).

32. Swindell JS. Facial allograft transplantation, personal identity and subjectivity. J Med Ethics. 2007;33(8):449–53 (PubMed PMID: 1766430).
33. Caplan A, Katz D. In brief: about face. Hastings center report. 2003;33(1)1–8.
34. Baylis F. A face is not just like a hand: pace Barker. Am J Bioethics 2004;4(3):30–2. PubMed PMID: 16192132.
35. Siemionow M, Sonmez E. Face as an organ. Ann Plast Surg. 2008;61(3):345–52. PubMed PMID:18724140.
36. Gordon CR[1], Siemionow M, Papay F, Pryor L, Gatherwright J, Kodish E, et al. The world's experience with facial transplantation: what have we learned thus far? Ann Plast Surg. 2009;63 (5):572–8. PubMed PMID 19806039.
37. Sheridan RL, Hinson MI, Liang MH, Nackel AF, Schoenfield DA, Ryan CM, et al. Long-term outcome of children surviving massive burns. JAMA. 2000;283(1):69–73 (PubMed PMID: 10632282).
38. Landolt MA, Grubenmann S, Meuli M. Family impact greatest: predictors of quality of life and psychological adjustment in pediatric burn survivors. J Trauma. 2002;53(6):1146–51 (PubMed PMID: 12478042).
39. Noronha DO, Faust J. Identifying the variables impacting post-burn psychological adjustment: a meta-analysis. J Pediatr Psychol. 2007;32(3):380–91 (PubMed PMID: 16882947).
40. Aitken ME, Tilford JM, Barrett KW, Parker JG, Simpson P, Landgraf J, et al. Health status of children after admission for injury. Pediatrics. 2002;110(2):337–42 (PubMed PMID:12165587).
41. Veatch RM. Transplantation ethics, Chapter 19. Washington, DC: Georgetown University Press; 2000.
42. Mill JS. Utilitarianism. Dover thrift edition 2007. In: Waldrop MC, Rogers TNR, editors; 1861.
43. Rawls JA. Theory of justice. Cambridge: Belknap Press of Harvard University Press; 1971.
44. Gordon CR, Siemionow M, Coffman C, Alam D, Eghtesad B, Zins JE, et al. The cleveland clinic FACES score: a preliminary assessment tool for identifying the optimal face transplant candidate. J Craniofac Surg. 2009;20(6):1969–74.
45. Kumnig M, Jowsey SG, DiMartini AF. Psychological aspects of hand transplantation. Curr Opin Organ Transpl. 2014;19(2):188–95.
46. Cendales LC, Rahmel A, Pruett TL. Allocation of vascularized composite allografts: what is it? Transplantation. 2012;93(11):1086–7 (PubMed PMID: 22406818).
47. Angelos P. Ethics and surgical innovation: challenges to the professionalism of surgeons. Int J Surg. (2013) 11(51), 52–55.
48. Babu MA, Heary RF, Nahed BV. Device innovation in neurosurgery: controversu, learning, and future directions. Neurosurgery. 2012;(4)70:789–795.
49. Dixon JB, Logue J, Komesaroff PA. Promises and ethical pitfalls of surgical innovation: the case of bariatric surgery. Obes Surg. 2013;23:1698–702.
50. Johnson J, Rogers W. Innovative surgery: the ethical challenges. J Med Ethics. 2012;38:9–12.
51. Knight JL. The dark side of surgical innovation innovations 2012;(5)7: 307–313.
52. Sagar SP, Law PW, Zlotnik Shaul R, Heon E, Langer JC, Wright JG, Hey, I. Just did a new operation! Ann Surg. 2014 1–2.
53. Johnson J, Rogers W. Innovative surgery: the ethical challenges. J Med Ethics. 2012;38(1):9–12 (PubMed PMID: 21697295).
54. Riskin DJ, Longaker MT, Krummel TM. The ethics of innovation in pediatric surgery. Semin Pediatr Surg. 2006;15(4):319–23 (PubMed PMID: 17055963).
55. Chagpar AB. Innovation in surgery: from imagination to implementation. Am J Surg. 2011;202(6):641–5.
56. Dahm P. The IDEAL stages of surgical innovation: distinguishing the development and exploration stages. Contemp Clin Trials. Letter to the Editor. 2014; 37(2):176–77.
57. Sagar SP, Law PW, Zlotnik Shaul R, Heon E, Langer JC, Wright JG. Hey, I just did a new operation! Introducing innovative procedures and devices within an academic health center. Ann Surg. 2014; 00(00).

58. Coffman KL, Siemionow MZ. Ethics of facial transplantation revisted. Curr Opin Organ Transpl. 2014;19:181–7.
59. Errico M, Metcalfe NH, Platt A. History and ethics of hand transplant. JR Soc Med Sh Rep. 2012;3:74.
60. Evans LA, Experiences of healthcare team members involved in facial transplant surgery and patient care. Nurs. Res. 2013;62(6):372.
61. Siemionow M, Ozturk C, Face transplantation: outcomes, concerns, controversies and future directions. J Craniof Surg. 2012;23(1):254.
62. Rosenthal R, Schafer J, Briel M, Bucher HC, Oertli D, Dell-Kuster S. How to write a surgical clinical research protocol: literature review and practical guide. Am J Surg. 2014;207 (2):299–312.
63. Rogers WA, Johnson J. Addressing within-role conflicts of interest in surgery. J Bioeth Inq. 2013;10(2):219–25 (PubMed PMID: 23519587).
64. Dixon JB, Logue J, Komesaroff PA. Promises and ethical pitfalls of surgical innovation: the case of bariatric surgery. Obes Surg. 2013;23(10):1698–702 (PubMed PMID: 23907325).
65. Zlotnik Shaul R, McDonald M, Langer JC. Facilitating innovation in the clinical setting: a pathway for operationalizing accountability. Healthc Q. 2009;12(3):60–65 (PubMed PMID: 19553766).
66. Evans LA. Experiences of healthcare team members involved in facial transplant surgery and patient care. Nurs. Res. 2013;62(6):372–82 (PubMed PMID: 24165213).
67. Institute of Medicine. 2013. Interprofessional education for collaboration: learning how to improve health from interprofessional models across the continuum of education to practice: workshop summary. Global Forum on Innovation in Health Professional Education, Board on Global Health, 1Y150.
68. Bueno EM, Diaz-Siso JR, Pomahac B. A multi-disciplinary protocol for face transplantation at Brigham and women's hospital. J Plast Reconstr Aesthet Surg. 2011;64(12):1572–9.
69. Singhal D, Pribaz JJ, Pomahac B. Face transplant program: a look back. Plastic Reconstr Surg 129:1.

Chapter 11
The Ethics of Managing Conflicts in the Era of Social Media

Sandra Amaral, Pamela Nathanson and Chris Feudtner

Abstract Increasingly, patients and families are engaging the media to influence transplant eligibility decisions for themselves or their family members. This chapter starts by reviewing the history of social media and how social media works as a means of persuasion, particularly in healthcare cases. The chapter then provides an overview of so-called social media crises, and presents an ethical framework to inform and assist policymakers, hospital centers and healthcare providers in responding to public outcry, while managing legal constraints to maintain patient confidentiality and ethical constraints to promote equity and utility in transplant access. In conclusion, the chapter addresses preventive strategies to mitigate both patient harm and public misperceptions.

Keywords Social media · Crisis management · Ethics · Privacy

S. Amaral (✉)
Division of Nephrology, Department of Pediatrics, The Children's Hospital of Philadelphia, Philadelphia, PA, USA
e-mail: amarals@email.chop.edu

S. Amaral
The Center for Clinical Epidemiology and Biostatistics, Perelman School of Medicine, The University of Pennsylvania, Philadelphia, PA, USA

P. Nathanson · C. Feudtner
Department of Medical Ethics, The Children's Hospital of Philadelphia, Philadelphia, PA, USA
e-mail: nathansonp@email.chop.edu

C. Feudtner
e-mail: feudtner@email.chop.edu

C. Feudtner
Department of Pediatrics, Perelman School of Medicine, The University of Pennsylvania, Philadelphia, PA, USA

C. Feudtner
Department of Medical Ethics and Health Policy, Perelman School of Medicine, The University of Pennsylvania, Philadelphia, PA, USA

R.A. Greenberg et al. (eds.), *Ethical Issues in Pediatric Organ Transplantation*, International Library of Ethics, Law, and the New Medicine 66,
DOI 10.1007/978-3-319-29185-7_11

11.1 Introduction

The era of social media has created new forces that lean upon key stakeholders in transplant decision making. These forces must be understood and managed to assure ethically appropriate care from transplant evaluation through management. Although the introduction of social media into the transplant process seems to have raised public consciousness of ethical challenges in transplantation, the tenets of ethical decision making and ethical responses to public concerns have not changed. Public awareness of conflicting opinions with regard to organ allocation and other aspects of transplant care heightens scrutiny of medical decision making but should not fundamentally change the way providers approach the practice of achieving ethical processes and outcomes for managing competing interests in transplant. Assuring patient privacy and confidentiality, weighing considerations of a just distribution of resources, making treatment decisions that are fair and unbiased, and reckoning with dire clinical situations where stopping or withholding life-sustaining interventions is justified because of diminishing prospects of any benefit and mounting harm: all these are pertinent ethical concerns.

This chapter provides a broad overview of social media and the ways in which social media influences society. We will focus on a few social media crises that have occurred in pediatric healthcare and then turn our attention to examining an ethical framework for managing social media crises regarding healthcare. We will then discuss how healthcare organizations can prepare to manage these crises in an ethically appropriate manner.

11.2 Social Media in Broader Historical Context

In this chapter, social media is defined as “websites and other online means of communication that are used by large groups of people to share information and to develop social and professional contacts” [1]. In the current era, social media enables messages to spread almost instantaneously across a tremendously wide audience. Patients and families are leveraging social media as a mechanism to garner support and advocate for their interests in transplant decision making.

In general, media communication influences social and political culture by shifting who controls, creates and distributes information. The advent of social media changed the landscape of information exchange by shifting greater control to the general public across all phases of communication from creation to dissemination.

In 1991, the World Wide Web was invented and the concept of a communication network was born. List-servs and email services fostered the creation of online community groups. By the late 1990s, several web platforms were re-defining how we communicate. Weblog publishing tool Blogger allowed internet users to post personal opinions and disseminate information on their own websites. The information was completely unvetted and unverified [2]. In 2001, Wikipedia emerged as a cooperative encyclopedia in which historical content is written and edited by a 'crowd' of volunteer amateur contributors. A few years later, platforms such as MySpace and Facebook expanded opportunities for the growth of online interpersonal social connections. In 2004, Mark Zuckerberg, the founder of Facebook, promised to make the Web more social with a goal to "make the world more open and connected" [2].

Although free speech and privacy had been long-standing tenets of media in the past, societal equity became another important aspect with the rise of social media. It is commonly perceived that the progression to electronic media liberated people from social constraints based on physical appearance or location. Zuckerberg and others promoted their media tools as enhancing ethics of transparency and serving as tools for democracy. Yet, the assertion that Facebook makes the world transparent and democratic is somewhat controversial since who controls and distributes individual and collective information is often unclear. Further, critics of high-speed communications have argued that public and private informed decision making relies on reliable and accurate accounts of events and, consequently, some control, or gatekeeping, of information is essential to promote constructive public discourse [3].

Nevertheless, social media has provided a voice for a variety of groups who have struggled to be heard [4]. For example, Twitter started in 2006 and is a free service that supports 140-character length messages to be 'tweeted' to a site where anyone can read them and then pass the information along further in a 're-tweet' [5]. Twitter reports are generally unverified and uncontrolled. Twitter enables the rapid dissemination of information during a crisis. During terrorist attacks in Mumbai, India in 2008, Twitter and Flickr, an online photo and video sharing community, permitted first-hand views of events to be spread widely and quickly. In 2009, after the Iranian election, Twitter became an agency for citizens to express dissent and expose their political oppression.

These political examples highlight the shift in media from a culture of distribution to one of active participation. In the present day, media reports are often created by citizens and then reshaped and reframed along the pathway of electronic communication, for example via Twitter. Wikipedia is another excellent example of the living, moving nature of history reported through community participation. This participatory culture acknowledges that individuals are not isolated, but rather belong to large social networks. Another forum which is used to solicit public support is Change.org [6]. Started in 2007, Change.org is a free, online petition tool intended to promote social change. Individuals can post petitions to support a wide range of causes, from criminal justice to health and education. Although the precise impact of Change.org is difficult to quantify, the site has been cited as influential in several high profile public controversies; for example, it may have contributed to criminal charges being brought against George Zimmerman, the neighborhood watch volunteer who shot and killed Trayvon Martin in Florida in 2012 [7].

There has also been a major shift in the imprint social media communication leaves behind. With social media, even seemingly inconsequential comments can reach large audiences and have long-lasting effects. Bobby Prasad suggests considering interactions via social media as similar to a conversation in an elevator —"anything you say will likely be heard and may well be used as evidence against you" [8]. This shift toward greater sharing of information challenges long-standing legal and ethical frameworks of privacy. For example, what are the legal and ethical implications when someone discloses personal information about others on the internet? Van Dijck argues in "The Culture of Connectivity" that the social, economic, cultural and legal definitions of sharing must be better defined in order to understand the legal values of privacy and control of information in the current era [2].

Long before the Internet, Harold Lasswell, an American political scientist (1902–1978), framed communication as "who says what to whom in which channel and with what effect" [3]. He argued that there are three equally crucial elements of communication—content, source and audience. This framework of communication provides an excellent conceptual model for managing social media in the healthcare setting. In transplantation, patients and caregivers are leveraging the ability to (1) share information about loved ones in need of organs, and (2) garner social support to promote access to transplantation for themselves or loved ones. One of the many challenges faced by healthcare teams in these cases is that patients and caregivers serve as the source of information, control the content of the information and, to some extent, target a specific audience of individuals with shared concerns. One puzzling aspect of privacy in healthcare is the disconnect between the value the healthcare system places on preserving patients' private health information and the amount of information patients themselves post on the internet publicly about the intimate details of their medical condition [2].

11.3 How and Why Do Messages Spread?

Although social media permits opportunities for information to spread widely and quickly, not every message is spread actively by its audience, or "goes viral". For example, in 2014, there were approximately 6000 tweets per second [9]. At best, about 25 % were forwarded on to others as re-tweets [10, 11]. The two main reasons for re-tweeting are to share information or to show support [12]. Similarly, in other media, the messages that tend to "go viral" are those that allow people to feel like they are helping others by educating them or providing social support. Additionally, "viral" messages tend to be more likely to evoke a strong emotional response (positive or negative) from their audience [13].

There are several recent examples of pediatric healthcare controversies that have gone "viral". Two of the first cases to reach widespread media attention occurred in Canada. In 2009, Jason Wallace and Crystal Vitelli introduced the story of their daughter, Kaylee, to the world [14]. Kaylee was born with a genetic disorder associated with multi-system organ failure, including severe neurologic impairment. Her initial prognosis was poor, and death was anticipated imminently.

Her parents learned that there was an infant with heart failure awaiting heart transplant in the same hospital, The Hospital for Sick Children in Toronto. The father held regular news broadcasts in front of the hospital to advocate for the directed deceased donation of Kaylee's heart to the designated infant and he approached the other family directly. The hospital emphasized that decisions regarding end-of-life care must be based on the interests of the child and not by the interests of a third party. They also stressed that organs could not legally be offered outside Canada's national transplant system. Later, the hospital supported the directed donation, however Kaylee survived after ventilator support was withdrawn and the heart transplant was not performed.

Once Kaylee survived the neonatal period, her parents focused on her survival. They continued to engage with social media but their message was reframed and they stressed that they should be the ones to decide Kaylee's quality of life. They claimed that the hospital staff suggested that too many healthcare resources were being used to support their daughter's life. The father was described in the news as "eager to face the cameras" and his character was questioned when he was charged with assault and robbery [14]. He later hired a public relations spokesperson as media messages turned against the family. Some reports questioned whether the family could provide effective care for the child. Kaylee's mother posted a disturbing video of Kaylee struggling to breathe. The video prompted public concerns of invasion of the child's privacy and dignity [15, 16]. Over a two year period, the parents continued to draw media attention to their daughter's medical condition. In 2011, they appealed to the public to advocate for Kaylee to be listed for kidney transplant and they appealed for living donors [17].

When Kaylee's story first garnered attention, it held strong emotional appeal for parents of children with disabilities and for those who had lost a child. The story focused on how quality of life is defined and who defines it. However, as the story unfolded, the parents were depicted as attention-seeking and self-serving. The audience remained engaged but the emotional response turned into anger and frustration toward the family and the story evoked ethical concerns regarding the child's best interest.

In 2011, another case in Canada drew attention to ethical dilemmas surrounding quality of life. The parents of Joseph Maraachli, an infant born with a terminal neurologic disorder, shared their story on Facebook, YouTube and on a website, SaveBabyJoseph.com [18–20]. The parents reported that London Health Sciences Centre in London, Ontario was denying Joseph a tracheostomy. Although the parents acknowledged their son's terminal diagnosis, they wished to take their son home to die after the tracheostomy procedure. They had previously had another child with the same diagnosis who survived at home for several months after receiving a tracheostomy. The hospital, with support from the Ontario Consent and Capacity Board and Ontario's Superior Court of Justice, deemed that the tracheostomy procedure was not in the child's best interests and the court directed Joseph's parents to consent to withdrawal of his breathing tube without tracheostomy placement [21–23]. The parents aligned with a right-to-life advocacy group, Priests For Life, who paid for the child to be flown to Cardinal Glennon

Children's Hospital in the United States where Joseph did receive a tracheostomy. Joseph subsequently returned to his home and died several months later. The social media debate included stories with titles such as, "Canadian Doctors Playing God?" and "Why Does Health Canada Want to Execute This Family's Infant?" [18, 20]. The message propagated by Priests For Life spokesman Father Frank Pavone was that doctors were placing a value judgment on the child's life and that the parents should have the "last word" in quality-of-life matters [20].

In the United States, the case of Amelia Rivera, a five-year-old with Wolf-Hirschhorn syndrome, drew widespread attention to transplant eligibility surrounding similar concerns of quality of life. In early 2012, Chrissy Rivera, Amelia's mother, posted a blog entry entitled "Brick Walls" in which she detailed her negative impressions of an encounter with transplant team members at The Children's Hospital of Philadelphia (CHOP) [24]. She also provided a detailed physical description of the doctor who allegedly denied her daughter a transplant. Mrs. Rivera was part of an online community at www.wolfhirschhorn.org. The blogpost received over 62,000 hits on the website within the first 24 h of the posting. Her story triggered follow-up stories in the New York Times and the Huffington Post, among other media outlets. A petition entitled "Allow the kidney transplant Amelia Rivera needs to survive" was started on Change.org and over 50,000 people signed it [25]. Mrs. Rivera's message was clear. She felt that her daughter was being discriminated against and denied access to transplant due to her cognitive disabilities and her quality of life as perceived by the transplant team. Her target audience consisted primarily of parents and caregivers of other children with physical and mental disabilities, but the story also garnered interest among the general public as well as legislators in the state of New Jersey where Amelia resided. Partially in response to Amelia's story, the state of New Jersey passed a bill that explicitly prohibits discrimination against a prospective organ transplant recipient solely on the basis of a physical or mental disability [26].

The following year, several other families used social media as tools to advocate for their children in need of solid organ transplants. The first case to garner public interest involved Sarah Murnaghan, a ten-year-old girl with cystic fibrosis in the United States. Her family engaged social networking sites and the media, sharing Sarah's story with video footage of the sick child with her oxygen mask on her face in her hospital bed [27]. The family approached members of congress and hired a lawyer to permit an exception for Sarah to be allocated lungs from an adult donor. The lung allocation policy at the time restricted adult lung allocation to pediatric patients 12 years of age and older [28]. Sarah was 10 years of age. A Change.org petition entitled "Allow Transplants of Adult Lungs to Children" received over 373,000 signatures [29]. The courts intervened to override the allocation policy and the Organ Procurement and Transplantation Network eventually modified its policies to permit exceptions [30]. This story gained widespread media and social media attention. The family's message was that the current organ allocation system was unfair. On one side, many members of the public expressed enthusiastic social support to prevent this particular child from dying. On the other hand, many professional societies, healthcare organizations and other members of the public expressed concerns of social injustice and inequity when the legal system intervenes to place the welfare of one person over that of the society [31]. Many argued that the legislative debate threatened to create a system where those with the most resources to create an emotionally appealing story would gain priority in organ allocation over those most disadvantaged. The Organ Procurement and Transplantation Network released a statement emphasizing the discrepancies between organ supply and demand, outlining specific numbers of people in need and lungs available, clarifying how policies are developed and stressing that individual need must be balanced with that of the greater good [32]. They later also published their response letter to the US Secretary of Health and Human Services with an action plan to address the court order [30].

In another case, in August 2013, 15-year-old Anthony Stokes was allegedly denied access to a heart transplant at The Children's Healthcare of Atlanta in Atlanta, Georgia [33]. Anthony's family posted the formal transplant denial letter on the internet [34]. The letter stated noncompliance as the rationale for denial. The family claimed that the real reason for denial was based on Anthony's poor school performance and run-ins with the law [35]. The patient was under house arrest with an ankle monitor at the time of his hospital admission. An interview was conducted with the patient in his hospital bed and distributed on YouTube [36]. The video named the doctor who allegedly denied the patient his heart transplant and included the individual's office phone number. The interviewer asked the audience to call the number directly to speak with the doctor to express concerns. The interviewer also made a plea for financial support for the family.

Anthony's story garnered interest from civil rights groups who spread the message that the case was evidence of society criminalizing black youth and devaluing their social worth [35]. Nevertheless, a Change.org petition "We want Anthony Stokes to be on the transplant list to receive a new heart" garnered only 45 supporters [37]. Although the message circulated nationally, the depth and breadth of the spreadability was much less than that of the Amelia Rivera or Sarah Murnaghan stories. Just as baby Kaylee's story lost momentum when her father was faced with criminal charges, Anthony's story may not have gained as much traction due to his history of encounters with the legal system.

Later the same year, the mother of Maverick Chenkus, an infant born with congenital heart disease and Coffin-Siris syndrome, went to CNN with allegations that her child was denied a heart transplant at New York Presbyterian for medical reasons that were invalid [38]. In a video featuring Ms. Chenkus with Maverick in their home, Ms. Chenkus reported that she had contacted researchers studying Coffin-Siris syndrome directly and that she had been told by these experts that the genetic disorder should not preclude transplant. She reportedly presented these medical data to the transplant team, however the child was still deemed ineligible for transplant. Ms. Chenkus reportedly contacted three other hospitals to attain transplant evaluations. Two declined evaluation. The third hospital accepted Maverick as their patient, treated him medically and, according to Ms. Chenkus, transplant was no longer necessary. The family filed a complaint against New York Presbyterian for a civil rights violation, claiming discrimination. They stated that they felt the real reason for denial was Maverick's disability and their main media message was that, in transplant, patients with disabilities are discriminated against, echoing the sentiments from the Rivera case. Ms. Chenkus was also quoted as saying, "The more I fought, the less they listened" [38]. Her quote highlights the significant role that poor communication between families and healthcare providers plays in invoking families to leverage media as a tool to be heard.

These cases played on the public's perception of social and political inequities in the transplant process. The organ allocation system is a complex and mysterious ranking system, and there is a long-standing history of distrust of the medical community's ability to make fair and unbiased decisions in determining transplant eligibility. Several studies have highlighted subjective bias in the transplant evaluation process. Most of these studies have focused on the psychosocial evaluation and have demonstrated an inconsistent nature of assessments of nonadherence and moral judgments of social worth [39–41]. Although the public worries about judgments of social worth, they express hesitation in allocating organs to patients with a history of noncompliance or an unhealthy lifestyle [42]. The judgment of social worth was a key issue in the Anthony Stokes case. Several studies of the general public's views of transplant decision making acknowledge the challenges faced by the transplant team. In general, the public has expressed preferences for fairness in the context of optimizing the public vs. individual good. Nevertheless, in social media, when faced with an emotionally appealing story, such as Sarah Murnaghan's, the public has a difficult time weighing the needs of society above those of the individual. Even medical criteria are somewhat inconsistent across centers [43, 44]. These inconsistencies leave room for patients and caregivers to claim discrimination—as occurred in the Chenkus case. Thus, there is widespread recognition that more robust, systematic guidelines are needed to inform consistency in transplant decision making [45]. In addition, it has been suggested that transplant centers should visibly post their transplant eligibility criteria on their websites to promote transparency [46].

11.4 Crisis Management in the Media

When patients or caregivers use social media to influence health outcomes, their portrayal of their healthcare providers and their healthcare system is often distorted. Importantly, as information spreads on the web or through social media, the information is further reframed. For instance, in several of the cases described above, specific health providers were named and depicted as singularly responsible for transplant denial. Most of the social media coverage grossly neglected to clarify that the transplant evaluation process is mandated as a multidisciplinary, multi-step process that requires input from a wide variety of healthcare professionals.

Thus, an important element of responding to a social media crisis is taking control of the information and serving as the source so that information is accurate. Social media responses must include three elements: (1) the public's concerns must be taken seriously; (2) there must be direct communication with the distributors of the information, in most cases, the patients or caregivers; and (3) the organization must acknowledge its role and responsibility in the crisis, demonstrating willingness to learn from its mistakes. Responding "no comment" is not likely to win support. Patients and caregivers want organizations to be transparent. Although, healthcare privacy laws severely restrict any reference to specific patients, responses that reinforce an organization's mission and principles are broadly encouraged. In the Anthony Stokes case, the Children's Healthcare of Atlanta Facebook message reinforced their dedication to the care of all their patients: "Dear friends, … you know [us]. Please know we are continuing to partner with the family as part of our ongoing commitment to all patients" [47].

In the Amelia Rivera case, The Children's Hospital of Philadelphia (CHOP) stressed their policies against discrimination [48]. Both hospitals communicated their ongoing commitment to partner with the displeased families to identify a solution. Nevertheless, the public often feels dissatisfied with what may be perceived as generic responses. Thus, it is important to emphasize the legal and ethical boundaries within which a response is constructed.

A media response must also convey that the patient's or family's concerns are valued. Responding swiftly helps reinforce the message that family concerns are highly regarded. In the case of Amelia Rivera, CHOP's initial Facebook response was posted three days after the blog entry and focused on their policies against discrimination and their general transplant evaluation procedures [48]. Many of the public's comments were negative and suggested the response was delayed and too carefully crafted. The institution posted a second response two days after that, starting with, "We hear you…" [49]. The second statement emphasized their commitment to patients and the recognition that caring for children is a "privilege" that is taken seriously. The second response was received more favorably. Approximately one month later, the hospital and family partnered in a joint statement expressing shared goals to optimize Amelia's health and to learn from their experience [50]. Specifically, the hospital acknowledged the need to improve the clarity of their policies and communications with families.

In the case of Amelia Rivera, the family and health providers moved forward together as Amelia received a kidney transplant from her mother at CHOP in July 2013 [51]. Since trust between patients, caregivers and health providers is often eroded when social media events occur, there must be concerted effort by the institution to work with the key stakeholders involved to forge a path forward. Institutions should provide crisis teams, who may include patient advocates, other healthcare providers and ethicists, to help mediate meetings between families and healthcare team members and to develop plans to work together, improve communication and rebuild confidence. Under some circumstances, it may be in the best interests of the patients, caregivers or healthcare team for the patient to transfer care elsewhere. This decision should ideally be reached mutually. When transfer of care is the selected next step, the institution must maintain its integrity and conscientiously limit the communication of biased information to the new care team.

In some circumstances, it is not possible for an institution to remain silent on the facts of the case. In the case of Baby Joseph, hospital officials from the London Health Sciences Centre chose to speak out publicly to defend themselves after security threats and defamatory comments were made against the hospital and healthcare providers. The hospital, nevertheless, was prudent in its response. The institution focused its response on the facts of the case which had already been publicly reported [21, 22]. Their statement was limited to clarifying the medical condition of the patient, defining tracheostomy, and explaining the hospital and court decisions related to substitute decision making. The institutional statement was also firm and clear that "legal action to protect the reputations of those against whom [outrageous and defamatory] allegations have been made are under consideration."

The support of providers who may be directly named in media reports is important. In several of the cases mentioned, specific healthcare providers were named and direct contact information was released to the public. The healthcare institution must partner with their information technology services to handle increases in phone calls, emails, etc. in response to media crises. Additionally, any healthcare providers or hospital leaders who might be involved in the case should receive specific guidance from the institution regarding how to respond and to whom to defer media inquiries. Affected healthcare providers may feel traumatized. They may struggle with how to manage their reputation and, in some cases, may need support in acknowledging their own mistakes. Providers should be offered psychological counseling. On occasion, heightened security may be warranted for the hospital and for individuals.

There is often an impulse by staff and employees at the institution to seek more information about the case. In general, it is advisable to alert all hospital staff and employees to practice restraint in acknowledging Facebook posts, public accusations, etc. Staff and employees must also be reminded not to access the medical record of the patient of interest unless they are directly involved in that patient's care.

The instantaneous nature of social media often misleads organizations into believing that social media crises are unavoidable. However, in many cases, families and caregivers go to the media when, as Ms. Chenkus suggested, the healthcare team is not listening. One approach to minimizing media scandals is to create a healthcare culture in which dissatisfied patients or caregivers are routinely referred to a patient advocate. Many healthcare organizations now provide support service teams that include patient and family representatives to listen to and mediate patient or caregiver grievances. These support teams can play an important role in the primary prevention of media crises by allowing opportunities for patients or caregivers to voice their concerns in a less public forum. Additionally, patient advocacy teams can benefit healthcare providers by offering a different viewpoint and clarifying the key concerns of the family.

Although social media crises are challenging situations, they also offer growth opportunities for healthcare organizations and professionals to learn from their mistakes. In the Amelia Rivera case, the crisis prompted recognition of ambiguity in CHOP's institutional transplant policies and procedures. In response, CHOP modified its organizational tools and operations to clarify the transplant evaluation process for families and caregivers as well as transplant team members. With the Sarah Murnaghan case, the United Network for Organ Sharing revisited its allocation policies and it continues to evaluate its processes for balancing equity and utility in an era of inadequate organ supply.

No healthcare organization or healthcare provider is invulnerable to being portrayed negatively in social media. The best approach to mitigating harm to the institution and individual healthcare professionals is proactive. Social media response teams, including healthcare providers, information services, legal/risk management, and public relations staff, should be engaged a priori. Response plans must align with an institution's vision, mission and goals, but also must be patient-centered and acknowledge collaboration with patients and caregivers to the extent possible.

11.5 Ethical Framework for Managing Social Media Crises

The ethical challenges related to organ transplantation in the age of social media are closely tied to the same ethical challenges faced in organ transplantation prior to the rise of social media. Organ transplantation raises significant ethical dilemmas regardless of the media environment. Issues such as privacy and confidentiality of patient information, distributive justice, fairness in allocation and eligibility determination, and limitations of treatment are daily challenges faced in healthcare and in the transplant arena [52]. The firestorm that can result when these issues are presented by the family of a patient in desperate need of an organ transplant and then propagated through social media is what has changed the landscape in the recent past.

11.5.1 Privacy and Confidentiality

When families publicize their stories on social media, they are making a decision to give up their privacy in exchange for public support for their cause. The difficulty for healthcare providers and institutions in this situation is that the obligation to protect their patients' privacy and confidentiality still exists in spite of their desire to respond to what may be perceived to be an unfair or inaccurate portrayal of the circumstances surrounding a particular case. Some of the ethical challenges involve protecting the privacy of other patients, of the providers working with this family, and of the child patient himself. Disclosing additional clinical details in order to clarify the hospital's position is not possible and commonly illegal, unless specific permission is granted from the family to disclose personal health information (PHI).

The duty to protect patient confidentiality can feel even more difficult when a particular provider or group of providers is being portrayed unfairly—or even as cruel or heartless—by the patient and family or by others commenting on social media. While providers have a natural instinct to defend themselves and their practice of medicine under these circumstances, it is important for the institution to maintain a calm presence and provide enough support for their providers to resist this impulse to respond to criticism. In the case of Joseph Maraachli, the hospital did issue a public statement reviewing the key facts of the case, but the institution was very careful to restrict its statement to facts that had been previously publicly disclosed by the family.

In pediatric transplant, an additional complication in these cases is the fact that most of the time it is the parents who are choosing to go public and give up their child's privacy as well as their own. Since information posted on social media will be part of an individual's virtual footprint forever, there may be concerns about how this information may affect the child in the future. Will friends or acquaintances learn information that the child would choose not to share? Could they be denied a job in the future due to concerns about potential illness? Parents may not be considering these factors, and even if they did, may not find them important enough to outweigh their wish to gain widespread support for their cause.

11.5.2 Distributive Justice

One of the main purposes of the United Network for Organ Sharing (UNOS) system of organ allocation is to avoid circumstances in which one patient's heart-rending story is pitted against another in vying for scarce organs. Some families have more resources and media savvy—or simply have characteristics such as youth or physical attractiveness that appeal to a particular constituency—to be able to marshal social and traditional media to pick up their cause and garner widespread support. This concept has been called "pushethics" and refers to the way families are pushing their own story into the "moral spotlight" [31]. While it is understandable that patients and families want to use their resources to advocate for themselves, it is important to ensure that the heightened attention to one individual does not unfairly disadvantage others equally in need. Specifically, social media attention should not influence an individual's priority in receiving a transplant [53]. This concept was nicely illustrated in coverage of the baby Kaylee story in which the parent of another infant with heart failure "worried other families might decide they've been waiting too long for an organ and go to the media, and that might mean another baby could get a heart before her infant" [54].

11.5.3 Fairness in Eligibility Determination

Even with well-defined and publicly available listing criteria, the process for determining individual patient eligibility for listing can still be perceived as unfair by a family that has been turned down for reasons that may seem discriminatory. The Rivera and Chenkus cases are examples. In these cases, disability and perceived quality of life were perceived by the families to be the sole reason for the determination of unsuitability for transplant. While these factors are frequently considered in the decision-making process for transplant, they are rarely the sole criteria considered. Cognitive disability in children in need of transplant often presents as part of a syndrome with other significant medical complications. In these cases, it is difficult to ascertain whether objective medial criteria alone would have prohibited listing for transplant. In other cases, issues surrounding projected difficulties with compliance or social support are determined subjectively and are open to interpretation as discriminatory. These types of issues were central to the Stokes case. Once these types of cases gain media attention, the fine details that may have influenced the listing determination are lost in the uproar. Since providers are unable to disclose details about how these decisions are made, they may be perceived to be hiding something.

With each case that goes viral, the public is exposed to stories that may portray the organ allocation system and transplant providers as unfair, unjust, biased or uncaring. The transplant community must be more proactive and transparent about how the organ allocation system works, and hospitals and transplant programs should provide more upfront information on the evaluation process to determine eligibility for listing, including the more subjective criteria like past and projected compliance and social support. It is difficult to know whether controversial cases spurred on by social media affect organ donation rates, and in what direction. On the one hand, these cases widely publicize the scarcity of organs available for donation and the plight of those on the waiting lists. On the other hand, stories that include perceived bias or unfairness in the allocation system may discourage some individuals from donation.

Fairness may also come into question when a provider publicly advocates for a particular patient to receive a transplant. For example, in the Sarah Murnaghan case, her pulmonologist testified at an emergency hearing on Sarah's behalf [55]. While that advocacy may benefit all of his patients, his other patients could perceive that their own case is not receiving the same level of concern or attention. Providers must use caution when advocating for particular patients, and ensure that they are also making a case for all of their patients waiting for transplant.

11.5.4 Limitations of Treatment

One of the most common ethical challenges faced by patients, families and providers surrounds coming to agreement around limiting interventions when the chance of success is small and the inherent risks are great. Organ transplantation may fall into this category, as patients walk a fine line between sick enough to be listed and being too sick to benefit from the intervention. Families who start on the path toward organ transplant have high hopes of their loved one being saved by the transplant, and if they are denied acceptance to a program, are forced to face their lack of options. Perhaps earlier referral to palliative care services, even when transplant is still a viable option, would help open the door to earlier communication about what might happen if a transplant does not occur [56].

11.5.5 Autonomy

During a social media crisis related to transplant, conflicts between patients, families, individual clinicians, or healthcare institutions can limit or threaten the ability of any of these agents to act in an autonomous manner to pursue what they perceive to be their own best interests. Ethical solutions to these conflicts are likely to not only respect autonomy but also attempt to restore autonomy as much as possible. Ideally, none of the parties should feel that they were forced—an admittedly high ideal, but ideal nonetheless. If conflicts cannot be resolved, patients and family caregivers should be offered second opinions or transfers of care. Respect for autonomy, though, goes beyond allowing for transfers of care. If either patients (or their families) or clinicians believe that the process of resolving the dispute is not handling their concerns and best interests fairly, those individuals have a right to seek legal or external professional counsel.

11.6 Planning for a Social Media Crisis

Healthcare providers and institutions that want to minimize the negative impact of a social media crisis on their practice must both embrace social media as a force to promote their mission and engage with their patients and families, while at the same time promote care and caution when managing a social media crisis. While social media has raised the stakes of negative publicity in terms of speed and scope of impact, organizations have long had policies in place regarding handling negative publicity in the traditional media. These policies can be adapted to the setting of social media [57].

Institutions must develop an overall social media policy for staff at all levels of the organization. That policy should acknowledge the widespread use of social media by families and caregivers and also provide guidance to staff for proper use of social media in order to prevent staff from overstepping their roles in terms of information sharing when a social media crisis arises. Just as most hospitals have media policies in which media requests are funneled through proper channels, staff must be educated in how to appropriately engage with social media in order to prevent missteps that could potentially "go viral".

11.6.1 Identify When a Crisis Is Happening

Since social media crises can expand dramatically in a short period of time, organizations must put in place systems to monitor social media and identify initial indicators that a social media crisis may be brewing. A social media plan should identify the individuals who will be tasked with monitoring online "chatter" for mentions of the institution's name, and set a threshold beyond which a determination of crisis is made. Since social media crises affecting healthcare organizations often arise in response to patient or family dissatisfaction, providers should identify those situations most likely to trigger such a response and alert the proper resources when they sense an impending crisis. Of course, a provider's first response should be continuing communication with the dissatisfied patient and family in order to prevent the social media complaint in the first place.

11.6.2 Elements of a Social Media Response Plan

Prior to any crisis, hospitals need to develop a social media crisis response plan, with input from all relevant stakeholders. While institutional differences exist, the following roles would most likely be included in a social media response planning process: clinical staff, family relations/family services, information services, internal communications, public relations, and legal/risk management. These stakeholders should identify the roles and responsibilities that each will play in the event of an actual crisis. Once a social media crisis is identified, a single leader should be identified who will be tasked with managing the crisis. Internal communications staff should develop and distribute communication to all staff with instructions on how to respond (or not respond) to inquiries about the situation, and where to turn for additional support [58]. Public relations staff will need to develop appropriate external statements, with review at the highest level of the organization. These external statements should be factual without disclosing protected patient information, and reiterate the institutional mission and values. Finally, institutions must put in place a coordinated social media effort of posting responses to complaints and comments, and posting of a series of ongoing proactive messages that are in line with the other external statements being delivered to traditional media.

An effective crisis management process should include the development of a crisis management plan long in advance of untoward events and should outline a framework for information dissemination, considering potential responses and reactions from key stakeholders and establishing a priori metrics for outcomes evaluation so that any crisis is framed as an opportunity for growth of the organization. In the Rivera case, the Children's Hospital of Philadelphia's joint statement with the family noted: "a strong desire to learn from this experience" [50].

Note that the extent of an effective response may vary depending on the specific circumstances of the case. In the case of the teenager who was reportedly denied a heart transplant at the Children's Healthcare of Atlanta, the hospital said very little. The institution pledged its ongoing commitment to all patients and reported that they were working with the family to resolve the conflict [47]. In the case of the Joseph Maraachli, the London Health Sciences Centre was compelled to issue a strong statement regarding the facts of the case and threatened legal action [21].

Any social media response plan should be reviewed during and after a crisis to determine whether the plan has met the needs of the organization and if any changes to the plan are needed. As social media continue to evolve, institutions and providers will need to keep current with trends and adapt response plans as necessary.

11.7 Conclusion

Ethical issues in organ transplantation will continue to raise challenges for providers both in practice and in terms of media or social media attention. The speed and reach of social media responses have raised the intensity and scope of these challenging situations, but the basic ethics issues remain the same. Institutions and providers must remain calm and present their case without disclosing patient information. Social media responses must be thoughtful and consistent with institutional values. The best approaches to social media management include the development of proactive social media policies that are formulated by relevant stakeholders, including patients and families. Engagement with social media requires active monitoring for early warning signs, a planned system for information dissemination and a priori metrics for outcomes assessment.

References

1. Dictionary.com. Definition of social media. Random House, Inc. 29 Jan 2015. Available from: http://dictionary.reference.com/browse/socialmedia.
2. Van Dijck J. The culture of connectivity: a critical history of social media. New York: Oxford University Press; 2013.
3. Briggs AB, Burke P. A social history of the media: from Gutenberg to the internet. 3rd ed. Malden, MA: Polity Press; 2013.

4. Jenkins HFS, Green J. Spreadable media: creating value and meaning in a networked culture. New York: New York University Press; 2013.
5. Twitter. Welcome to Twitter. Twitter, Inc. 29 Jan 2015. Available from: https://www.twitter.com.
6. Change.org. The world's platform for change. Change.org, Inc. 29 Jan 2015. Available from: https://www.change.org.
7. Wikipedia. Change.org. Wikimedia Foundation, Inc. 29 Jan 2015. Available from: http://en.wikipedia.org/wiki/Change.org.
8. Prasad B. Social media, health care, and social networking. Gastrointest Endosc. 2013;77 (3):493.
9. Internetlivestats.com. Twitter usage statistics. Internetlivestats.com. 29 Jan 2015. Available from: http://www.internetlivestats.com/twitter-statistics/.
10. Lui Y Chloe K-S, Mislove A. The tweets they are a changin': evolution of twitter users and behavior. 2014. Available from: http://www.ccs.neu.edu/home/amislove/publicatons/Profiles-ICWSM.pdf.
11. Crum C. Wonder what percentage of tweets are retweets? 2009. Available from: http://www.webpronews.com/wonder-what-percentage-of-tweets-are-retweets-2009-06.
12. Recuero R, Araujo R, Zago G. How does social capital affect retweets? In: Proceedings of the fifth international AAAI conference on weblogs and social media (serial on the Internet). 2011. Available from: www.aaai.org.
13. Konnikova M. The six things that make stories go viral will amaze, and maybe infuriate, you. New Yorker (serial on the Internet). 2014. Available from: http://www.newyorker.com/tech/elements/the-six-things-that-make-stories-go-viral-will-amaze-and-maybe-infuriate-you.
14. 16×9 onglobal. Baby Kaylee's heart transplant battle. In: 16×9, editor. Global News; 2009.
15. Blatchford C. Praying that baby Kaylee is comforted, not filmed. The globe and mail (serial on the Internet). 2010. Available from: http://license.icopyright.net/user/viewFreeUse.act?fuid=MTg4MTY5NTc%3D.
16. DiManno R. Discord darkens baby Kaylee's final days. Toronto star (serial on the Internet). 2010. Available from: http://www.thestar.com/news/gta/2010/04/03/dimanno_discord_darkens_baby_kaylees_final_days.html.
17. Kuitenbrower P. New drama for Kaylee and parents. 2011. Available from: pkuitenbrouwer@nationalpost.com?subject=newdrama for kaylee and parents.
18. FoxNews. Why does health Canada want to execute this family's infant? America's Newsroom: ill318; 2011.
19. LIfeSiteNews. Baby Joseph NOT in vegetative state. In: News LS, editor; 2011.
20. FoxNewsInsider. Canadian doctors playing god? 2011.
21. LHSC. Statement regarding diagnosis and treatment of Baby Joseph. LHSC media release (serial on the Internet). 2011. Available from: http://www.lhsc.on.ca/About_Us/LHSC/Media_Room/Media_Releases/2011/2011_03_23.html.
22. LHSC. Just the facts: medical, legal and ethical issues. LHSC media release (serial on the Internet). 2011. Available from: http://www.lhsc.on.ca/About_Us/Baby_Joseph/index.htm.
23. Weeks C. Hospital confronts social media uprising over care of dying baby. The globe and mail (serial on the Internet). 2011, Available from: http://license.icopyright.net/user/viewFreeUse.act?fuid=MTg4MTYzMDc%3D.
24. Rivera C. Brick Walls. 2012. Available from: http://wolfhirschhorn.org/2012/01/amelia/brick-walls/.
25. Change.org. Allow the kidney transplant Amelia Rivera needs to survive. Change.org. 2012 (cited 2015 29 Jan 2015). Available from: https://www.change.org/p/children-s-hospital-of-philadelphia-allow-the-kidney-transplant-amelia-rivera-needs-to-survive.
26. NJStateAssembly. PL2013, chapter 80 senate, no. 1456. In: Jersey SaGAotSoN, editor. Senate and General Assembly of the State of New Jersey; 2013.
27. Murnaghan J. Janet-Sarah Murnaghan lung transplant. 2013 (29 Jan 2015). Available from: https://www.facebook.com/janet.murnaghan?fref=ts.

28. Lupkin S. MK. Sarah Murnaghan, 11, home after controversial lung transplant. 2013. Available from: http://abcnews.go.com/Health/sarah-murnaghan-11-home-controversial-lung-transplant/story?id=20080946.
29. Change.org. Allow transplant of adult lungs to children. Philadelphia: family and friends of Sarah Murnaghan. 2013. Available from: https://www.change.org/p/allow-transplants-of-adult-lungs-to-children.
30. OPTN. OPTN response to court order regarding Sarah Murnaghan. United Network for Organ Sharing. 2013. Available from: http://www.unos.org/about/index.php?topic=newsroom&article_id=2795.
31. Savulescu JEB. Pushethics: how to get a lung for your child. Practical ethics (serial on the Internet). 2013 (19 Jan 2015). Available from: http://blog.practicalethics.ox.ac.uk/2013/06/pushethics-how-to-get-a-lung-for-your-child/.
32. OPTN. OPTN statement regarding lung transplantation and pediatric priority. Organ procurement and transplantation network. 2013. Available from: http://optn.transplant.hrsa.gov/news/optn-statement-regarding-lung-transplantation-and-pediatric-priority.
33. Park A. Teen denied access to transplant list receives new heart. TIME (serial on the Internet). 2013. Available from: http://healthland.time.com/2013/08/22/teen-denied-access-to-transplant-list-receives-new-heart/.
34. Michelson B. Anthony stokes denied transplant: 5 facts you need to know. 2013. Available from: http://heavy.com/news/2013/08/anthony-stokes-heart-transplant-enlarged/.
35. Culp-Ressler T. Dying teen is being denied a heart transplant because he's had trouble with the law. Think progress (serial on the Internet). 2013. Available from: http://thinkprogress.org/health/2013/08/12/2453941/dying-teen-heart-transplant/.
36. Anthoni R. Anthony stokes denied a heart transplant by Dr. Rahn Anthoni Show; 2013.
37. Change.org. We want Anthony stokes to be on the transplant list to receive a new heart. Change.org. 2013 (21 Jan 2015). Available from: https://www.change.org/p/petition-for-15-year-old-anthony-stokes-heart-transplant-we-want-anthony-stokes-to-be-on-the-transplant-list-to-receive-a-new-heart.
38. Cohen E. Disabled baby denied heart transplant. CNN health (serial on the Internet). 2013. Available from: http://www.cnn.com/2013/11/30/health/disabled-transplants/.
39. Dobbels F, Hames A, Aujoulat I, Heaton N, Samyn M. Should we retransplant a patient who is non-adherent? A literature review and critical reflection. Pediatr Transpl. 2012;16(1):4–11 (Review).
40. Furth SL, Hwang W, Neu AM, Fivush BA, Powe NR. Effects of patient compliance, parental education and race on nephrologists' recommendations for kidney transplantation in children. Am J Transpl. 2003;3(1):28–34 (Research Support, Non-U.S. Gov't Research Support, U.S. Gov't, P.H.S.).
41. Volk ML, Biggins SW, Huang MA, Argo CK, Fontana RJ, Anspach RR. Decision making in liver transplant selection committees: a multicenter study. Ann Intern Med. 2011;18, 155 (8):503–8 (Multicenter Study Research Support, N.I.H., Extramural Research Support, Non-U.S. Gov't).
42. Irving MJ, Tong A, Jan S, Wong G, Cass A, Allen RD, et al. Community preferences for the allocation of deceased donor organs for transplantation: a focus group study. Nephrol Dial Transpl. 2013;28(8):2187–93 (Research Support, Non-U.S. Gov't).
43. Secunda K, Gordon EJ, Sohn MW, Shinkunas LA, Kaldjian LC, Voigt MD, et al. National survey of provider opinions on controversial characteristics of liver transplant candidates. Liver Transpl. 2013;19(4):395–403 (Research Support, Non-U.S. Gov't).
44. Tong A, Hanson CS, Chapman JR, Halleck F, Budde K, Papachristou C, et al. The preferences and perspectives of nephrologists on patients' access to kidney transplantation: a systematic review. Transplantation. 2014;15, 98(7):682–91 (Research Support, Non-U.S. Gov't).
45. Batabyal P, Chapman JR, Wong G, Craig JC, Tong A. Clinical practice guidelines on wait-listing for kidney transplantation: consistent and equitable? Transplantation. 2012;15, 94 (7):703–13 (Research Support, Non-U.S. Gov't Review).

46. Bramstedt KA, Young JB. Use of the internet by United States heart transplant centers to promote transparency in the process of patient selection. Telemed J E Health. 2006;12(3):359–62.
47. CHOA. Dear friends… The children's healthcare of Atlanta. 2013 (updated posted 12 Aug 2013). Available from: https://www.facebook.com/ChildrensHealthcareofAtlanta.
48. CHOP. We have watched with great concern… The children's hospital of Philadelphia. 2012 (updated 13 Jan 2012). Available from: https://www.facebook.com/Childrens HospitalofPhiladelphia.
49. CHOP. To our Facebook community: we hear you; 2012.
50. CHOP-Rivera. Joint statement of the Rivera family and the children's hospital of Philadelphia. 2012. Available from: http://wolfhirschhorn.org/2012/02/amelia/joint-statement-of-the-rivera-family-and-the-childrens-hospital-of-philadelphia/.
51. Painter K. HN. Amelia Rivera gets kidney after transplant debate. USA Today (serial on the Internet). 2013. Available from: http://www.usatoday.com/story/news/nation/2013/07/30/amelia-rivera-kidney/2600551.
52. Cohen E. Organ transplantation: defining the ethical and policy issues. Staff discussion paper. President's Council on Bioethics([serial on the Internet). June 2006. Available from: https://bioethicsarchive.georgetown.edu/pcbe/background/staff_cohen.html.
53. de Sante J, Caplan A, Hippen B, Testa G, Lantos JD. Was Sarah Murnaghan treated justly? Pediatrics. 2014;134(1):155–62 (Case Reports).
54. Journal Pioneer. Publicity around infant girls donor drama angers other parents at sick kids. The Journal Pioneer (serial on the Internet). 2009. Available from: http://www.journalpioneer.com/Living/2009-04-09/article-1393430/Publicity-around-infant-girls-donor-drama-angers-other-parents-at-Sick-Kids/1.
55. Dale M. Pa. girls gets on lung donor list; boy also sues. Big story (serial on the Internet). 2013. Available from: http://bigstory.ap.org/article/judge-backs-dying-pa-girl-who-needs-new-lungs.
56. Fowler A, Freiberger D, Moonan M. Palliative and end-of-life care in pediatric solid organ transplantation. Pediatr Transplant. 2015;19(1):11–7.
57. Morrison M. Hospital crisis management: keep your social media audience involved. 2011. Available from: http://www.hospitalimpact.org/index.php/2011/05/25/hospital_crisis_management_keep_your_soc.
58. Rivkin SS, Seitel F. Taming a hospital crisis: 7 rules of the road. Becker's Hospital Review; 2012.

Part IV
Allocating Organs

Chapter 12
Neurodevelopmental Status as a Criterion for Solid Organ Transplant Eligibility

Aaron Wightman, Jodi Smith and Douglas S. Diekema

Abstract In this chapter we will review existing policies and laws that speak to whether children with neurodevelopmental disabilities can be considered as candidates for solid organ transplant. We will highlight the differences in policy that exist for different organs (heart, liver, kidney). We will review studies of provider attitudes and the published literature of outcomes of these children. Finally we will perform a normative analysis to consider whether current policies and practices are justified, and offering a framework to assist in determining whether, when, and how consideration of neurodevelopmental status should be used when making decisions about candidacy for solid organ transplantation.

Keywords Intellectual disability · Social value · Rationing · Justice

12.1 Introduction

Solid organ transplantation in children with intellectual disability remains a contentious issue [1]. Historically, children with intellectual disabilities were denied access to transplantation [2, 3]. Reasons for exclusion included the lack of sufficient cognitive ability to understand transplantation and comply with the required

---

A. Wightman · J. Smith · D.S. Diekema
University of Washington School of Medicine, Seattle, WA, USA
e-mail: Jodi.Smith@seattlechildrens.org

D.S. Diekema
e-mail: diek@u.washington.edu

J. Smith
University of Washington School of Public Health, Seattle, WA, USA

A. Wightman (✉) · D.S. Diekema
Treuman Katz Center for Pediatric Bioethics, Seattle, WA, USA
e-mail: Wightman@pediatrics.wisc.edu

R.A. Greenberg et al. (eds.), *Ethical Issues in Pediatric Organ Transplantation*,
International Library of Ethics, Law, and the New Medicine 66,
DOI 10.1007/978-3-319-29185-7_12

post-transplant therapy, a belief that transplantation would not improve quality of life in these children, and the reduced life expectancy that accompanied at least some of the conditions that afflicted these children [1–3]. In this chapter, we will review existing policies that consider whether children with neurodevelopmental disabilities are and should be considered candidates for solid organ transplant, highlight the differences in listing policy that exist for different organs, and review studies of provider attitudes and the literature of outcomes of transplanted children with intellectual disability. Finally, we will consider whether current policies and practices are justified and offer a framework to assist in determining whether, when, and how consideration of neurodevelopmental status should be used when making decisions about candidacy for solid organ transplantation.

12.2 Historic Cases

In 1996 Sandra Jensen, a 35-year-old woman with Down syndrome, was denied listing for heart-lung transplantation by the University of California San Diego and Stanford University after referral from her primary care physician [4, 5]. Both institutions claimed that patients with Down syndrome lacked sufficient intelligence to follow the complicated protocol that accompanies transplant and would suffer poorer outcomes as a result [6]. In the rejection letter from the University of California San Diego, Cecilia Smith, the medical director of the heart transplant program, wrote, "that it is the recommendation of the transplant committee that Ms. Jensen not be a candidate for heart-lung transplantation at our program… Ms. Jensen is limited in her ability to have recall and memory… We are left with great concerns of her ability to not only adhere to the medical regimen required of her, but to understand the complexities of the transplant procedure and the complications she will face after the transplantation…" [6, p. 2]. Stanford's rejection letter stated transplant was not "appropriate" for a person with Down syndrome [7, p. 4]. Ms. Jensen had graduated high school, worked as a waitress and disability rights advocate, and lived independently. In response to public outcry and a warning from the United States Department of Justice that rejection for transplantation on the basis of disability could represent a violation of the Americans with Disabilities Act, both hospitals agreed to reconsider. Ms. Jensen ultimately underwent successful transplant at Stanford University, but died 18 months later due to a complication of her immunosuppressive medications [5].

In 2000 18-year-old Joanne Harris became the first person with Down syndrome to be placed on the United Kingdom's National Health Service (NHS) waitlist for heart and lung transplant. Ms. Harris was also initially denied listing. However, her case became highly publicized following a feature in the news program *Inside Out* [8, 9] and after the winner of a popular game show offered his winnings to help pay for a heart and lung transplant in America [10]. Ms. Harris's mother described the medical community's argument against transplant as "there are so few organs and they're not going to waste one on my child" [10, p. 2000]. The medical director of

Harefield Hospital, interviewed as part of the news program, stated "it's not just Down's we are discriminating against, it's anyone with a disability. We take, when faced with a choice, the person most likely whole" [9, p. 499]. Following a national debate, Ms. Harris was ultimately removed from the NHS wait list and died from an infection in 2005 [11].

In 2010, the Veneto region of Northern Italy became the first region in the world to explicitly declare intellectual disability a contraindication to solid organ transplantation [12]. Legislation in that region defines an IQ <50 as an absolute and IQ <70 as a relative contraindication to solid organ transplantation [12].

In 2012 Amelia Rivera, a three-year-old girl with Wolf-Hirschhorn syndrome and failing kidneys was allegedly denied acceptance as a candidate for renal transplantation by the Children's Hospital of Philadelphia. That evening Amelia's mother, Chrissy Rivera, wrote on a blog of Wolf-Hirschhorn parents:

> The doctor begins to talk and I listen intently on what he is saying… He places two sheets of paper on the table. I can't take my eyes off the paper. I am afraid to look over at Joe because I suddenly know where the conversation is headed. In the middle of both papers, he highlighted in pink two phrases. Paper number one has the words, "Mentally Retarded" in cotton candy pink right under Hepatitis C. Paper number two has the phrase, "Brain Damage" in the same pink right under HIV. I remind myself to focus and look back at the doctor. I am still smiling. [13, p. 1]

Chrissy Rivera's post triggered dramatic public outcry and a national discussion about whether children with intellectual disability should be considered candidates for solid organ transplants [14]. In the course of press coverage, Dr. Kurt Hirschorn, a geneticist and pediatrician who first identified Wolf-Hirschorn syndrome, was interviewed. When asked if Amelia should be considered for a transplant, Hirschorn responded, "no, what our ethics committee would tell the parents is that we can't give her a kidney, because there is a shortage of kidneys, especially for little kids, and her impairments are too significant" [13, p. 1]. Hirschorn did state that Amelia could be considered for living donor transplantation because in that situation she would not be taking a kidney from the donor pool. Ultimately, Amelia received a living donor transplant from her mother at the Children's Hospital of Philadelphia. However, the national discussion of transplant candidacy in children with intellectual disabilities has not been resolved [14, 15].

12.3 Definitions-Legal-Americans with Disabilities Act

12.3.1 Definitions

Historically there has been a lack of consensus on a definition of intellectual disability amongst medical providers. This heterogenous population has also been referred to as "mentally delayed," "mentally retarded," "learning disabled," "intellectually handicapped," "mentally deficient," and "mentally delayed." The historic lack of consensus has resulted in the grouping of a broad, diverse population that ranges from those with mild limitations to those with profound disabilities and made rigorous study difficult. Various studies have used differing definitions based upon historic American Psychiatric Association (APA) criteria [16–18] or cutoffs based upon formal standardized intellectual quotient (IQ) or developmental quotient (DQ) testing [16, 17, 19].

Recent publications by the APA and the American Association on Intellectual Disabilities have moved closer toward creating a consensus definition for intellectual disability. The APA has updated their definition of intellectual disability for the Diagnostic Statistical Manual of Mental Disorders 5th Edition (DSM-V) [20]. The new definition characterizes intellectual disability as involving "impairments of general mental abilities that impact adaptive functioning in three domains or areas." These domains include the conceptual (e.g. language, reading, math, memory), social (empathy, interpersonal communication skills), and practical (e.g. personal care, job money management) [20, p. 2]. The DSM-V also acknowledges the limitations of overemphasizing IQ testing and no longer considers IQ testing in the definition [20]. The American Association on Intellectual and Developmental Disabilities defines intellectual disability as "a disability characterized by significant limitations in both intellectual functioning and in adaptive behavior, which covers many everyday social and practical skills… and originates before age 18" [21, p. 1]. Both organizations highlight that diagnosis requires both individualized clinical assessment and standardized testing. These new definitions should aid in consistently identifying this population allowing for generalizability in future research and guidelines.

Currently, there are no standardized mechanisms for transplant centers to assess for intellectual disability. In the remainder of this chapter, we will use intellectual disability to describe the diverse population of children and infants who meet the DSM-V definition of intellectual disability with or without other comorbidities.

12.3.2 Legal-Americans with Disabilities Act

Article 25 of The United Nations Convention on the Rights of Persons with Disabilities holds that "persons with disabilities have the right to the enjoyment of the highest attainable standard of health without discrimination on the basis of disability" [22].

In the United States, The Americans with Disabilities Act (ADA) of 1990 prohibits discrimination and ensures equal opportunity for persons with disabilities in employment, state and local government services, public accommodations, commercial facilities, and transportation [23]. A leading ethical and legal scholar on the ADA, David Orentlicher has written a thorough analysis of the application of the ADA to healthcare and transplantation [24]. The ADA defines a disability as existing when an individual has an illness, physical, or psychological, that substantially limits a major life activity such as walking learning, breathing, working, or participating in community events [24]. By this definition, all those with end-stage organ failure qualify are considered disabled. The ADA prohibits discrimination on the basis of disability "in the full and equal enjoyment of the services of any place of public accommodation" operated by a private entity, including hospitals [25, p. 12182a]. This suggests that intellectual disability cannot be a sufficient reason to deny access to organ transplant programs. The ADA also prohibits the use of eligibility criteria that tend to screen out persons with disabilities. This rule mandates individual assessments of candidates and makes it difficult for centers to use either an intellectual disability or it's psychosocial effect as criteria in evaluation of candidates for transplantation.

Orentlicher believes the ADA offers room for centers to consider degree of disability in their decisions about candidacy. It allows hospitals and insurers to engage in traditional risk classification as long as the practice is not designed to evade the ADA [24]. A transplant center could argue that allocation may occur based upon the traditional practice of prioritizing those who would gain the greatest benefit from transplant, either in terms of longevity, or quality of life. On this basis, centers may use eligibility criteria to screen out persons with intellectual disability for transplantation as long as the criteria are *necessary* for the operation of the program. This means that under the ADA it may be acceptable to include psychosocial criteria for transplantation as long as criteria help distinguish among different candidates in terms of their likelihood of benefiting from the transplant [24].

12.4 Unique Features and Provider Attitudes

The following section will briefly review unique features associated with each transplanted organ and highlight the available literature on provider attitudes. It should be cautioned that surveys of transplant provider attitudes are completed by a wide range of providers (nurse coordinators, medical specialists, social workers, transplant surgeons) and may not reflect the widespread beliefs or practices of an individual center, country, or specialty.

A recent review by Martens et al. summarized a 2004 Internet survey of 205 members of the ARC of the United States (an intellectual disability family organization) in which 80 % of responders reported they believed people with intellectual disability are discriminated against with regard to access to organ transplant operations [26]. A 2006 survey of pediatric transplant centers revealed that 38 % of centers had denied listing of a child solely on the basis of a coexisting neurodevelopmental disability [19]. In the same survey, 43 % of pediatric centers "usually" or "always" considered neurodevelopmental status in listing decisions; however, only 19 % of centers required a formal evaluation of neurodevelopmental status, and no centers reported using formal guidelines by which neurodevelopmental status was used in the listing decision [19]. On the other hand, 40 % of responding pediatric transplant centers reported that neurodevelopmental status alone was never considered an absolute contraindication to solid organ transplantation [19].

12.4.1 Kidney

Kidney transplantation differs from other forms of solid organ transplantation in important ways. Transplantation is generally held as the goal of treatment for a child with End Stage Kidney Disease (ESKD) due to considerations of quantity and quality of life; however treatment with the artificial kidney (dialysis) is also available. In addition to deceased donor transplantation, most healthy individuals with two kidneys can donate a kidney relatively safely, allowing for parents or others to act as living donors to potential recipients [27]. There is a worldwide shortage of kidneys available for transplantation. The National Kidney Foundation reports that 100,602 people in the U.S. are currently waiting for a kidney transplant and 2500 names are added to that list each month [28]. In 2013, 3381 individuals died while on the deceased donor kidney wait list [28]. The 2012 Organ Procurement and Transplant Network (OPTN) and Scientific Registry of Transplant Recipients (SRTR) Annual report indicates that 1275 children were on the active deceased donor wait list on 12/31/2012 and that 27 children died awaiting kidney transplantation in 2012 in the United States [29].

Provider attitudes regarding renal transplantation in children with disabilities have shifted in the last 25 years. In a 1992 survey of 217 adult and pediatric kidney transplant centers, responders reported that an IQ of 50–70 was considered a relative contraindication for kidney transplantation by 51.3 % of centers and an absolute contraindication by 2.6 % of centers. In the same survey an IQ <50 was considered a relative contraindication by 52 % of centers and as an absolute contraindication by 24 % of centers [30]. A 2005 survey found that 30 % of kidney transplant programs (n = 23) "always or usually" consider neurodevelopmental disability in their decisions [19]. While no centers considered mild or moderate cognitive disability (IQ 35–70) a relative contraindication to transplant, 40 % of transplant centers consider an IQ <35 an absolute contraindication to kidney transplantation, and 56 % consider it a relative contraindication [19]. In the same survey responders also indicated they would be more open to transplantation if a parent or family member was willing to act as a living donor. One respondent stated they would "prefer" and even "pressure" for living related donation, so that the child with neurodevelopmental disability would not "take" an organ from the deceased donor wait list [19, p. 849]. These studies indicate a shift among transplant centers to increasingly consider children with intellectual disabilities as potential recipients of solid organ transplants.

Goldberg et al. [15] recently proposed a framework for the evaluation of kidney transplant candidacy of children with multiple comorbidities. The authors recommend individualized assessments and therapeutic plans that are based upon an analysis of the benefits and burdens of therapy for the particular child, as opposed to criteria based upon a certain level of intelligence.

12.4.2 Heart

Patients who require heart transplantation do not have the long-term option of an artificial organ nor the possibility of living donor transplantation. It has been reported that those with intellectual disability are less likely to be referred for heart transplantation. In 2000, a 14-year review of a high-volume British heart transplant center found that of over 800 referrals to the center, only one was made for a patient with Down syndrome [31]. The OPTN and SRTR Annual Data Report indicates that on 12/31/2012 there were 330 children on the cardiac transplant wait list in the United States [29]. In the year 2012, 67 children died while on the cardiac transplant wait list [29].

A 1991 international survey of 204 adult and pediatric heart transplant centers reported that an IQ of 50–70 was considered a relative contraindication for consideration by 59 % of US centers and 40.5 % of non-US centers, and an absolute contraindication by 26 % of US centers and 54 % of non-US centers [32]. In the same survey, an IQ <50 was considered a relative contraindication by 19.7 % of US centers and 3 % of non-US centers, and an absolute contraindication by 74 % of US centers and 92 % of non-US centers [32]. A 2005 survey of 15 pediatric heart transplant centers found that 71 % of heart transplant programs "always or usually" consider mental retardation in their decisions and 46 % of heart transplant programs considered mild or moderate mental retardation to be a relative contraindication to transplantation [19].

12.4.3 Liver

Liver transplantation offers the opportunity for deceased donation, living donation, or split donation, but lacks the option of a long-term artificial organ. In a 1992 survey of 72 adult and pediatric liver transplant centers, responders reported that an IQ of 50–70 was considered a relative contraindication for transplant by 70 % of centers and an absolute contraindication in 11 % of centers. In the same survey an IQ <50 was considered a relative contraindication by 41 % of centers and an absolute contraindication by 46 % of centers [30]. A 2005 survey of pediatric transplant programs (n = 12) found that 33 % of centers reported that cognitive disability was "always or usually" considered in their decisions. However, no pediatric centers considered mild or moderate cognitive disability alone (IQ 35–70) to be a relative or absolute contraindication to transplantation.

The 2012 OPTN and SRTR Annual Data Report indicates that there were 585 children on the pediatric liver transplant wait list in 2012 and that 37 children died while waiting for a liver transplant [29].

12.4.4 Lung

The attitudes and beliefs of pediatric and adult lung transplant providers toward the eligibility of children with intellectual disabilities have not been assessed.

12.5 Outcomes

Outcome studies have largely been limited to single center case series and are further limited by the fact that they only include children who received an organ transplant, excluding children listed for transplant who remain on the wait list, those who were refused listing by a transplant center, and those with organ failure not referred to a transplant center. Children with intellectual disability who receive solid organ transplant may be the ones with the most familial resources and support, which may skew towards more favorable outcomes.

12.5.1 Kidney

Several small case series and case-control studies from Brazil [33], the United States [17, 18], Japan [34], and Ireland [35] have demonstrated excellent graft and patient survival in children with intellectual disabilities (see Table 12.1). Where examined, there was no difference in graft survival compared to a control group without intellectual disability [33, 34]. In one study involving children with intellectual disability, the severity of intellectual disability was not associated with graft or patient survival [34]. In the same study, the primary caregiver of each recipient with intellectual disability reported improvement in their perception of their child's quality of life and 60 % of caregivers reported improvement in their own quality of life [34].

A recent analysis of the UNOS dataset of all children receiving a first kidney-alone transplant from 2008 to 2011 identified 332 children with definite (117) or probable (215) intellectual disability based upon center reporting, accounting for 16 % of all first pediatric renal transplants. Children with definite intellectual disability were not significantly different from children without intellectual disability with respect to sex, or ethnicity. Intellectual disability was not significantly associated with rejection, short-term graft failure, or decreased patient survival at three years post-transplant. Contrary to the sentiment noted in the surveys reviewed above, a higher percentage of recipients with intellectual disability received kidneys from deceased donors rather than from living donors.

Table 12.1 Outcomes

Author	Year	Country	N	Graft Survival	Patient Survival	Comment
Kidney						
Galante et al. [33]	1997–2006	Brazil	16 (4 LRRT, 12 DDRT)	100 % at 5 years	75 % at 5 years (deaths due to sepsis)	Defined ID as IQ<70
Ohta et al. [16]	1998–2004	Japan	25 (23 LRRT 2 DDRT)	100 % at 5 years	87 % at 5 years	
Benedetti et al. [17]	1968–1996	USA	8 (7DDRT, 1 LRRT)	100 % at 5 years	100 % at 5 years	
Baqi [18]	1987–1995	USA	14 (6LRRT, 8 DDRT)			Only Down syndrome Analysis of NAPRTCS
Hamdi Kamel et al. [35]	1997	Ireland	DDRT	100 % at 9 years	100 % at 9 years	Kabuki Syndrome
Heart						
Orr et al. [36]	1991	USA	1		100 % at 10 years	8p deletion
Samuelson-Jones et al. [38]	1978–2010	USA	5		80 % at 5 years	Adult patients
Irving et al. [37]	1985–2006	UK	2		Reported as successful	2/821 infants with Down syndrome
Liver						
Gentile et al. [40]	1990	Italy	1		100 % at 5 years	COACH syndrome
Herzog et al. [41]	1996	Canada	1		100 % at 5 years	COACH syndrome
Ewart Toland et al.	1998	USA	1		"Successful" at 3 months	Kabuki syndrome

12.5.2 Heart

Limited data exist on outcomes of heart transplantation in children with intellectual disabilities. Reasons for lack of data could include poor outcomes, limited reporting, or a lack of referral for children with intellectual disability for heart transplant [31]. Orr reported the outcome of a boy with 8p deletion (expected to experience moderate to severe mental retardation as a result of his underlying genetic condition) who received a transplant at eight months of age [36]. At follow-up at 10 years of age, the boy was noted to be "mildly delayed" and have a functioning heart graft [36, p. 220]. In a review of 821 infants born with Down syndrome in the North East of England from 1985–2006, two underwent heart transplantation as young children [37]. Samelson-Jones et al. [38] reviewed the cases of five heart transplants in adults with intellectual disability. They reported excellent outcomes in 4 of 5 patients, with survival times ranging from 8 to 16 years at time of publication. The patient with a poor outcome died four years after transplant, likely related to complications of medication non-adherence due to functional decline in the primary caregiver.

12.5.3 Liver

Reports of liver transplantation in children with intellectual disability are similarly limited. One reason for the limited publication may be under recognition of intellectual disability. Kennard et al. [39] reported long-term (>3 years) outcome of 50 pediatric liver transplant recipients and found that 18 % had IQ scores <70 [39]. The authors did not assess for a correlation between IQ and graft or patient survival [39]. Gentile et al. [40] and Herzog et al. [41] each reported single cases of successful liver transplantation in a child with COACH syndrome (an autosomal recessive disorder characterized by mental retardation, ataxia due to cerebellar hypoplasia, oligophrenia, ocular coloboma, and hepatic fibrosis). Both cases were considered successful although one child died 4 years later as a result of Hepatitis C infection. Ewart-Toland et al. [42] described a child with Kabuki syndrome who received a liver transplant at eight months of age and was doing well at the three-month follow-up. Toker and Salzer described a case of retransplantation in an 11-year-old who developed blindness, muteness, deafness, and severe mental retardation following complications of the original liver transplantation [43]. After developing chronic graft failure, the patient met medical criteria for retransplantation, but the decision to pursue transplant was questioned by the medical team due to concern for the patient's ability to comply with the complex post-transplant regiment, potentially reduced life expectancy compared to a child without intellectual disability, and ambiguity regarding the potential for improved quality of life in a child with severe intellectual disability. The issue was referred to the ethics committee. The ethics committee recommended education of the potential donor (an uncle) regarding the risks and potentially limited benefits of organ donation and the patient's poor overall prognosis. Ultimately the potential donor declined to donate.

12.6 Current Guidelines

12.6.1 General

The American Medical Association Code of Medical Ethics Opinion 2.03 addresses allocation of limited resources such as organs for transplant and agrees that use of criteria such as social value, implicitly or explicitly, by transplant centers is unjustified [44, 45]:

> Decisions regarding the allocation of limited medical resources among patients should consider only ethically appropriate criteria relating to medical need. These criteria include likelihood of benefit, urgency of need, change in quality of life, duration of benefit, and, in some cases, the amount of resources required for successful treatment… Non-medical criteria, such as ability to pay, age, social worth, perceived obstacles to treatment, patient contribution to illness, or past use of resources should not be considered…. The allocation procedures of institutions controlling scarce resources should be disclosed to the public as well as subject to regular peer review from the medical profession. [45, p. 35)

Similarly, the draft 2010 Transplantation Society of Australia and New Zealand Eligibility Guidelines And Allocation Protocols (http://www.tsanz.com.au/organallocationprotocols/) state that "decisions regarding eligibility and allocation will not be made on the basis of race, religion, gender, marital status, sexual orientation, social status, disability or age" [46, p. VII]. These guidelines do allow consideration of cognitive disability only insofar as they allow denial of transplantation on the basis of inability to comply with complex medical therapy, such as might occur with an individual with cognitive deficits in the absence of a caregiver able to take on this role.

12.6.2 Organ-Specific Guidelines

In many countries, guidelines have evolved independently for each organ system and will be discussed separately below.

12.6.2.1 Kidney

The Patient Care and Education Committee of the American Society of Transplant Physicians 1995 guidelines state that mental retardation should only be considered a contraindication to transplantation when the cognitive impairment is so severe that it impairs compliance with essential medical support regimens [47]. The guidelines acknowledge that support from family or caregivers may compensate for an individual's inability to comply with medical regimens on their own. The 2000 European Best Practice Guidelines for Renal Transplantation do not include intellectual disability as a contraindication transplantation [48]. The 2005 Canadian Society of Transplantation consensus guidelines on eligibility for kidney transplantation state "children who would otherwise be considered for renal replacement therapy should not be excluded from consideration for transplantation solely on the basis of diminished cognitive or physical capacity" [49, p. s4]. The guideline highlights the potential benefit in quality of life for children and their families and recommends that decisions be made according to the best interests of the child. The UNOS/Organ Procurement Transfer Network (OPTN) 2007 Ethics Committee report states that patients should not be excluded from renal transplantation because of mental retardation [12]. The European Association of Urology 2010 Guidelines on Renal Transplantation states that mental retardation is not necessarily a contraindication for transplantation if the patient "is able to understand the procedure and can adhere to the procedures and medication required" [50, p. 35]. The guidelines of Italian Society of Nephrology hold that mental retardation may be considered a contraindication if the patient is unable to understand the complexity of the transplant process [12]. These latter two guidelines remain silent on whether a caregiver who is willing and able to assist the patient as needed would compensate for the patient's inability to do so.

12.6.2.2 Heart

The 2006 International Society for Heart and Lung Transplantation Guidelines for the Care of Cardiac Transplantation Candidates [51] includes mental retardation as a relative contraindication to transplantation. The authors write, "there is general agreement that heart transplantation should be reserved for those most likely to benefit both in terms of quality of life and survival" [51, p. 1034]. The authors recommend that each patient be evaluated in four areas: compliance, comprehension, quality of life, and social evaluation. However, they also state "care must be taken to ensure that psychosocial factors predictive of outcome are not confused with judgments of an individual's social worth" [51, p. 1035]. Intellectual or physical disability is not mentioned in either the 1995 American Heart Association Statement on Selection and Treatment of Candidates for Heart Transplantation [52] or the 2008 Joint Guidelines for Heart Transplantation from the Netherlands Society of Cardiology and the Netherlands Association for Cardiothoracic Surgery [53]. The draft 2010 Transplantation Society of Australia and New Zealand Eligibility Guidelines and Allocation Protocols considers inability to comply with complex medical therapy, such as chronic cognitive deficits in the absence of a caregiver able to take on this role, as a relative contraindication to all organ transplantation [46, p. 2].

12.6.2.3 Liver

The 2005 American Association for the Study of Liver Diseases Practice Guidelines holds that "children with mental retardation pose significant logistical and ethical challenges," but does not directly comment on whether intellectual disability is a relative contraindication [54, p. 7].

12.6.2.4 Lung

The 2006 Consensus report from the pulmonary scientific council of the International Society for Heart and Lung Transplantation International Guidelines for the Selection of Lung Transplant Candidates does not explicitly consider intellectual disability as an allocation criterion, but acknowledges that "considerations of quality life in addition to survival should be used to assess the effectiveness of the procedure" [55, p. 745]. However, "owing to the shortage of available organs, it is not currently possible to support transplantation solely for quality of life purposes" [55, p. 746].

12.7 Ethical Considerations

12.7.1 Intellectual Disability as Medical Criteria

The existing medical literature does not support the conclusion that children with intellectual disability have poorer outcomes, poorer compliance, a greater number of complications, or poorer improvement in quality of life than other potential transplant recipients. Children with intellectual disability appear to derive medical benefit at an equal or greater likelihood, quality, and duration when compared to other transplant recipients. The principle of justice requires that we avoid discriminating between individuals on morally irrelevant grounds and apply criteria to all in an equal manner [56]. This concept of fairness is rooted in the principle of formal justice which holds that "equals must be treated equally and unequals must be treated unequally" [57, p. 250]. If the only relevant factors in transplant receipt are need, likelihood of success, and duration of success, and if intellectual disability does not have prognostic medical significance with regard to measurable outcomes for transplant such as graft survival, length of survival, compliance, and quality of life, then presence and degree of disability should not be considered in decisions about eligibility or allocation. Bourgois, offering a different argument, notes, that if individuals with cognitive disability are considered eligible to donate organs, then the principle of reciprocity would suggest they should also be considered recipients [58].

Physicians at the bedside, or in the conference room, should not make rationing decisions regarding access to solid organ transplantation in children solely on the basis of intellectual disability; either implicitly as demonstrated in the Richards [19] paper or explicitly as in the Rivera case. In these cases, physicians are considering intellectual disability as a social value criteria, even if unaware that they are doing so, rather than medical criteria in justifying exclusion. The next question to be asked is how may such inclusion or exclusion of persons with cognitive disability from transplantation be justified by appeals to justice?

12.7.2 *Social Justice Arguments for Inclusion of Children with Intellectual Disabilities*

The philosopher Norman Daniels suggests an adaptation of Rawlsian theory of justice that we feel may be applied justify inclusion of children with intellectual disabilities as candidates for solid organ transplantation. Daniels believes that disease and disability result in "natural disadvantages" that limit individuals' species and societal-relative functioning, resulting in denial of equal opportunity to the normal range of life plans that a rational individual would choose based upon their own conception of the good [59, p. 166]. Two individuals with equal indices of wealth and income will not be equally well off nor have equal opportunities if they differ in healthcare needs [59]. Healthcare allows individuals to achieve or maintain species-typical functioning. While health itself is not a primary good (something every rational being is presumed to want e.g. income, wealth, power, authority), denial of health results in a loss of opportunity. Daniels argues that opportunity is a primary good, therefore healthcare should be governed by the Principle of Fair Equality of Opportunity [59]. Daniels interprets Rawls concept of the "natural lottery" (the advantages that we are born with e.g. race, class, gender, ethnicity) to include advantages that play a role in health such as genetics and available economic resources to pay for preventative and therapeutic healthcare [59, p. 166]. An extended Principle of Fair Equality of Opportunity would require society to provide resources to counter the advantages in opportunity some receive in the "natural lottery." Society would also be required to consider disease and disabilities acquired in life as well, because these also have a limiting effect on normal functioning, particularly in those less advantaged, in order to allow members of society the "normal opportunity range for a given society" [59, 60].

Daniels acknowledges that his focus on "normal functioning" does not provide guidance for individuals with severe disabilities such as children with intellectual disability or those who require burdensome, invasive therapies such as transplant [59, p. 154]. Such patients may never be able to enjoy "normal species-typical functioning" even at the lowest end of the spectrum for society and an attempt to provide treatment to achieve that goal may be futile and fall prey to bottomless pit problems. In our view, this weakness need not prove fatal. In Daniels' model, children with intellectual disability may be candidates for solid organ transplantation. Organ failure represents a natural disadvantage that denies opportunity to children with and without intellectual disabilities. Solid organ transplantation lessens this disadvantage. In the instance of kidney disease, solid organ transplant has been demonstrated to improve the health and quality of life for these children and their families compared to dialysis [34]. This change reflects an improvement in opportunity even if that opportunity remains limited. Children with severe intellectual disabilities may never achieve a narrow interpretation of "normal functioning," however; they are human and could be included in a broader interpretation of "species functioning" that includes all human beings. Interpreted in this way, solid organ transplant functions to spare recipients and their families the impediments to opportunity caused by organ failure.

12.7.3 Social Justice Arguments for Exclusion of Children with Intellectual Disabilities

Some have used arguments based on the Principle of Utility to promote exclusion in settings of resource scarcity, such as solid organ transplantation. A utilitarian approach to solid organ distribution could, for example, focus on maximizing cost-effectiveness and quality adjusted life years (QALY) as surrogate measures for overall social utility. This approach would favor those whose transplant could be accomplished most efficiently and whose graft would last longest and with the highest measured quality of life.

Savulescu has argued that considerations focused solely on the principle of equal access fall short when resources are scarce, and applying that principle results in "fewer people surviving and those who do will live shorter and worse quality lives" [8, p. 875]. Relying only on the principle of equality of access precludes "value-based" considerations such as "lifespan, ability to get the most out of life, and not being a burden upon others or upon society" [8, p. 876]. Savulescu argues that value-based considerations are incorporated in denying transplant to those with severe dementia, not because it is against that individual's interests, but because of a utilitarian appeal that scarce organs could be better used elsewhere. According to Savulescu, the relevance of Down syndrome in allocating scarce organs depends on how the associated disability detracts from a "good life." He further argues that the quality of life considerations are used when matching organ transplant recipients by blood type or body size. Given the shortage of available organs it would be wrong to transplant a child "barely conscious with a severe intellectual disability or a child with congenital myotonia dystrophica dependent on a ventilator" [8, p. 876] although he acknowledges that such a practice would likely be illegal in Australia, Europe, and the United States.

Others appear to support Savulescu's argument to exclude those children with most severe intellectual disability. Santiago-Delpin argued in favor of exclusion but, "only in the case of extreme limitation and dependency, or when medical considerations dictate otherwise can exclusions be ethically valid" [61, p. 259]. Similarly Arciniegas and Filley [62] argue for individualized consideration of individuals with cognitive impairment due to static encephalopathy may not be considered to be candidates depending on their degree of impairment and presence of social supports. Each of these arguments seems to be based on judgments regarding quality of life.

12.7.3.1 Quality of Life

Quality of life may be generally defined as an "individual's perceptions of their position in life in the context of the culture and value systems in which they live, and in relation to their goals, expectations, and concerns" [63, p. 153]. For children, QOL includes, but is not limited to, the child's perception of their social, physical, and emotional well-being [64]. Quality of life measurements tools are inherently limited, not designed for comparisons between patients, and incorporate implicit judgments based upon the social value of "normal function" [45]. Many tests of quality of life rely on functional abilities as a primary determinant of QOL and, especially for children and those with intellectual disabilities, require assessments of the quality of an individual's life through the eyes of an assessor who does not have that disability [64]. These assessments may in turn reflect perceptions of what life with a disability would be like rather than the subjective experience of that child. Physicians and parents consistently rate the quality of life experienced by children with medical conditions and disabilities as worse than the child rates their own quality of life [64, 65].

We feel a criterion focused on societal perceptions of quality of life will inherently favor those without intellectual disability. While all transplant candidates likely have a relatively low quality of life, those without a preexisting intellectual disability are more likely to return to "normal functioning," something not possible for someone with an intellectual disability, as solid organ transplant is unlikely to effect significant preexisting intellectual disability [66]. In effect, as Harris argues, we are counting the disabled patient's "desire to live the life that was available to him as less value than his rivals" [67]. The individual without disability is judged to have higher value.

12.7.3.2 Limitations of Social Value in Rationing Decisions

The use of social values in rationing decisions may reflect the values of those choosing. Application of a social value criterion can result in healthcare being used as a form of double punishment or compounded discrimination or unfairness related to societal position. To children with intellectual disability a social value criterion focuses on their limitations and discounts the contributions that these children make to their families and society. The literature identifies contributions made by children with intellectual disabilities through teaching others compassion, love, responsibility, and respect [68–70].

For social value criterion to be employed implicitly as done by transplant centers in their consideration of the candidacy of children with intellectual disabilities adds additional concern. Implicit rationing occurs in a haphazard, unsystematic manner. It is not part of a publicly stated policy or plan and no attempt is made to justify it by appealing to ethical standards. The varied practice of considering neurodevelopmental status by transplant centers provides evidence that this consideration is applied implicitly, and depending on the particular centers involved, similar children could be considered in very different manners [19]. Implicit rationing criteria may be applied inconsistently and be subject to biases. As demonstrated by the Seattle "God Committee" the use of implicit social value criteria led to centers selecting patients most like the decision-makers.[1] Children with intellectual disabilities will never be the patients most similar to the transplant center physicians who select candidates for transplantation.

Social value judgments do not reflect the pluralistic values of Western Society; they often reflect the values and biases of the decision-makers, and make unwarranted assumptions about the homogeneity of groups [45]. In discussing the Amelia Rivera case, Frader has gone further and argued, that when determining transplant eligibility, other than medical success, "we have no universally accepted moral basis for settling these disputes" [75, p. 136]. Physicians are not trained to assess or weigh social value and to do so violates the trust inherent in the fiduciary relationship between physician and patient [76]. The same holds true when transplant centers consider the social value of children with intellectual disability as candidates for solid organ transplantation.

---

[1]In the early 1960s it was estimated that there were 50–100 adults needing dialysis for every available position at the Seattle Artificial Kidney Center. Introduction of strict geographic, medical, and psychological criteria, and a requirement to guarantee payment for 3 years ($10,000 annually in 1962) decreased the ratio to 4:1 [72, 73]. To perform the final patient selection an anonymous, lay committee was created consisting of a labor leader, businessman, housewife, minister, lawyer, and two physicians [74]. Members of the Admissions and Policies Committee were purposely given no explicit guidelines for selection [72]. In their deliberations members considered markers of social value including age, sex, marital status, number of dependents, income, net worth, religious activity, occupation, past contributions to society, potential future contributions to society, educational background, emotional stability, and names of individuals who could serve as references to the candidate's character [73]. With no explicit guidelines, personal biases intruded upon judgments. Deliberations included "for the children's sake we've got to reckon with the surviving parent's opportunity to remarry, and a woman with three children has a better chance to find a new husband than a very young widow with 6 children." [74, p. 115]. "If we are looking for the men with the highest potential of service to society then I think we must consider that the chemist and the accountant have the finest educational backgrounds." [74, p. 115]. In effect, committee members, without explicit guidelines and selecting patients based on social value, selected those most like themselves. Other biases present with social value rationing include preferences of English speakers, male sex, and white race [75]. The net effect was summarized by Shana Alexander in her seminal article, "A candidate who plans to come before this committee would seem well-advised to *father* a great many children, then to throw away all his money, and finally to fall ill in a season when there will be a minimum of competition from other men dying of the same disease" [74, p. 125]. Those who did not meet the implicit social value criteria of the lay committee were not favored, "The Pacific Northwest is no place for a Henry David Thoreau with bad Kidneys." [73, p. 378].

Beauchamp and Childress [57] and Kilner [74] argue that, barring for exceptional circumstances, comparisons based upon social value should be avoided. They are difficult to apply in practice, incorporate societal biases and discrimination, and can create a tendency to view people as only a means to produce benefits for society. Kilner concluded that "People's values are so diverse and intangible that they simply cannot be quantified or otherwise compared" [74, p. 38]. Further, even if social-value standards could be identified and persons could be accurately assessed in relation to those standards, the resulting selection of patients might be far from optimal in terms of social value produced [74]. A society that saves only those of highest social value, such as cardiac transplant surgeons, may be less vibrant than one with a diverse population.

12.7.4 Power Differential

Finally, there is a concern regarding power differential. Transplant consideration decisions are made by committees usually made up by transplant surgeons and medical specialists, with occasional input from allied health professions such as social work and nursing [77]. Family members are usually not present at these meetings and do not have an effective means of appealing a center's decision. This scenario places all decision-making power with the transplant team rather than sharing with the family and excludes an important stakeholder from the discussion. There is no check upon the use of implicit rationing criteria such as social value by transplant centers. Procedural justice could provide an answer to this concern by making the transplant consideration evaluation process similar to that employed by UNOS. Once a center accepts a patient as a transplant candidate, if it is for a deceased donor transplant, they are placed into the UNOS database and their relative candidacy is determined by a published and publicly available algorithm that does not consider intellectual functioning [25]. Similarly, the decision-making at the center level could be made open, with explicit criteria, and with a procedure in place for patients or their families to appeal similar to processes suggested by Daniels and Sabin [78].

12.8 Conclusions and Recommendations

The issue of consideration of children with intellectual disabilities as candidates for solid organ transplantation remains contentious. Children with intellectual disability do not appear to have poorer outcomes, poorer compliance, greater complications, or poorer improvement in quality of life than other potential transplant recipients. The principles of equal access and fairness suggest that this population should be considered the same as other children. Further ethical debate focuses on the social value placed on children with intellectual disabilities.

Recommendations:

1. Consistent with the recommendations of the American Society of Transplantation, we recommend individual formal assessments of candidacy. Intellectual disability should only be considered if the disability is found to be relevant to patient or graft survival or improvement in the patient's quality of life or if no caregiver is available to provide support.
2. Considerations should be made in a consistent and explicit manner based upon as objective criteria as possible.
3. Further research is needed to examine outcomes of this population and patient, family, and provider attitudes.
4. Further research is needed to understand the complex issues that arise in transition [79].

References

1. Veatch RM. Transplants and mental disability: the meaning of discrimination. Ethics Intellect Disabil. 2001;6(1):1–5.
2. Silber T, Batshaw ML. Ethical dilemmas in the treatment of children with disabilities. Pediatr Ann. 2004;33(11):752–61.
3. Collins TL, Wayne Holden E, Scheel JN. Cognitive functioning as a contraindication to organ transplant surgery: dilemmas encountered in medical decision making. J Clin Psychol Med Settings. 1996;3(4):413–22.
4. Goldberg C. Her survival proves doubters wrong: retarded woman recovers from transplant she'd been denied. New York Times. 1996;3:1996.
5. Shurkin J. The, "right" to a heart. Nat Med. 1996;2(3):264.
6. Fremon C. "…we do not feel that patients with Down syndrome are appropriate candidates for heart-lung transplantation.": these words were a death sentence for Sandra Jensen. And that, she decided, just wasn't going to happen. Los Angeles Times Mag. 1996;1996:1–7.
7. Down-syndrome woman gets transplant. Eugene Register-Guard. 1996 1/24/1996.
8. Savulescu J. Resources, down's syndrome, and cardiac surgery. BMJ. 2001;322(7291):875–6.
9. Sommerville A. A view from the back of the queue: channel 4—inside out: a heart for Jo. BMJ. 1996;313(7055):499.
10. Scott H. Are people with down's syndrome denied care? Br J Nurs. 2000;9(18):2000.
11. My special little lady: Big Brother's Craig pays tribute to Down's syndrome sufferer 2008 [7/22/2014]. Available from: http://www.dailymail.co.uk/news/article-560472/My-special-little-lady-Big-Brothers-Craig-pays-tribute-Downs-Syndrome-sufferer.html.
12. Panocchia N, Bossola M, Vivanti G. Transplantation and mental retardation: what is the meaning of a discrimination? Am J Transpl (official Journal of the American Society of Transplantation and the American Society of Transplant Surgeons). 2010;10(4):727–30.
13. Belkin L. Denying a transplant to a 'retarded' child. The Huffington Post [Internet]. 2012 1/16/2012. Available from: http://www.huffingtonpost.com/lisa-belkin/denying-transplant_b_1207630.html?view=print&comm_ref=false.
14. Painter K, Hellmich N. Amelia Rivera gets kidney after transplant debate. USA Today [Internet]. 2013 3/31/14. Available from: http://www.usatoday.com/story/news/nation/2013/07/30/amelia-rivera-kidney/2600551/.
15. Goldberg AM, Amaral S, Moudgil A. Developing a framework for evaluating kidney transplantation candidacy in children with multiple comorbidities. Pediatr Nephrol. 2014.

16. Ohta T, Motoyama O, Takahashi K, Hattori M, Shishido S, Wada N, et al. Kidney transplantation in pediatric recipients with mental retardation: clinical results of a multicenter experience in Japan. Am J Kidney Dis. 2006;47(3):518–27. PubMed PMID: 16490632. Epub 2006/02/24.
17. Benedetti E, Asolati M, Dunn T, Walczak DA, Papp P, Bartholomew AM, et al. Kidney transplantation in recipients with mental retardation: clinical results in a single-center experience. Am J Kidney Dis (official Journal of the National Kidney Foundation). 1998;31 (3):509–12. PubMed PMID: 9506689.
18. Baqi N, Tejani A, Sullivan EK. Renal transplantation in Down syndrome: a report of the North American pediatric renal transplant cooperative study. Pediatr Transplant. 1998;2(3):211–5. PubMed PMID: 10084745.
19. Richards CT, Crawley LM, Magnus D. Use of neurodevelopmental delay in pediatric solid organ transplant listing decisions: inconsistencies in standards across major pediatric transplant centers. Pediatr Transplant. 2009;13(7):843–50.
20. Association AP. Intellectual disability definition, American Psychiatric Association; 2013 [cited 2014 6/1/2014]. Available from: http://www.dsm5.org/documents/intellectualdisabilityfactsheet.pdf.
21. Disabilities AAoIaD. Definition of intellectual disability [cited 2014]. Available from: American Association on Intellectual and Developmental Disabilities.
22. Convention on the Rights of Persons with Disabilities, UN General Assembly; 2006 [2014].
23. Americans with Disabilites Act; 1990.
24. Orentlicher D. Psychosocial assessment of organ transplant candidates and the Americans with Disabilities Act. Gen Hosp Psychiatry. 1996;18(6 Suppl):5S–12S.
25. Americans with Disabilities Act; 1990.
26. Martens MA, Jones L, Reiss S. Organ transplantation, organ donation and mental retardation. Pediatr Transplant. 2006;10(6):658–64.
27. Clancy MJ, Geddes CC. Acceptable mortality after living kidney donation. Kidney Int. 2014;85(5):1240–1.
28. Foundation NK. Organ donation and transplantation statistics; 2014 [cited 2014 9/9/2014]. Available from: http://www.kidney.org/news/newsroom/factsheets/Organ-Donation-and-Transplantation-Stats.
29. Scientific Registry of Transplant Recipients Annual Data Report 2012 2012 [cited 2014]. Available from: http://srtr.transplant.hrsa.gov/annual_reports/2012/Default.aspx.
30. Levenson JL, Olbrisch, ME. Psychosocial evaluation of organ transplant candidates: a comparative survey of process, criteria, and outcomes in heart, liver, and kidney transplantation. Psychosomatics. 1992;34:314–23.
31. Leonard H, Eastham K, Dark J. Heart and heart-lung transplantation in Down's syndrome. The lack of supportive evidence means each case must be carefully assessed. BMJ. 2000;320 (7238):816–7.
32. Olbrisch ME, Levenson JL. Psychosocial evaluation of heart transplant candidates: an international survey of process, criteria, and outcomes. J Heart Lung Transpl Off Publ Int Soc Heart Transpl. 1991;10(6):948–55.
33. Galante NZ, Dib GA, Medina-Pestana JO. Severe intellectual disability does not preclude renal transplantation. Nephrol Dial Transpl Off Publi Eur Dial Transpl Assoc Eur Renal Assoc. 2010;25(8):2753–7. PubMed PMID: 20215391.
34. Ohta T, Motoyama O, Takahashi K, Hattori M, Shishido S, Wada N, et al. Kidney transplantation in pediatric recipients with mental retardation: clinical results of a multicenter experience in Japan. Am J Kidney Dis (official Journal of the National Kidney Foundation). 2006;47(3):518–27.
35. Hamdi Kamel M, Gilmartin B, Mohan P, Hickey DP. Successful long-term outcome of kidney transplantation in a child with Kabuki syndrome. Pediatric transplantation. 2006;10(1):105–7. PubMed PMID: 16499597.
36. Orr RD, Johnston JK, Ashwal S, Bailey LL. Should children with severe cognitive impairment receive solid organ transplants? J Clin Ethics. 2000;11(3):219–29. PubMed PMID: 11127636.

37. Irving CA, Chaudhari MP. Cardiovascular abnormalities in Down's syndrome: spectrum, management and survival over 22 years. Arch Dis Child. 2012;97(4):326–30. PubMed PMID: 21835834.
38. Samelson-Jones E, Mancini DM, Shapiro PA. Cardiac transplantation in adult patients with mental retardation: do outcomes support consensus guidelines? Psychosomatics. 2012;53 (2):133–8. PubMed PMID: 22424161. Epub 2012/03/20.
39. Kennard BD, Stewart SM, Phelan-McAuliffe D, Waller DA, Bannister M, Fioravani V, et al. Academic outcome in long-term survivors of pediatric liver transplantation. J Dev Behav Pediatr JDBP. 1999;20(1):17–23.
40. Gentile M, Di Carlo A, Susca F, Gambotto A, Caruso ML, Panella C, et al. COACH syndrome: report of two brothers with congenital hepatic fibrosis, cerebellar vermis hypoplasia, oligophrenia, ataxia, and mental retardation. Am J Med Genet. 1996;64(3):514–20. PubMed PMID: 8862632. Epub 1996/08/23.
41. Herzog D, Martin S, Yandza T, Alvarez F. Hepatic insufficiency and liver transplantation in a patient with COACH syndrome. Pediatr Transplant. 2002;6(5):443–6. PubMed PMID: 12390436. Epub 2002/10/23.
42. Ewart-Toland A, Enns GM, Cox VA, Mohan GC, Rosenthal P, Golabi M. Severe congenital anomalies requiring transplantation in children with Kabuki syndrome. Am J Med Genet. 1998;80(4):362–7.
43. Toker A, Salzer L. Pediatric liver transplantation—ethical dilemmas in a disabled patient. Pediatr Transpl. 2012;16(6):E257–60.
44. Workman JK, Myrick CW, Meyers RL, Bratton SL, Nakagawa TA. Pediatric organ donation and transplantation. Pediatrics. 2013;131(6):e1723–30.
45. Council on Ethical and Judicial Affairs, American Medical Association. Ethical considerations in the allocation of organs and other scarce medical resources among patients. Arch Intern Med. 1995;155(1):29–40.
46. Zealand TSoAaN. Transplantation Society of Australia and New Zealand Eligibility Guidelines and Allocation Protocols 2010 [cited 2014 06/01/2014]. Available from: http://www.tsanz.com.au/organallocationprotocols/.
47. Kasiske BL, Ramos EL, Gaston RS, Bia MJ, Danovitch GM, Bowen PA, et al. The evaluation of renal transplant candidates: clinical practice guidelines. Patient Care and Education Committee of the American Society of Transplant Physicians. J Am Soc Nephrol JASN. 1995;6(1):1–34.
48. Ebpg, European Renal A, European Society for Organ T. European best practice guidelines for renal transplantation (part 1). Nephrol Dial Transpl Off Publ Eur Dial Transpl Assoc Eur Renal Assoc. 2000;15 Suppl 7:1–85.
49. Knoll G, Cockfield S, Blydt-Hansen T, Baran D, Kiberd B, Landsberg D, et al. Canadian Society of transplantation: consensus guidelines on eligibility for kidney transplantation. CMAJ Can Med Assoc J (journal de l'Association medicale canadienne). 2005;173(10):S1–25.
50. Kälble T, Alcaraz A, Buddle K, Humke G, Karam M, Lucan G, et al. European Association of Urology Guidelines on Renal Transplantation 2010 [7/21/14]. Available from: http://www.uroweb.org.
51. Mehra MR, Kobashigawa J, Starling R, Russell S, Uber PA, Parameshwar J, et al. Listing criteria for heart transplantation: International Society for Heart and Lung Transplantation guidelines for the care of cardiac transplant candidates—2006. J Heart Lung Transpl Off Publ Int Soc Heart Transpl. 2006;25(9):1024–42.
52. Costanzo MR, Augustine S, Bourge R, Bristow M, O'Connell JB, Driscoll D, et al. Selection and treatment of candidates for heart transplantation. A statement for health professionals from the Committee on Heart Failure and Cardiac Transplantation of the Council on Clinical Cardiology, American Heart Association. Circulation. 1995;92(12):3593–612.
53. de Jonge N, Kirkels JH, Klopping C, Lahpor JR, Caliskan K, Maat AP, et al. Guidelines for heart transplantation. Neth Heart J (monthly journal of the Netherlands Society of Cardiology and the Netherlands Heart Foundation). 2008;16(3):79–87.

54. Murray KF, Carithers RL, Jr., Aasld. AASLD practice guidelines: evaluation of the patient for liver transplantation. Hepatology. 2005;41(6):1407–32.
55. Orens JB, Estenne M, Arcasoy S, Conte JV, Corris P, Egan JJ, et al. International guidelines for the selection of lung transplant candidates: 2006 update—a consensus report from the Pulmonary Scientific Council of the International Society for Heart and Lung Transplantation. J Heart Lung Transpl (official publication of the International Society for Heart Transplantation). 2006;25(7):745–55.
56. Jecker NS. The problem with rescue medicine. J Med Philos. 2013;38(1):64–81.
57. Beauchamp TL, Childress JF. Principles of biomedical ethics, 6th edn. Oxford University Press, New York.
58. Bourgeois N. Liver transplantation in the mentally retarded. Acta gastro-enterologica Belgica. 2002;65(2):135.
59. Daniels N. Health-care needs and distributive justice. Philos Public Aff. 1981;10(2):146–79.
60. Daniels N. Fair equality of opportunity and decent minimums: a reply to Buchanan. Philos Public Aff. 1985;14(1):106–10.
61. Santiago-Delpin EA. Ethical dilemmas: transplantation in prisoners and the mentally disabled. Transpl Proc. 2003;35(5):2057–9.
62. Arciniegas DB, Filley CM. Implications of impaired cognition for organ transplant candidacy. Curr Opin Organ Transpl. 1999;4(2):168.
63. Group W. Study protocol for the World Health Organization project to develop a quality of life assessment instrument (the WHOQoL). Qual Life Res. 1993;2:153–9.
64. Waters E, Davis E, Ronen GM, Rosenbaum P, Livingston M, Saigal S. Quality of life instruments for children and adolescents with neurodisabilities: how to choose the appropriate instrument. Dev Med Child Neurol. 2009;51(8):660–9.
65. Saigal S. Quality of life of former premature infants during adolescence and beyond. Early Human Dev. 2013;89(4):209–13.
66. Icard P, Hooper SR, Gipson DS, Ferris ME. Cognitive improvement in children with CKD after transplant. Pediatr Transplant. 2010;14(7):887–90.
67. Harris J. QALYfying the value of life. J Med Ethics. 1987;13(3):117–23.
68. Nussbaum MC. Frontiers of justice: disability, nationality, species membership. Cambridge: Belknap Press; 2007, 512 p.
69. Hastings RP, Beck A, Hill C. Positive contributions made by children with an intellectual disability in the family: mothers' and fathers' perceptions. J intellect Disabil JOID. 2005;9(2):155–65.
70. Blacher J, Baker BL. Positive impact of intellectual disability on families. Am J Mental Retard AJMR. 2007;112(5):330–48.
71. Murray JS, Tu WH, Albers JB, Burnell JM, Scribner BH. A community hemodialysis center for the treatment of chronic uremia. Trans Am Soc Artif Intern Organs. 1962;8:315–20.
72. Sanders D, Dukeminier JJ. Medical advance and legal lag: hemodialysis and kidney transplantation. UCLA Law Rev. 1968;14:366–80.
73. Alexander S. They decide who lives, who dies: medical miracle and a moral burden of a small committee. Life. 1962;9:102–25.
74. Kilner J. Who lives? Who dies? Ethical criteria in patient selection, 359 p. Yale University Press; 1990.
75. Frader JE. Re-evaluating the recipient criteria for organ transplants. Pediatr Ann. 2012; 41(4):135–6.
76. Kilner JF. Selecting patients when resources are limited: a study of US medical directors of kidney dialysis and transplantation facilities. Am J Public Health. 1988;78(2):144–7.
77. Volk ML, Biggins SW, Huang MA, Argo CK, Fontana RJ, Anspach RR. Decision making in liver transplant selection committees: a multicenter study. Ann Intern Med. 2011;155(8):503–8.
78. Daniels N. Accountability for reasonableness. BMJ. 2000;321(7272):1300–1.
79. Kaufman M. Transition of cognitively delayed adolescent organ transplant recipients to adult care. Pediatr Transplant. 2006;10(4):413–7.

Chapter 13
Should Transplantation Be Offered to Pediatric Foreign Nationals?

Marie-Chantal Fortin and Rebecca A. Greenberg

Abstract The provision of medical care, and particularly organ transplantation, to foreign nationals (i.e. non-citizens and non-residents) is a matter of heated debate in many countries. Worldwide, the practices regarding the provision of transplantation services to foreign nationals vary enormously. Developed countries with transplant programs often receive requests to list foreign nationals for transplantation. This chapter will explore (1) international practices and policies, (2) arguments for and against transplantation for foreign nationals, (3) whether or not there are morally relevant distinctions between children and adult foreign nationals, and (4) whether or not there are morally relevant distinctions between life-saving (e.g. heart) and life-enhancing organ (e.g. kidney) transplants. We will suggest guidance and possible options for policy makers on listing children who are foreign nationals for transplantation.

Keywords Organ transplant · Foreign national · Pediatrics · Public policy · Health care rationing

13.1 Introduction

The provision of medical care, particularly organ transplantation, to foreign nationals (FNs) (i.e. non-residents and non-citizens) is a matter of heated debate in many countries [1, 2]. This is especially true when the organ recipient is a child. The issue of allocation of organs for transplantation is particularly important, since there is an absolute scarcity of organs available for transplantation and some

M.-C. Fortin (✉)
Centre de recherche du CHUM, Nephrology and Transplant Division Centre hospitalier de l'Université de Montréal, Montreal, Canada
e-mail: Marie-chantal.fortin@umontreal.ca

R.A. Greenberg
Bioethics Department, The Hospital for Sick Children, Toronto, Canada
e-mail: rebecca.greenberg@sickkids.ca

R.A. Greenberg et al. (eds.), *Ethical Issues in Pediatric Organ Transplantation*, International Library of Ethics, Law, and the New Medicine 66,
DOI 10.1007/978-3-319-29185-7_13

patients die without having the chance to receive an organ. In recent years, stories of FNs and their difficulty in accessing organ transplantation services have made the headlines around the world [2–4].

Our definition of FNs children includes children claiming refugee status, undocumented immigrants, travelers, students and temporary workers. It is difficult to get official data on the numbers of undocumented children residing in various countries. In Canada, for example, it is estimated that the number of undocumented immigrants and their children is between 200,000 and 500,000, with the majority living in urban areas [5]. In the United States (US), in January 2011, the population of undocumented immigrants was estimated to be 11.5 million. In some countries, children may present illegally without their parents, as is currently the case at the southern US border [6].

We have chosen to include in our definition of FNs both undocumented immigrants and those who do not reside in the country but are seeking transplantation. An argument could be made that these two groups require different analysis, primarily because there is a conception that those who travel to seek transplantation are wealthy while undocumented immigrants are marginalized, generally disadvantaged and contribute to the society for which they are seeking transplantation. In the case of children, contribution is not a relevant consideration as the majority of children are not contributing financially to the country for which they reside. In countries with universal medical coverage, the primary concern with this presumed discrepancy is that treating the wealthy contributes to widening the health gap between socio-economic classes which is fundamentally incongruent with the ideology of universal coverage. However, whether the majority of non-residents seeking transplantation are very wealthy has yet to be substantiated. There have been media reports in the US of wealthy FNs purchasing organs [7]. However, there is only weak empirical data to substantiate this claim. A study by Schold et al. described the demographic characteristics of FNs who received kidney and liver transplants in the US and compared them to nationals. The study shows that FNs were less likely to have a high school degree, but more likely to have a graduate degree as compared to US citizens. The population of FNs was both those who were self-paying (23 % of kidney recipients and 38 % of liver recipients) and those who had government health coverage. The percentage of those self-paying was significantly higher than US citizens which suggests a disproportionate number of affluent recipients. However, it does not suggest that this group is predominantly made up of the later, but rather that it is a diverse group [8]. Further research is needed to better study the population of FN recipients and their socio-economic status. In the absence of conclusive information about the wealth disparity of recipients we decided to include both groups in our analysis. Additionally, as this paper pertains specifically to FN children, we did not think that much of the reasons to differentiate these populations applied as they predominantly centre on the ability to pay or contribute to society. Further discussion of the relevance of contribution to society will be discussed in Sect. 13.3.1.

When a child presents at a hospital with a life threatening condition, many countries recognize an ethical and legal obligation to provide treatment, regardless of the child's residency or immigration status [9–11]. This obligation could be interpreted differently in the arena of organ transplantation in pediatric patients. The issue of organ transplantation for FN children raises numerous, multi-level conflicts between clinicians' duty of care, the financial constraints of institutions, a societal duty to fairly allocate organs and the obligations of the state toward FNs.

Developed countries with transplant programs often receive requests to list FNs for transplantation. This chapter will explore (1) international practices and policies, (2) arguments for and against transplantation for FNs, (3) whether or not there are morally relevant distinctions between children and adult FNs, and (4) whether or not there are morally relevant distinctions between life-saving (e.g. heart) and life-enhancing organ (e.g. kidney) transplants. We will suggest guidance and possible options for policy makers on listing children who are FNs for transplantation.

13.2 International Practices and Policies

13.2.1 Canada

In Canada, there are no guidelines at either the federal or provincial level regarding the provision of organ transplantation services to FNs [10, 12]. Some provincial organ procurement organizations have begun to take steps towards implementing a policy on FNs [13]. However, transplant centres are in general left to deal with this thorny issue on their own, and some have agreed to put a limited number of FNs on their transplant waiting lists.

13.2.2 EuroTransplant

EuroTransplant (ET) is responsible for the allocation of donor organs in Austria, Belgium, Croatia, Germany, Hungary, Luxembourg, the Netherlands and Slovenia.

Within the ET countries there is, with the exception of Belgium, Croatia and Slovenia, no legislation in the area of transplantation for FNs. In Belgium, transplantation for FNs is possible based on a Royal decree and individual decisions made by the Ministry of Health. FNs can be listed for an organ transplant after approval by the Belgian Ministry of Health [14]. Also, in the Netherlands, citizenship or residency are not criteria for allocation of deceased organ [15].

It is of interest to note that the criteria that qualifies a person a "resident" varies within the ET countries. Based on a European Union Council Directive, a non-European Union national can be granted long-term residency status after 5 years of continuous legal residence [16]. However in Belgium, Croatia and Slovenia a person is considered a "resident" after 6 months of legal residence in the respective country. The definition of residency is important given that a legal resident has access to the organ transplantation wait list as would any national citizen.

ET previously had a quota policy for transplantation, according to which FNs could not exceed 5 % of the number of patients on the waiting list (renal, pancreas and intestine transplantations were denied for FNs) [17]. ET opposes transplant tourism as transplantation for non-residents may undermine any particular ET country's ability to provide transplant services for its own population. According to the ET manual of 2014, ET will no longer carry a 5 % policy and will aim at achieving the best possible transparency about practices related to FNs. Transplantation activities will be reported on an annual basis per transplant centre according to the different categories of residency status. Living donor transplantation for non-residents will not be included in the Annual Report since living donation is the responsibility of the transplant centres. ET makes no specific mention of children in their publicly accessible documents.

13.2.3 United Kingdom

In the United Kingdom (UK) organs are preferentially allocated to citizens and residents [18]. The National Health Service Blood and Transplant (the authority that manages blood, tissue and organs) recognizes two groups of patients. Group 1 is comprised of patients who qualify for National Health Service (NHS) by virtue of ordinary UK residence, reciprocal international health agreement, or employment in an armed force or Crown service performed abroad. Group 2 is comprised of individuals who do not fall into Group 1 categories [19].

The NHS *Patient Selection and Organ Allocation Policy* stipulates that organs donated by deceased donors should be allocated first to Group 1 patients and to Group 2 patients only if there is no suitable Group 1 patient in the UK [20]. This is reflected in their liver allocation policy but is not spoken to in their kidney allocation policy [21].

In the UK, the Policy was modified to require transplant centres to report any organ transplantation of non-citizens and non-residents [18]. Following concerns about the number of organs from deceased UK donors being transplanted into non-UK resident EU nationals each year, an independent report to clarify the rules on organ transplants for NHS patients and non-UK EU residents was commissioned by the Health Secretary in 2009 [18]. The report's recommendations, which were broadly accepted by the Department of Health, aimed to ensure public confidence in the fairness and transparency of the organ allocation system.

The report recommended that organs from deceased donors should only be transplanted into patients who are entitled to transplants on the NHS (Group 1 patients). Patients in Group 2 would not be eligible for transplant with organs from deceased donors in the UK. These recommendations have not yet been implemented but were accepted. The Department of Health, which has broadly accepted the report's recommendations, is currently considering the implications for each of them and has established an implementation group consisting of representatives from the transplant community, NHS Blood and Transplant, commissioners and others governmental departments to consider each recommendation and to advise on implementation. The report's recommendations include: (1) an agreement to ban all private clinical practice in the UK involving solid organs donated after death within the NHS, (2) establishing an implementation group to work with NHS Blood and Transplant and commissioners of transplantation to monitor referrals from overseas, (3) developing Department of Health guidance for transplant centres to provide clarification on the eligibility criteria for people from abroad, and (4) raising the need to build both capacity and expertise in developing transplant programs in EU Member States, as well as the building of reciprocal agreements between neighbouring countries [18]. The UK makes no specific mention of children in their publicly accessible documents.

13.2.4 United States

The United Network for Organ Sharing (UNOS) is responsible for managing the US organ transplant system. The Organ Procurement and Transplantation Network (OPTN), operated under contract with the US Department of Health and Human Services by UNOS, maintains the only national patient wait list in the country.

As of 2012, according to the OPTN policy *Transplantation of Non-Resident Aliens*, deceased donor organ allocation will not differ on the basis of a candidate's citizenship or residency status [22]. Any non-US citizen/non-US resident on the organ transplant wait list will not be discriminated against and will be allocated organs in the same way as US citizens. Each candidate is referred on a case-by-case and physician-to-physician basis and there is no limit to the number of non-US citizen/non-US residents who can be referred. There is no specific mention of children in their non-resident alien policy.

Previous policies included an audit provision that stipulated that a committee would review the activities of transplant centres where non-resident alien recipients constituted more than 5 % of recipients of any particular type of deceased organ. This was widely misunderstood as a 5 % cap on a transplant centre's ability to list and transplant non-US citizens. However, since the newly formed OPTN Final Rule mandates that OPTN policies be based on medical criteria, the 5 % audit rule was removed. In its place, there is a policy that allows the OPTN to review listings and transplants of non-US citizens/non-US residents so that they can report on transplant-by-citizenship data to the public [23]. The Ad Hoc International Relations Committee prepares and provides public access to an annual report of Member transplant centre activities related to the listings and transplantation of non-US citizens/non-US residents.

Certain states allow access to FNs via state run medical aid programs. For example, as of October 2014, a law in Illinois provides access to kidney transplantation for non-citizens with end stage renal disease who meet certain residency and financial requirements, even if they are residing in Illinois as undocumented residents [24].

13.2.5 France

France allows FNs to be transplanted under certain conditions: provide proof from the Ministry of Health in the FN's country of origin that transplantation is not available, patients should not be on waiting lists in other countries and patients should be able to afford the transplantation procedure [25]. FNs, whether resident or not in France, can register on the national waiting list under administrative and financial conditions. An analysis of the national kidney registry from 2004 to 2008 [25] found that 15.3 % are of non-French nationality; of which 3.4 % are other European, 5.9 % North African, 3.9 % sub-Saharan African, 2.9 % other [26]. None of the papers reporting the French experience explicitly mentions children.

13.2.6 Scandiatransplant (Denmark, Finland, Iceland, Norway, Sweden)

According to their published guidelines, non-Nordic nationals, whose status in a Nordic country entitles them to receive treatment under the public health care system on the same terms as national subjects of that country, are eligible for organ transplantation within Scandiatransplant. Non-Nordic nationals without such status are not eligible for organ transplantation in the Nordic countries. Prior to 2014 there was an exception for people who fulfilled the urgent criteria for liver transplantation. Under certain conditions Scandiatransplant member nations can arrange cooperation agreements that include transplantation for non-Nordic nationals and the use of organs from non-Nordic countries (e.g. an attempt to establish a balance between the number of organs made available to Scandiatransplant, and the number of organs transplanted into non-national patients) [27]. Scandiatransplant makes no specific mention of children in their publically accessible documents.

13.2.7 Australia and New Zealand

Citing reasons of the Declaration of Istanbul (i.e. treatment of patients from outside the country or jurisdiction is only acceptable if it does not undermine a country's ability to provide transplant services for its own population.) [28], and the fact that the demand for organs far exceeds the supply, The Transplantation Society of

Australia and New Zealand (TSANZ) considers it inappropriate for international patients to receive transplantation, except under exceptional circumstances. In these exceptional cases, such as acute organ failure, transplantation may be permissible if the recipient will be able return to an environment that permits appropriate ongoing post-transplant surveillance and treatment [29]. TSANZ makes no specific mention of children in their publically accessible documents.

13.2.8 Council of Europe Report: Transplantation of Non-nationals and Non-residents in the Countries of the Council of Europe

In 2008 the European Committee on Organ Transplantation of the Council of Europe, coordinated by the Italian National Transplant Centre, conducted a study of practices related to transplantation for FNs [30]. A questionnaire was circulated to 37 health authorities of which 29 replied. For 77 % of the countries, residency is used as a restricting criterion and in three countries nationality or citizenship is used as an additional restricting element. However, 88 % of countries with restrictions on access to transplantation for FNs did not apply this restriction to pediatric cases. Additionally, 67 % of countries with restriction for transplantation would consider transplants in the case of a life-threatening medical emergency.

13.3 Ethical Arguments Regarding Transplantation for Foreign Nationals

Various arguments and considerations are used to support or oppose the provision of organ transplants to FNs, which we will summarize in the following section [12].

13.3.1 Arguments Against Transplantation for Foreign Nationals

In this section we will review arguments that have been raised against transplantation for FNs. Additionally, we will discuss some of the shortcomings of these arguments.

Disadvantaging Nationals Organs are an absolute scarcity. Performing transplantations on FNs will directly affect nationals as every organ allocated to a FN will translate into one less organ in the pool for citizens. There is a great discrepancy between the supply and demand for organs across the globe. In Canada in 2012, 4,612 patients were waiting for an organ and 2,225 transplants were performed. Among the patients on the waiting list 230 died while waiting for an organ [31].

Membership and Citizenship Given the organ shortage, it can be reasonably argued that children who are nationals should be prioritized for transplantation or even treated exclusively. Citizenship is a concept with multiple definitions. In the political and social spheres, citizenship confers special rights, such as the right to vote as well as access to certain collective resources such as health care. It could therefore be argued that citizenship ought to be a necessary requirement to have access to scarce resources such as organs, which are managed by the state [32]. Some might also argue that, from a liberal realism point of view, given the limited resources available to countries with social security systems, it is acceptable to pursue self-interest and not to provide specialized medical care to FN children in order to secure the resources necessary for specialized medical care to citizens and residents [33].

Lack of Contributions by FNs It has been argued that FNs who receive organ transplants are "free riders", since they benefit from a service (transplantation) and a resource (organs) without having contributed to the donor pool [32]. This may not hold true as non-nationals act as living donors and also deceased donors. A review of data in the US over a 19 month period commencing in March 2012 reveals that the 0.3 % of all deceased donors were FN and 0.3 % of recipients of deceased organs donors were FNs [34]. Furthermore, prior willingness to donate and contribute as a deceased donor is not a prerequisite for listing of nationals. Deceased organ donation is a generous act and there is no direct reciprocation (it is however beyond the scope of this chapter to discuss the nature of gift exchange in organ donation); requiring reciprocation or balancing of donors and recipients contradicts the supererogatory nature of organ donation [35, 36]. Similarly there have been arguments that FNs do not contribute economically to the host country and therefore should not be eligible for transplant listing [37]. This argument is not all that applicable to children as most national children do not contribute economically to their country's economy. Nonetheless, they have potential to contribute to the economy and society in the future. It should be noted that some adult patients waiting for an organ are also not working or contributing economically. To base eligibility for transplantation solely on contributions, financial participation or productivity is therefore arbitrary and unfair [38].

The Risk of Becoming a Transplant Destination It has been argued that if countries permit transplantation for FN children they risk being regarded as a destination for children who do not have access to transplantation in their home country [1]. This could potentially perpetuate the current organ shortage, overwhelm host countries' health care systems and remove the incentive for other countries to develop their own transplantation and organ procurement programs, which is recommended by the Declaration of Istanbul [39]. Yet this risk appears to be low. In the U.S., where transplant centres are permitted to list FNs for deceased donor organ transplantation, a 2005 study found that only 0.9 % of patients on the kidney waiting list and 1.5 % of patients on the liver waiting list were FNs [8]. Another US study on undocumented residents requiring hemodialysis treatment showed that these patients had spent more than 35 % of their life in the US before receiving hemodialysis, and only 3 % were aware of their renal condition prior to moving to the US, contradicting the widely held belief that people are moving to the US to receive dialysis treatment [40]. An Israeli study showed similar results with respect to Israel [41]. Some countries have implemented quotas for FNs as a means to ensure their system is not overly burdened.

Reduction in Donation Rates Offering transplants for FN children could have a negative impact on public trust and, in turn, on organ donation rates. While there is scant empirical data on this issue, one survey showed that 38 % of American respondents would be less willing to donate their organs if people were allowed to travel to the US to be transplanted. However, children were seen as a special exception, with 28 % of respondents agreeing that it was acceptable to travel to the US to receive a transplant in this situation [42]. While public opinion should be taken into account, it is obviously not the only measure of what is ethically acceptable. It might also be possible to lessen negative public perceptions by presenting clear and transparent data about the actual demand by FNs for organ transplantation, and the associated costs and benefits of treating or not treating children in need.

13.3.2 Arguments for Transplantation for Foreign Nationals

Transplantation of FNs Exclusively With Organs From Living Donors One option to circumvent the problem of the organ shortage might be to consider organ transplantation for FN children who have a potential living organ donor, so as not to deplete the deceased donor organ pool. However, this could only be an option in renal and liver transplantation where one organ or part of an organ can be procured from a living donor. In the case of renal transplantation, we can reasonably assume that a high proportion of FN children requiring a kidney would have a potential living organ donor, most commonly a parent. A survey conducted with adult undocumented immigrants receiving hemodialysis treatment in a hospital in Queens, New York, showed that 60 % had identified a living organ donor [43]. If a transplant centre were to consider living donor transplants for FN children, it would have to ensure that the transplant associated costs would be reimbursed by the state and follow-up care is accessible for both the organ recipient and the living organ donor. Alternatively, if the recipient has financial means and is able to pay for the medication and follow-up long term it may be acceptable to proceed with transplantation. The challenge is for the physician to feel confident that the patient will be able to cover the funds for follow-up and medical care long term. This challenge is not unique to the care of FNs as many health care systems do not provide coverage or full coverage for the cost of medication for nationals requiring transplantation. The physician should feel confident that the costs can be covered but it is not reasonable to expect the physician to research the patient's and family's financial situation exhaustively. As the relationship between patient and provider is founded on trust, it follows that the physician needs to be able to trust the information provided by the patient and family, unless there is a founded reason to think otherwise. If the medications and follow-up care are not available for the recipient this will compromise the graft and patient survival and will annul the benefits of the transplantation.

Physicians' Duties One of the primary duties of physicians is to consider, promote and protect the health and well-being of their patients [44, 45]. In addition, physicians should not refuse to treat patients for unjust or discriminatory reasons. This requirement is enshrined in medical codes of ethics [44–46]. It can therefore be argued that a physician facing a child with organ failure should consider listing the patient for a transplant on medical suitability alone, regardless of their citizenship, since transplantation offers the best medical outcome and, in the case of non-renal transplantation, the only chance of survival. Physicians should not be concerned with issues of payment or insurance; they should focus on providing the best possible standard of care.

Transplant physicians should be concerned with patients' access to post-transplant follow-up care and to immunosuppressive drugs necessary for graft survival [47]. (A brief discussion of physician's role in ensuring follow-up care is discussed in the previous section.) Proceeding with transplant without ensuring access follow-up would cause harm to the patient (a fundamental principle that physicians are not to breach) and would waste a viable organ. The waste of an organ would go against the role of transplant physician as stewards of societal resources. Ensuring adequate support post-transplant is necessary in order not to jeopardize the graft and the patient's survival.

13.3.2.1 Cost-Saving for FNs on Dialysis

FN children receive dialysis as emergency treatment in some countries, like Canada and the US. It is exceedingly uncommon for pediatric patients to stop dialysis, once it is initiated, since discontinuation would be fatal in almost all cases. Since transplantation, in the medium- and long-term, is less expensive than dialysis, it has been argued that offering transplantation for FNs who are receiving dialysis would save the system money [43, 48, 49].

Goldberg et al. [49] have argued for kidney transplants for non-documented residents in the US. They argue that non-documented residents receiving dialysis in the US should be offered transplants as doing so would open a dialysis spot for documented residents and would save money after as little as 2.7 years post-transplant [49].

13.3.2.2 Alignment with Humanitarian Efforts

Providing FN children access to organ transplantation is in keeping with same ideology that underpins humanitarian work. In our increasingly globalized world humanitarian aid is commonly supported by federal governments. It is not uncommon for countries to provide support to those more disadvantaged in various forms which include peacekeeping, foreign aid and military support. These efforts can be at significant cost to the country providing aid in the both financially and to the health and well-being of nationals. Offering FN children access to transplant waitlist aligns with the values behind humanitarianism. Additionally, countries who have universal health care believe in the concept of helping the most disadvantaged. This sentiment is also consistent with providing FNs access to transplant wait lists.

13.4 Pediatric Exceptionalism—Should FN Children Be Treated Different Than Adults?

In many jurisdictions children are given special status for organ transplant listings and are prioritized. In this section we will explore the arguments as to whether or not children should have special status in deceased organ transplantation compared to adults. These arguments are relevant in assessing whether FN children should have different treatment than FN adults.

13.4.1 Arguments for Pediatric Exceptionalism

The Medical Argument Children are generally, and in the arena of transplantation, regarded as a vulnerable population—a group in need of protection. Physical growth is stunted in children with end stage kidney disease, and neurodevelopment may also be delayed. The longer children wait for transplant, the less opportunity there is to catch up on these delays [50]. Consequently, it can be argued that it is ethically justifiable to give children different consideration in allocation policies for deceased donor organs because of their medical vulnerability. Several jurisdictions including the US, all Canadian provinces, Europe, Australia and New Zealand give children extra priority for kidneys allocation [51]. Children are prioritized in transplantation both for the time they are on the wait list and for the quality of organs in order to maximise the graft survival. For extra-renal organs, other criteria such as the size of the organs and the patients are taken into consideration.

Human Rights and the Child The belief that children should be afforded special status and rights has been enshrined in the United Nations Convention on the Rights of the Child, a human rights treaty which sets out the social, health, economic, political and cultural rights of children. 192 countries have ratified the Convention [52]. The preamble states that "childhood is entitled to special care and assistance." Article 24 states that "States Parties recognize the right of the child to the enjoyment of the highest attainable standard of health and to facilities for the treatment of illness and rehabilitation of health. States Parties shall strive to ensure that no child is deprived of his or her right of access to such health care services"

[53]. The Article further declares that there should be international co-operation to achieve the realization of this right, with particular attention to the needs of developing countries. Providing access to organs for FN children is in keeping with this well accepted treaty.

The Right to an Open Future A central guiding principle in pediatric ethics is the 'right to an open future'—the idea that a child's opportunities in the future should not be unduly restricted [54]. This principle is invoked when arguing that every child ought to have access to public education, no matter their residency or immigration status. Without access to organ transplants this principle would not be upheld, as on account of morbidity and mortality, FN children would not be able to maximize future opportunities. For example, children who are on renal dialysis have difficulties in school and maximizing scholarly opportunities [55].

Age-based Justice and Children Building further on the special circumstances for affording children special consideration when allocating organs is the concept of 'Fair Innings', which maintains that children should be given priority over adults as they have had fewer healthy life years [56]. This resource allocation argument has been used specifically in the context of allocating kidneys. Ross et al. have discussed the 'Fair Innings' argument; specifically, whether children should be given priority in allocation schemes as those developing end stage renal disease (ESRD) at a younger age are worse off than those developing ESRD when older. This concept of 'Fair Innings' is in keeping with Veatch's concept of justice over a lifetime, where everyone should have the chance to enjoy the same amount of medical well-being [57]. For instance, a 65-year-old patient who was lived 60 healthy years should be given less priority than a 10-year-old patient who has not had the opportunity to enjoy the same amount of well-being. Veatch's position is congruent with Norman Daniels' argument on prudential lifespan, which takes into account the different needs we have at different stages of our lives. Daniels [58] proposes that resources should be allocated so as to ensure that everyone can attain the normal age-relative opportunity range and justifies treating people of different ages differently. Finally, prioritization of children is in accordance with Rawlsian principles of justice [59]. Specifically, the second principle stipulates that inequalities are to be arranged so that they afford the greatest benefit to the least-advantaged members of society. Indeed, sick children are generally more vulnerable than sick adults and could be seen as the worst off in society. It follows that it would be well reasoned that FN children be afforded different consideration than adults and be granted more liberal access to organ transplantation.

Capability Approach The capability approach to justice, developed by Martha Nussbaum and Amartya Sen, suggests that everyone should enjoy equal opportunities. According to this theory, children require special consideration. Capabilities, according to Nussbaum [60, p. 20], refer to "the abilities residing inside a person but also the freedoms or opportunities created by a combination of personal abilities and political, social, and economic environments". This approach to justice also requires accounting for the differences between groups of individuals, such as adults versus children, and acknowledges that the needs for care of children and their vulnerability differ from that of adults. Consequently, it is justified to provide different treatment to children compared to adults [61].

Public Support for Prioritizing Children Some studies have shown a public preference for allocation of organs to the youngest. A study conducted with the public in the US showed that respondents preferred to allocate organs to younger people (not necessarily children) [62]. However, patients' perspectives are divided. In the UK, where there is a pediatric priority, a study of dialysis patients and transplanted patients found that 50.4 % of respondents did not want to prioritize an 8-year-old child over a 35-year-old adult. According to these respondents, age should not be a criterion in the allocation scheme [63]. An Australian focus group study of patients found that some agreed that the age of the recipient should be considered in the allocation of benefits to the youngest [64].

13.4.2 Arguments Against Pediatric Exceptionalism

Despite the many arguments in favour of prioritization of children, some scholars have questioned this approach. Some have argued that prioritization of children should not be permitted as it has decreased the rates of living organ donation in renal transplantation, Indeed, given many children now receive a deceased organ in a timely manner, potential living organ donors are less urged to proceed to donation [65]. Also, some have asserted there is an overestimation of the impact of a delayed transplantation on growth and cognitive development [66]. Adults also suffer if their transplantation is delayed, albeit in different ways. Another argument put forward is that prioritizing children could lead to increased graft loss given that adolescents have a high incidence of non-adherence leading to graft loss [67]. Finally, prioritisation of children could represent a form of ageism against older potential recipients [66].

In conclusion, as per the aforementioned reasons in the section on *Arguments for Pediatric Exceptionalism* many transplant programs, internationally, confer a special status to children for listing for a deceased donor organ. It follows that a FN child would also have a special status compared to an adult FN. Policies pertaining to the listing and allocation of organs for FN children requires different considerations than for FN adults.

13.5 Imminently Life-Saving Versus Life-Enhancing

There are a number of jurisdictions (i.e. European countries who are part of Eurotransplant) which offer transplantation to FNs but exclude non-life saving organs [17]. By life-saving we mean heart, lung, liver and intestinal transplantation where there is no widely used replacement therapy such as dialysis for kidney transplantation. (That being said, with the development of ventricular assisted device or hepatic dialysis, there will be alternatives to transplantation for these organs in the future.) It is understandable why jurisdictions which permit transplantation for FNs often limit it to life-saving organ transplantation, given that without transplantation patients are condemned to immediate death.

Life-enhancing transplants include kidney, pancreas, and vascularized composite allotransplantation. Intestinal transplantation could also be qualified as non-life-saving transplants when the primary indication is to improve the quality of life (see Sect. 13.5.3). Should non-imminently life-saving transplants they be offered to FN children? In this section, we will look closer at the distinction between imminently life-saving and life-enhancing organs in the debate of transplantation for FN children.

A consideration in assessing the ethical permissibility of offering life-enhancing transplantation to FNs is the gravity of the consequences of not having access to transplantation listing. This factor can be considered on both sides. To some the compelling reason for listing FNs for imminently life-*saving* organs is that without transplantation the consequences are extremely grave and would otherwise result in death. However, by listing FNs the corollary is that national children are put at greater risk, as wait times are likely to increase. In contrast, the consequences to national children with the allocation of life-*enhancing* transplants are far less severe. As such, some may argue that the offering of life-*enhancing* transplants to FNs is more ethically justified as it will not result in the imminent death of national children or adults. Similarly, one could state that since the consequences are not grave there is not enough justification to warrant a transplant to FNs.

13.5.1 Kidneys

Although kidney transplantation is not imminently life-saving, the harm suffered by children denied a transplant is significantly greater than that suffered by adults and may warrant exceptional considerations. Children who remain on dialysis will suffer substantial long-term consequences such as growth and neurodevelopmental delays [68, 69]. It has been well established that kidney transplant is superior to dialysis in terms of morbidity, mortality, quality of life, and cost [70, 71]. Children who receive kidney transplants, versus dialysis, will do better and will live longer [72]. Correspondingly some may say kidney transplant should be considered life-saving in the long term. Kidney transplantation also provides children with the greatest opportunities to maximize their health and well-being. As such kidney transplantation for FN children can be ethically defensible.

13.5.2 Vascularized Composite Allotransplantation (VCA)

VCA refers to limb, hand and face transplantation. In children it is still a novel concept, with the first pediatric limb transplant performed in 2015 [73] and face transplantation being considered experimental with none performed in children [35]. Further, the debate has yet to be settled about the ethical viability of VCA in children. The ethics of pediatric VCA generally are discussed in this book in Chap. 10 by Zlotnik et al. and more specifically of face transplant in a paper by Flynn et al. [35]. In order to discuss the ethics of VCA and FNs we must first operate under the assumption that, in some circumstances, VCA is ethically viable as there is still a debate in the literature pertaining to the ethics about many aspects of VCA for children.

There are a number of unique considerations with VCA transplantation as to why it is ethically defensible to offer to FNs. Although VCA may not be considered imminently life-saving it can drastically improve quality of life and prolong life. Pediatric burn patients may benefit from face transplants immensely as there is a correlation with primary psychiatric diagnosis that include anxiety and depressive disorders in this population [74]. Transplant may offer therapeutic benefits and reduce the risk of these disorders [35]. Of more significance, research shows that there is a correlation with young adults suffering from pediatric burn injuries and with suicide potential [75]. VCA may be the optimal treatment for some indications (i.e. patients with severe burns). Nonetheless, there are limits to currently available reconstructive surgery such that both cosmetic and functional impairments can never be completely corrected [76].

VCA requires a highly specialized skills set that is only available in a very limited number of centres worldwide. Without offering transplantation to FN only a very small number of children worldwide would be able to access VCA transplantation. Because VCA is in its infancy, with few centres offering transplants the ability to individualize allocation is noncompetitive and offering transplantation to FNs would not overwhelm the waitlists [48]. However, this will likely change over time.

Considering (1) the potential overall positive impact of VCA on a child's health, and in theory particularly with face transplants, (2) the numbers of VCA are so small that transplantation for FNs will not significantly have negative effect on access to transplant for nationals, and (3) care is highly specialized and not available in many countries, in some circumstances listing FNs for VCA would be ethically permissible.

13.5.3 Isolated Intestine Transplant

Most children listed for an intestinal transplant are also listed for other organs and listed for imminently life-threatening indications. It is difficult to justify intestinal transplant for poor quality of life because incidence of morbidity and mortality following transplant is high with long-term survival rates of 50–60 % of those who receive a transplant, but it is possible in some cases. For example, children with short bowel syndrome have sufficiently high and erratic fluid losses that make it impossible to provide appropriate replacement outside of the hospital. As such the children are unable to participate in age appropriate activities, including, but not limited to, school [77]. Intestinal transplantation could, in these cases, increase the quality of life of patients with short bowel syndrome. The wait list for isolated intestine transplant has been steadily decreasing due to the advances in intestinal care and rehabilitation. Additionally, the number of deaths on the wait list for isolated intestine transplants is less than other solid organ transplants. [78] Considering the significant impact on quality of life and the availability of intestines—relative to other organs—it may be permissible to list FNs children for transplantation. An in-depth exploration of the ethical issues relevant to small bowel transplantation is discussed in Chap. 9 by Annie Fecteau.

13.6 Possible Policy Options

The various arguments for and against transplantation for FN children give rise to six possible policy options [12].

1. *Refuse to transplant any FN children.* This option could be justified because (i) there is an organ shortage; (ii) residency or citizenship ought to be a requirement to have access to a national resource such as organs; (iii) transplant physicians could harm FN children if they do not have access to immunosuppressive drugs and post-transplant follow-up care in their country of origin; physicians may violate the ethical principle of non-maleficence (to do no harm) and would not be acting as responsible stewards of a precious resource and; (iv) the possible negative public response which could lead to a reduction in donation rates. This option raises rights-based concerns, is not congruent with physicians' codes of ethics and duties, and fails to consider the unique considerations of children.
2. *Allocate organs to FN children that are not suitable for residents or citizens.* This option would be consistent with the view that citizens or residents should receive priority. This approach, in which FN children are treated as second-class patients, could be viewed as unjustified discrimination. Also, it is unlikely this approach would be practically feasible. With the current organ shortage, the criteria for acceptance of deceased donor organs have been broadened (e.g. marginal donors and the procurement of organs following cardiac death), so there are fewer "unwanted" organs than expected.
3. *Permit organ transplantation only for imminently life-saving organs for FN children.* This would preclude any transplantation for life-enhancing organs such as kidney, pancreas, VCA. The main argument against this option is that allocation of a life-saving organ to a FN child will lead to less organs being transplanted in residents or citizen and could be responsible for increased deaths on the wait list.
4. *Permit organ transplantation only for life-enhancing organs for FN children.* Transplantation of life-enhancing organ to a FN child will not be directly responsible for increased mortality citizen or resident children on the waiting

list. The main argument against this option is that the consequences of not offering transplantation do not result in imminent death and so there may not be enough justification to warrant a transplant to FN children.
5. *Permit transplantation for FN children only when there is a suitable and available living organ donor.* This solution would not remove organs from the deceased donor organ pool and would allow some access to transplantation for certain FN children. However, this is only a solution for children requiring a liver or kidney transplant and who have a viable living donor.
6. *Have a transparent, national policy that would allow the wait-listing of FN children for life-saving and life-enhancing organ transplantation using deceased or living donor organs* This option would be compatible with the view that entitlements are not only related to citizenship, but also to human rights. This option would also require wealthy developed countries to assume a certain responsibility toward children for transplantation. In listing FN children there are two options:
   (a) No quota policy.
   (b) A quota policy that needs to be determined. This quota policy could help to ensure that some countries do not become a transplant destination.

13.7 Conclusion

Transplantation for FN children is a thorny issue. This issue highlights the fundamental tension between physicians' duties towards their individual patients, the obligation of organizations to fairly allocate organs and the obligations of the State towards children. Policies on listing FN children for transplantation should not be left to individual transplant centres. This could lead to unfair and unjust allocations in which FN children in one centre have access to organ transplantation, whereas other clinically similar FN children presenting at another centre, in the same country, would be denied this service. It is essential that countries have a transparent, reasoned policy regarding the allocation of organs to FNs in order to ensure equitable access to transplantation and to preserve trust from wait-listed patients and their families, health providers, and the general public. Further research is needed to explore different stakeholders' views on the issues of transplantation for FN children.

References

1. Stevens D. Asylum seekers and the right to access health care. N Ir Legal Q. 2010;61(4):363–90.
2. Holtz M. Illegal immigrant gets kidney. Loyola covers costs of transplant. The Chicago Tribune. 9 Dec 2012;Sect. News.

3. Bernstein N. For illegal immigrant, line is drawn at transplant. New York Times. 2011 December 20;Sect. N.Y./Region.
4. Marsden S. NHS organs donated to foreign patients. The telegraph. 2013;Sect. Health News.
5. Miklavcic A. Canada's non-status immigrants: negotiating access to health care and citizenship. Med Anthropol. 2011;30(5):496–517.
6. Krogstad JM, Gonzalez-Barrera A, Lopez MH. Children 12 and under are fastest growing group of unaccompanied minors at U.S. border: pew research center; 2014 (updated July 22; cited 17 Sept 2014). Available from: http://www.pewresearch.org/fact-tank/2014/07/22/children-12-and-under-are-fastest-growing-group-of-unaccompanied-minors-at-u-s-border/.
7. Ornstein C, Glionna JM. After livers, cash to UCLA. LA Times. 2008 May 31.
8. Schold JD, Meier-Kriesche H-U, Duncan PR, Reed AI. Deceased donor kidney and liver transplantation to nonresident aliens in the United States. Transplantation. 2007;84(12):1548–56.
9. Gouvernement du Québec. Chapitre S-4.2 Loi sur les services de santé et les services sociaux 2013 [cited 2013 June 18]. Available from: http://www2.publicationsduquebec.gouv.qc.ca/dynamicSearch/telecharge.php?type=2&file=/S_4_2/S4_2.html.
10. Bruni R, Wright L. Sharing organs with foreign nationals. Prog Transplant. 2011;21(1):78–82.
11. ServiceOntario. Public HOSPITALS ACT R.S.O. 1990, CHAPTER P.40 1990 (updated 2010; cited 9 Sept 2013). Available from: http://www.e-laws.gov.on.ca/html/statutes/english/elaws_statutes_90p40_e.htm#BK20.
12. Fortin M-C, Williams-Jones B. Should we perform kidney transplants on foreign nationals? J Med Ethics. 2014;40:821–6.
13. Trillium Gift of Life Network. Organ transplantation for uninsured and non-residents. preliminary jurisdictional scan and review. In: Trillium gift of life network transplant ethics working group, editor; 2014.
14. Eurotransplant Foundation. Eurotransplant manual version 4.0. Leiden, The Netherlands: Eurotransplant Foundation; 2014.
15. Zuidema W, Tronchet N, Luchtenburg A, de Klerk M, IJzermans JNM, Weimar W. Nonresident living kidney donors. Transplant Proc. 2005;37:598–9.
16. The Council of the European Union. Council Directive 2003/109/EC of 25 November 2003 concerning the status of third-country nationals who are long-term residents. Official J European Union. 2004;47(L16):44.
17. Eurotransplant Foundation. Eurotransplant manual version 3.2. Leiden, The Netherlands: Eurotransplant Foundation; 2013.
18. Buggins E. Allocation of organs to non UK EU residents: department of health, NHS blood and transplant; 2009. Available from: http://www.bts.org.uk/Documents/Publications/Buggins%20Report%20-%20ALLOCATION%20OF%20ORGANS%20TO%20NON%20UK%20EU%20RESIDENTS.pdf.
19. NHS Blood and Transplant. Eligibility of non-UK residents to receive donated organs 2009 (cited 25 Sept 2014). Available from: http://www.organdonation.nhs.uk/newsroom/statements_and_stances/statements/non_uk_eligibility.asp.
20. Zalewska K. Introduction to patient selection and organ allocation policies 2013 [cited 2015 January 7]. Available from: http://www.odt.nhs.uk/pdf/introduction_to_selection_and_allocation_policies.pdf.
21. NHS Blood and Transplant. Guidance and policies 2014 (cited 17 Sept 2014). Available from: http://www.odt.nhs.uk/transplantation/guidance-policies/.
22. Organ Procurement and Transplantation Network. Policy 6.0 deceased donor organ transplantation of non-US residents/non-US citizens, and the importation of deceased donor organs from foreign sources; 2013.
23. Organ Procurement and Transplantation Network. Proposed revisions to and reorganization of policy 6.0 (Transplantation of Non-Resident Aliens), which include changes to the non-resident alien transplant audit trigger policy and related definitions 2012 [cited 2013 January 9]. Available from: http://optn.transplant.hrsa.gov/PublicComment/pubcommentPropSub_290.pdf.

24. Illinois General Assembly. Bill status of SB0741 2014 [cited 2015 January 5]. Available from: http://www.ilga.gov/legislation/BillStatus.asp?DocTypeID=SB&DocNum=741&GAID=12&SessionID=85&LegID=70138#top.
25. Cantrelle C, Laurens C, Luciolli E, Loty B, Tuppin P. Access to kidney transplantation in France of non-French patients and French patients living in overseas territories. Transplantation. 2006;81(8):1147–52.
26. Cantrelle C, Pessione F, Macher M-A, Thuong M. Évolution de l'accès à la greffe rénale en France des patients étrangers ou résidant outre-mer. Bulletin de la Société de Pathologie Exotique. 2012;105:115–22.
27. Scandiatransplant. Guidelines for deceased organ transplantation of non-nordic nationals within Scandiatransplant and the use of non-nordic organs for Scandiatransplant recipients 2014. Available from: http://www.scandiatransplant.org/organ-allocation/GuidelinesnonNordicnationalsMay2014.PDF.
28. Steering Committee of the Istanbul Summit. Organ trafficking and transplant tourism and commercialism: the declaration of Istanbul. Lancet. 2008;372(9632):5–6.
29. The Transplantation Society of Australia and New Zealand. Organ transplantation from deceased donors: consensus statement on elibiligy cirteria and allocation protocols version 1.3. 2014.
30. Carella C, Cozzi E, Di Ciaccio P, Nanni Costa A. Transplantation of non-nationals and non-residents in the countries of the council of Europe: results of a survey conducted in the context of the initiatives of the European committee on organ transplantation. Newslett Transplant. 2012;17(1):83–7.
31. Canadian Institute for Health Information. Canadian organ replacement register annual report: treatment of end-stage organ failure in Canada, 2003 to 2012. Ottawa: CIHI; 2014.
32. Prottas JM, Jonasson O, Kleinig JI. In Organ transplants, American first? Hastings center report. 1986:23–5.
33. Cole P. Human rights and the national interest: migrants, healthcare and social justice. J Med Ethics. 2007;33:269–72.
34. Glazier AK, Danovitch GM, Delmonico FL. Organ transplantation for nonresidents of the United States: a policy for transparency. Am J Transplant. 2014;14(8):1740–3.
35. Flynn J, Shaul RZ, Hanson MD, Borschel GH, Zuker R. Pediatric facial transplantation: ethical considerations. Can J Plast Surg = Journal canadien de chirurgie plastique. 2014;22(2):67–9.
36. Simmerling M, Angelos P, Goldberg AM, Frader JE. Do gifts create moral obligations for recipients? Am J Bioeth. 2004;4(4):20–2.
37. Bramstedt KA. Supporting organ transplantation in non-resident aliens within limits. Eth Med Christ Perspect Issues Bioeth. 2006;22(2):75–81.
38. Annas GJ. The prostitute, the playboy, and the poet: rationing schemes for organ transplantation. Am J Publ Health. 1985;75(2):187–9.
39. Participants in the International Summit on Transplant Tourism and Organ Trafficking convened by The Transplantation Society and International Society of Nephrology. The declaration of Istanbul on organ trafficking and transplant tourism 2008 (cited 20 August 2008). Available from: http://www.sciencedirect.com/science/MiamiMultiMediaURL/B6T1B-4SWYBM2-5/B6T1B-4SWYBM2-5-1/4886/cad07b9aabfc1fbe9c50d5818806e7fc/mmc1.pdf.
40. Raghavan R, Sheikh-Hamad D. Descriptive analysis of undocumented residents with ESRD in a public hospital system. Dialysis Transpl. 2011;40:78–81.
41. Chernin G, Go-Oz A, Shashar M, Schwartz D, Weinstein T. Care of undocumented-uninsured immigrants in a large urban dialysis unit. BMC Nephrol. 2012;13:112.
42. Volk ML, Warren GJW, Anspach RR, Couper MP, Merion RM, Ubel PA. Foreigners traveling to the U.S. for transplantation may adversely affect organ donation: a national survey. Am J Transplant. 2010;10:1468–72.
43. Linden EA, Cano J, Coritsidis GN. Kidney tranpslantation in undocumented immigrants with ESRD: a policy whose time has come? Am J Kidney Dis. 2012;60(3):354–9.

44. Canadian Medical Association. CMA Code of Ethics 2004 (updated March 2012; cited 6 Sept 2013). Available from: http://policybase.cma.ca/dbtw-wpd/PolicyPDF/PD04-06.pdf.
45. Code de déontologie des médecins. Loi médicale 2013 (updated Sept 1; cited 6 Sept 2013). Available from: http://www2.publicationsduquebec.gouv.qc.ca/dynamicSearch/telecharge.php?type=3&file=/M_9/M9R17.HTM.
46. American Medical Association. Opinion 9.12—Patient-Physician Relationship: Respect for Law and Human Rights 2007 (cited 7 Sept 2014). Available from: http://www.ama-assn.org/ama/pub/physician-resources/medical-ethics/code-medical-ethics/opinion912.page.
47. Friedman AL, Friedman EA. Illegal aliens are not entitled to federally funded organ transplants. Transplantation. 2007;83(1):21–3.
48. Cendales LC, Rahmel A, Pruett TL. Allocation of vascularized composite allografts: what is it? Transplantation. 2012;93(11):1086–7.
49. Goldberg AM, Simmerling M, Frader JE. Why nondocumented residents should have access to kidney transplantation: arguments for lifting the federal ban on reimbursement. Transplantation. 2007;83(1):17–20.
50. Crafter SR, Bell L, Foster BJ. Balancing organ quality, HLA-matching, and waiting times: impact of a pediatric priority allocation policy for deceased donor kidneys in Quebec. Transplantation. 2007;83(11):1411–5.
51. Baran D. Deceased donor kidney allocation in the US, Europe, Australia and New Zealand. Montréal: Canadian Council for Donation and Transplantation; 2006.
52. UNICEF. Convention on the rights of the child- frequently asked questions 2005 (cited 9 Sept 2014). Available from: http://www.unicef.org/crc/index_30229.html.
53. United Nations General Assembly. Convention on the rights of the child. In: Office of the high commissioner for human rights, editor. 1989.
54. Feinberg J. A child's right to an open future. In: Aiken W, LaFollette H, editors. Whose Child? Parental rights, parental authority and state power. Totowa, NJ: Littlefield, Adams, and Co; 1980. p. 124–53.
55. Moser JJ, Veale PM, McAllister DL, Archer DP. A systematic review and quantitative analysis of neurocognitive outcomes in children with four chronic illnesses. Pediatric Anesthesia. 2013;23:1083–96.
56. Ross LF, Parker W, Veatch RM, Gentry SE, Thistlethwaite JR Jr. Equal Opportunity Supplemented by Fair Innings: equity and efficiency in allocating deceased donor kidneys. Am J Transplant. 2012;12(8):2115–24.
57. Veatch RM. Transplantation ethics. Washington D.C.: Georgetown University Press; 2000. 427 p.
58. Daniels N. Just health: meeting health needs fairly. Cambridge: Cambridge University Press; 2008.
59. Rawls J. A Theory of Justice. Cambridge: Harvard University Press; 1999.
60. Nussbaum MC. Creating Capabilities. Cambridge: The Belknap Press of Harvard University Press; 2011 237 p.
61. Dixon R, Nussbaum MC. Childrens' rights and a capabilities approach: the question of special priority. Cornell Law Rev. 2012;97:549–94.
62. Stahl JE, Tramontano AC, Swan JS, Cohen BJ. Balancing urgency, age and quality of life in organ allocation decisions—what would you do? A survey. J Med Ethics. 2008;34:109–15.
63. Geddes CC, Rodger SC, Smith C, Ganai A. Allocation of deceased donor kidney for tranpslantation: opinions of patients with CKD. Am J Kidney Dis. 2005;46(5):949–56.
64. Tong A, Jan S, Wong G, Craig JC, Irving M, Chadban S, et al. Patient preferences for the allocation of deceased donor kidneys for transplantation: a mixed methods study. BMC Nephrol. 2012;13(18).
65. Axelrod DA, McCullough KP, Brewer ED, Becker BN, Segev DL, Raog PS. Kidney and pancreas transplantation in the United States, 1999–2008: the changing face of living donation. Am J Transplant. 2010;10(Part 2):987–1002.
66. Capitaine L, Van Assche K, Pennings G, Sterckx S. Pediatric priority in kidney allocation: challenging its acceptability. Transplant Int. 2014;27(6):533–40.

67. Andreoni K, Forbes R, Andreoni R, Phillips G, Stewart H, Ferris M. Age-related kidney transplant outcomes: health disparities amplified in adolescence. JAMA Int Med. 2013;173 (16):1524–32.
68. Pape L, Ehrich JH, Zivicnjak M, Offner G. Growth in children after kidney transplantation with living related donor graft or cadaveric graft. Lancet. 2005;366(9480):151–3.
69. Qvist E, Pihko H, Fagerudd P, Valanne L, Lamminranta S, Karikoski J, et al. Neurodevelopmental outcome in high-risk patients after renal transplantation in early childhood. Pediatr Transplant. 2002;6(1):53–62.
70. Tonelli M, Wiebe N, Knoll G, Bello A, Browne S, Jadhav D, et al. Systematic review: kidney transplantation compared with dialysis in clinically relevant outcomes. Am J Transplant. 2011;11(10):2093–109.
71. Manns BJ, Mendelssohn DC, Taub KJ. The economics of end-stage renal disease care in Canada: incentives and impact on delivery of care in Canada. Int J Health Care Finance Econ. 2007;7:149–69.
72. McDonald SP, Craig JC. Long-term survival of children with end-stage renal disease. New Engl J Med. 2004;350(26):2654–62.
73. The Children Hospital of Philadelphia. World's first bilateral hand transplant on a child performed at the children's hospital of Philadelphia 2015 (cited 15 Sept 2015). Available from: http://www.chop.edu/news/world-s-first-bilateral-hand-transplant-child-performed-children-s-hospital-philadelphia#.Vfhpun2CTIl.
74. Stoddard FJ, Norman DK, Murphy JM. A diagnostic outcome study of children and adolescents with severe burns. J Trauma. 1989;29(4):471–7.
75. Rosenberg L, Robert R, Thomas C, Holzer CE 3rd, Blakeney P, Meyer WJ 3rd. Assessing potential suicide risk of young adults burned as children. J Burn Care Res Official Publ Am Burn Assoc. 2006;27(6):779–85.
76. Sheridan RL, Hinson MI, Liang MH, Nackel AF, Schoenfeld DA, Ryan CM, et al. Long-term outcome of children surviving massive burns. JAMA. 2000;283(1):69–73.
77. Kaufman SS, Atkinson JB, Bianchi A, Goulet OJ, Grant D, Langnas AN, et al. Indications for pediatric intestinal transplantation: a position paper of the American Society of Transplantation. Pediatr Transplant. 2001;5(2):80–7.
78. Mazariegos GV, Steffick DE, Horslen S, Farmer D, Fryer J, Grant D, et al. Intestine transplantation in the United States, 1999–2008. Am J Transplant Official J Am Soc Transplant Am Soc Transplant Surg. 2010;10(4 Pt 2):1020–34.

Part V
Recipient Issues

Chapter 14
The Ethics of Uncertainty, Risk, and Trust in Pediatric Heart Transplantation

Daniel Z. Buchman and Anne I. Dipchand

Life is short
And the art long;
The occasion instant,
Experiment perilous,
Decision difficult.

Hippocrates [1, p. 697]

I act with complete *certainty. But this certainty is my own.*
Wittgenstein [2, p. 174]

Abstract Pediatric heart transplantation occurs in a context characterized by high levels of risk and uncertainty, which necessitates a strong family-centered trust relationship with clinicians. Pediatric patients and their families are burdened with multiple difficult decisions, both medical and ethical, and require support as they go through the process of decision-making. The challenge for all parties is to determine the child's best interest from a medical, physical, emotional, developmental, and psychosocial perspective, in order to provide a child who is a high-risk transplant candidate not only with an option for survival, but also with options for an acceptable quality of life within their life circumstances, with or without transplantation. This can be achieved through a commitment to epistemic humility and a strong, family-centered trust relationship.

D.Z. Buchman (✉)
Bioethics Program and Multi-Organ Transplant Program, University Health Network, Toronto, ON, Canada
e-mail: daniel.buchman@uhn.ca

D.Z. Buchman
Joint Centre for Bioethics and Division of Clinical Public Health, Dalla Lana School of Public Health, University of Toronto, Toronto, ON, Canada

A.I. Dipchand
Labatt Family Heart Centre, Hospital for Sick Children, Department of Pediatrics, University of Toronto, Toronto, ON, Canada
e-mail: anne.dipchand@sickkids.ca

R.A. Greenberg et al. (eds.), *Ethical Issues in Pediatric Organ Transplantation*, International Library of Ethics, Law, and the New Medicine 66,
DOI 10.1007/978-3-319-29185-7_14

Keywords Uncertainty · Risk · Trust · Bioethics · Family-centred care · Pediatric heart transplant · Epistemic humility · Decision-making

14.1 Introduction

Innovations in medical science and technology have led to an increase in the possibility of pediatric heart transplantation for patients[1] with end-stage heart failure. While these developments have saved lives and enhanced the quality of life for many, novel technological interventions include complex procedures that have known, unknown, and poorly understood risks. In situations characterized by high levels of risk, uncertainty is common. In the pediatric heart transplant context, patients, their families, and the healthcare team, are often faced with a series of challenging, and often agonizing ethical decisions. These decisions occur in a context where the demand for organs far exceeds the supply. While many children are listed for transplant, many also die on the waiting list. When the child's best interests are at stake, families and the healthcare team may be unclear in how they ought to act. Trust is essential in situations characterized by risk and uncertainty, and so it is ethically necessary for clinicians and institutions to cultivate and maintain a trusting therapeutic atmosphere wherein treatment plans can be developed according to the child's best interests. The concept of best interests, or the best interest standard, is widely discussed in law, medicine, and bioethics, often in the context of pediatric substitute decision-making. Best interests are an umbrella term to describe decision-making that maximizes well-being for an incapable person, such as a child, who may not have had the opportunity to identify treatment preferences, establish their values, or ever had the ability to make them known [3].

In this chapter, we focus on the ethical tensions related to notions of uncertainty, risk, and trust in the context of pediatric heart transplantation; more importantly, in the context of managing a disease process that cannot be cured or when all treatment options are sub-optimal. Specifically, we are concerned with issues raised by congenital heart disease or end-stage heart failure in children, and heart transplantation. Disease management in this context comes with varying levels of uncertainty and risk, including difficulty estimating prognosis. We examine the question, how do we—the healthcare team, pediatric patients and their families, both separately and together—make tough, moral decisions in a context of high risk and uncertainty? Despite our cardiac-specific focus, we draw upon literature from other organ groups given the limited experience and data in this area as compared to what is available for adult transplantation. The chapter concludes with a proposal of how a collaborative family-centered approach to

[1]We use the terms "child" and "children" to refer to infants, children, and adolescents up to the age of 19, and use these terms interchangeably with "patient" or "pediatric patient".

decision-making based in epistemic humility can help clinicians attempt to earn and sustain the trust of children and families in situations characterized by high levels of risk and uncertainty.

14.2 Family-Centered Care in the Context of Pediatric Heart Transplantation

Most people are situated in and influenced by their families, so understanding the role of family in pediatric healthcare decision-making is important. We adopt Lindemann's [4, p. S98] definition of family as "the structure of intimacy in which most people live out their lives, characterized by relationships that are more or less ongoing and close, usually (but not always) involving long periods of living together in the same household". Family members, usually parents, are often called upon to make treatment decisions on behalf of their child who lacks treatment decision-making capacity by virtue of age or competence.[2] A family-centered perspective in health care views the psychosocial and developmental well-being of the patient as being enmeshed with the well-being of the family [5]. In this chapter we will not engage the question as to whether familial involvement in treatment decision-making is or is not morally good. We recognize that family members, parents in particular, are often involved in the healthcare decision-making for their children and may have differing preferences with respect to their degree of involvement. There are also limits to parental substitute decision-making, as factors such as incompetence, unavailability, abuse, neglect, or decisions contrary to the best interests of the child may be ethically and legally defensible grounds to relieve parents of their decision-making responsibilities [6].

Pediatric heart transplantation is characterized by acutely or chronically ill patients, increasingly complex technological interventions, and families with understandably high expectations of care. There is a tremendous diversity of values and beliefs both within and across families that influence the decisions that families make for their child. Indeed, there is a body of evidence that demonstrates how pediatric organ transplantation impacts the family in profound ways [7]. For example, studies have identified elevated parental stress, financial burden, caregiver burnout, and overall increased family stress following transplantation and continued for several months thereafter [8–10].

The path to end-stage organ failure is also variable for families; decisions and the values upon which these decisions are based can and will change over time. Sometimes changed decisions or values will be in relation to a change in medical

[2]Consent and capacity legislation differs depending on jurisdiction. For example, in the province of Ontario, Canada, there is no age of consent and capacity to consent to treatment is determined by specific criteria. In other jurisdictions, there is an age of consent (e.g., 18 years) and legal guardians of the child make treatment decisions for the child up until the legal age of majority.

situation or circumstance—for example, a new stroke in a child with an already uncertain neurologic prognosis that leads the family to reconsider their decision-making around listing for transplantation. Sometimes they will be more nebulous and multifactorial, but this variability needs to be appreciated, acknowledged, accepted and addressed by the healthcare team throughout the transplantation process.

Pediatric organ transplantation is further situated in the context of interprofessional healthcare teams that are highly motivated to succeed and who generally define success as graft and patient survival. This is in the context of a drive towards innovation and advancing knowledge, especially in academic health care centres. This gives rise to multiple duties which can often be in tension—the duty to alleviate suffering, the duty to respect the child's emerging autonomy, and the duty to act in the child's best interest [11]. The right balance between these duties is different for every child, and challenges may ensue when there are additional tensions between the clinician's primary duty—the patient—and the goals of family centered care. What parents believe is in their child's best interest may differ from the perspective of the clinician, especially where survival projections are at either end of the spectrum (very high or very low) and/or intervention or lack thereof would result in significant pain and suffering. Such differences may impact how the child's best interest is defined. How such discussions proceed and the decisions evolve within the context of pediatric heart transplantation is explored further below.

14.3 Uncertainty

Uncertainty is the state of being when something is not known or definite [12]. Uncertainty itself is neither good nor bad; yet, uncertainty often characterizes situations in which moral dilemmas arise. How uncertainty is managed is important ethically when the well-being of oneself or the best interests of the child is at stake.

14.3.1 Medical Forms of Uncertainty

The great Canadian physician Sir William Osler mused that medicine is a "science of uncertainty and an art of probability" [13, p. 129]. Indeed, medical decisions always involve some degree of uncertainty [14]. When medical forms of uncertainty arise, they tend to be expressed in terms of the unpredictable nature of treatment or the limits in medical knowledge [15]. Complexities of diagnosis, doubt regarding the veracity of some symptoms, limitations to evidence-based treatments, unclear courses of action, and challenges in communication all affect uncertainty in the clinical setting [14]. Chronic illness may produce uncertainty by disrupting daily routines leading to less predictability and therefore more uncertainty in everyday encounters and in overall well-being [16]. For example, in the context of heart failure, a real risk of sudden deterioration or sudden death exists. How that uncertainty is perceived and managed by individuals is subjective, as it depends on the individual's particular social context and lived experiences [17]. As we discuss in later sections, uncertainty is associated with notions of risk and therefore is central to the risk-benefit calculus and discussions about the child's best interest.

The literature on medical uncertainty and transplant is primarily focused on adult patients. Although evidence and practice from adult healthcare cannot always be applied to pediatric healthcare settings, situations of uncertainty fall all along the age spectrum. Furthermore, treatment and goals of care decision-making for both pediatric and adult transplantation is situated in a context where the demand for organs far exceeds the supply. This situation of absolute scarcity raises ethical questions about who should receive the organ, a scarce resource, and by what criteria. Although we do not address ethical issues associated with organ allocation in this chapter, we recognize the importance of macro-level factors and how they may shape notions of uncertainty and risk.

Martin et al. [18] explored notions of medical, personal, and social forms of uncertainty among adult transplant patients. Prior to transplantation, adult patients reported uncertainty with respect to many factors, including insufficient information about diagnosis, complex decisions about transplantation, unknown or unknowable organ availability, unclear expectations about medical procedures and outcomes, ambiguous financial consequences, and unclear implications on interpersonal relationships. Post-transplantation, uncertainty was related to complex medication regimens, unpredictable future health/prognosis, complex role and identity challenges, unclear financial consequences, possible stigmatizing reactions, unclear relational implications, and complex interactions with the deceased donor's family. Brown et al. [19] used a phenomenological approach to explore what it is like for patients who are awaiting a liver transplant. The authors found that the waiting period—which is burdened with uncertainty and associated with high rates of morbidity and mortality—is, among many factors, a time in which one both prepares for life and prepares for death. A grounded theory study by Scott et al. [20] indicated that social support helps transplant patients manage feelings of uncertainty, as the support provides relational stability during a very challenging time.

Qualitative research on pediatric treatment decision-making revealed that uncertainty is a common experience for parents [21]. Children with end-stage heart failure and their families who care for them live with uncertainty on a daily basis. Waiting for a heart transplant is a time of considerable anxiety and uncertainty for many pediatric patients and families. They face changes related to the child's physical functioning, medical and social needs, and daily routines. Families also deal with increasing financial issues while simultaneously trying to live a 'normal' life [22]. Although the transplant may provide hope for the future for children and families, it is also an experience that is described as uncertain because of the health risks and unclear long-term prognosis for the child [23].

Situations characterized by uncertainty may raise questions for pediatric patients and families about clinicians' abilities to promote the welfare of the child. This is especially true as transplant can only manage but not cure the disease. When making transplant-related treatment decisions, patients, families, and the health care team, must determine how far both the disease and the treatment options, including transplantation, compromise quality of life for the patient. In this way, informed judgments can be made about whether or not treatment is medically appropriate, and, critically, which might be the best option for the child. However, the 'best option' for the child might be difficult to identify when all options are sub-optimal, given the inherent likelihood of risks and potential benefits of each option. The impact that transplantation has on quality of life will vary by organ, pre-existing morbidities, and the ability of the child's family to mobilize and access resources necessary for recovery [11]. A typical example in a child with complex congenital heart disease is making a decision between (a) being listed for a transplant with the uncertainty of dying whilst waiting, the risk related to the transplant procedure itself, and uncertain long term outcomes post-transplant with an anatomically normal heart, and, (b) a staged surgical procedure with an upfront risk, outcomes related to the success of that surgical procedure, and subsequent procedures on a heart with remaining anatomical abnormalities. Equipoise between two pathways from a medical perspective can accentuate uncertainty from the patient and family perspective.

Uncertainty also contextualizes the transition from hospital to the community. Families may be concerned with the unknowns related to their ability to adhere with post-operatives plans and the impact the child's post-transplant condition will have on the family unit [24, 25]. Similarly, the transition to adult care may be a source of anxiety and uncertainty for patients and families as they often have to contend with a new healthcare team, new institutions, and diminished access to financial resources that are reserved for pediatric patients [26]. All told, these interconnected factors that are specific to the pediatric transplant context makes it challenging for the healthcare team, patients, and families to manage various forms of uncertainty.

14.4 Risk

The concept of risk has been studied from multiple perspectives, such as psychology, economics, sociology, and ethics. Risk is a concept with multiple definitions and meanings. Risks can be optionally accepted by, or imposed on, an individual [27]. In making decisions under risk, the outcome of a possible course of action may not be known. Rather, there are a range of possible outcomes based on known probabilities [28]. Accordingly, risk involves aspects of uncertainty as well. We consider risk as a "negatively-valenced likelihood assessment that an unfavorable event will occur" [29, p. 981]. Risk perceptions—subjective judgments individuals make about the severity of a risk—are central to theories of human health and behaviour.

14.4.1 Medical Forms of Risk

Risk is involved in almost every course of action or inaction in healthcare. For example, bruising is a risk associated with a needle injection for a vaccine; however, one increases their risk of developing measles—and infecting others—if they do not receive the measles vaccine. When patients, families, and clinicians make shared decisions based on unclear outcomes, the benefits of acting or not acting are weighed against their likely risks and benefits [30]. Some clinicians may discuss risk factors in the statistical sense, such as the probability of surviving without transplant versus mortality from surgery. However, in the setting of "high risk" heart transplantation, even risk factors may not be able to be quantified, can be theoretical, or can be based solely upon expert opinion or an educated "guess". The institution of newer mechanical support devices as advanced forms of life support is an example where a high risk procedure may never have been performed before, is discussed amongst a group of experts as a technical possibility that would "probably" work, but have no evidence base to provide risk factors or statistical estimates of probability of success.

Like uncertainty, judgments about risk also have a psychosocial dimension, in terms of how risk is perceived and understood by patients. For example, patients and families, who, for religious reasons believe in the sanctity of life and wish to prolong life at all costs, may be more accepting of higher risk interventions in order to achieve this outcome. Conversely, a familial underestimation of risk due to a belief that a child is going to "be cured" is a different framework within to think about decision-making and risk assessment. In this context, moral distress can arise within the medical team as a result of a concern that a family is making a decision not to embark upon a treatment path, for example transplantation, for reasons other than an understanding of the elements of risk and medical uncertainty. Risk assessment combines both objective and subjective information.

Framing effects also influence risk perceptions. Framing effects relate to the presentation of two logically equivalent situations where one is presented in positive terms and the other in negative terms [31]. For example, if an intervention (e.g., surgery) is framed positively (e.g., 90 % probability of survival) there is greater compliance with the recommendation than if framed negatively (e.g., 10 % probability of mortality) [32]. Regardless of the numerical ratio of the risk-benefit ratio, within the context of any given child-family-healthcare team triad is differing thresholds or tolerances for accepting risk. Communication of the risk by the healthcare team and actual understanding and appreciation of the risk by the patient and/or family is often discrepant—not due to lack of effort on either side, but in relation to the hierarchical nature of the clinician-family relationship as discussed below.

Bioethicists and clinicians may discuss risk from the perspective of a risk-benefit ratio, where the potential risks, benefits, and likely consequences of the intervention must be informed and understood, and the likely benefits of the intervention may or may not outweigh the potential risks for the patient. This calculus is based on the presumption of capacity, and the same criteria are applied to the legal substitute decision-maker. Yet, with so much uncertainty and risk associated with pediatric heart transplantation, it is worthwhile questioning whether parents or other legal decision-makers are making informed decisions as defined. There has been some effort to answer this question empirically. For example, Dellon et al. [8] examined the decision-making process for lung transplantation in cystic fibrosis from the perspective of caregivers. The authors found that many patients and caregivers were not well informed about transplantation, including alternatives to transplantation. Given the strong desire in this population to prolong life, issues of risk were not well addressed and thus patients were generally not well informed. However, from a medical perspective, it is challenging to fully assess the level of understanding of families and patients and their appreciation of risk; too often, the decision revolves around the slim "hope" (or "probability" from the medical perspective) of survival. For example, even when openly quoted a very low probability of survival (e.g. <10 %), some families will cling to that 10 % despite the objective risks outlined by the medical team, leading one to question the true level of understanding of the information and whether this constitutes true informed consent. The nature of question of the existence of true informed consent becomes most apparent in situations of high risk with a negative outcome (e.g. death) wherein the family behavior subsequently points to a lack of appreciation of the outcome as a possibility (in spite of whatever extensive discussion process preceded the decision). The opposite scenario, though less distressing, is in situations where there is a good outcome in a very high risk situation; lack of appreciation of the "medical miracle" can contribute to unrealistically high expectations in the future and impact future decision-making and thresholds for risk, both for that individual patient and for those exposed to the outcome—a thought-provoking concept.

Risk factor analyses help to facilitate the identification of high-risk patients, patient characteristics and patient groups from a medical risk perspective [33, 34]. Success needs to be considered in different terms than in a standard (i.e., not high-risk) risk group. Some decisions become straightforward based on risk factor analyses. For example, a specific patient characteristic (e.g., renal failure in infants) that heart transplantation would not be recommended, and the main discussions and decisions to make would be oriented toward end-of-life care. Indeed, clinicians are not legally or ethically obligated to inform patients or their substitute decision-makers of treatment options that the medical team considers to be clinically futile. In the context of pediatric solid organ transplantation and end-stage organ failure, the risks of transplant are substantial and prognosis tends to be guarded. For example, pediatric lung transplantation remains challenging with 5-year survival reported at 51 %; heart-lung even lower at <35 % [35]. Accordingly, additional considerations include allocating a scarce resource to a patient who is not likely to benefit in the short or long-term from the transplant, versus allocating the organ to a patient who is likely to have better outcomes. Yet, clinical futility alone, in some situations, may not be enough for a clinician to justifiably withhold transplant as a potential treatment option: patients and families may hold a broader notion of "futility" in which any benefit—physiological or psychological and no matter how statistically remote or the burdens that may be incurred—is worth pursuing. Accordingly, patient and family assessments of quality of life, best interests, and pain and suffering must be considered in the decision-making process.

There are many pre-transplant and post-transplant risk factors that may influence outcomes. These include medical risk factors such as renal function, infections, respiratory problems, amongst others, in addition to non-medical factors such as race, sex, and socioeconomic status. Knowledge of risk factors that allow for the prediction of these outcomes may inform the decision-making process, for example, renal failure and its significant negative impact on post-heart transplant outcomes. These factors are multiple and they may in some circumstances be hard to predict how they influence outcomes, for example, socioeconomic status or race. Every factor contributes toward a pediatric patient's risk of transplantation and is identified and shared in organ-specific publications, allowing for translation to practice in centres worldwide [36]. While strategies have been proposed for clinicians to help minimize post-transplant risks in their pediatric patients [37], how clinicians communicate the risks to patients and families is an ongoing challenge. As Lunde [38, p. 86] stated, "[i]t is sometimes difficult to explain the risks of the procedure to a person who has no concept of a phenomenon such as tissue rejection".

Risk as it is portrayed above relates to medical factors—some of which are modifiable and some of which are not. There are identifiable risk factors on the part of the patient and family behaviour that impact the therapeutic relationship and the dynamics around the decision-making process. Adherence to treatment is a challenge in pediatric populations, given the life-threatening consequences of failing to take medications, and several risk factors for non-adherence have been identified. A study of risk factors in pediatric liver transplant recipients found that a history of substance abuse, child abuse, not having two parents at home, having received public assistance, having been diagnosed with a psychiatric disorder, and history of school dropout, were all predictive of non-adherence [39]. Thus again, the family and social context is overlaid on the medical issues and intertwined in the trust issues as discussed below.

14.5 Trust

Trust is necessary when uncertainty exists and the tolerance for risk is low. Many different conceptualizations of trust are available. Trust has been defined as an affective or emotional attitude that depends on a person's good will and motivations toward another, with the belief that a trustee is competent and will follow through with an expected action or predicted set of behaviours significant to a trustor [40, 41]. Norms of trust suggest that the trustor will give discretionary powers to the trustee because of the expectation that the relationship will be of benefit. Accordingly, granting discretionary powers involves some degree of risk, as the trustor becomes vulnerable to the trustee [42].

Trusting is not always immediate and conscious and trust may be placed without critical reflection. Past experiences of placing trust may inform the individual whether or not placing trust again in similar contexts will be advantageous. Factors such as gender, age, social class, and ethnicity are also relevant in deciding whether to place trust [43]. In the setting of a rare occurrence such as high risk pediatric heart transplant, an element of trust may be imposed or forced upon families due to a limited number of medical team members with the required knowledge and expertise; adding yet another layer to the trust dynamics.

Parents or adult guardians, family members, and others are influential in crafting the moral attitudes of their children, and shaping a child's trusting and distrusting attitudes. In an ideal situation, adults who raise children are faced with a dilemma: on the one hand, they desire to raise the child in a climate characterized by good will, respect, and mutual trust, but on the other hand, they need to teach the child to be cautious about the world, the intentions of others, and how to place trust wisely. Distrusting some individuals may safeguard against harms rather than cause them [44].

14.5.1 Trust and Healthcare

Clinicians have historically enjoyed a high level of patient and public trust, owing to society's perception of medicine as a valued profession [45, 46]. Organ donation and transplantation could not exist without public support and trust in healthcare systems and in transplant programs. Indeed, the provision of healthcare is grounded in trust and the trustworthiness of its professionals and institutions, and yet trust may be taken for granted in therapeutic relationships. Healthcare has historically operated from a paternalistic model of the therapeutic relationship, where the clinician, practicing chiefly from the duty of beneficence, holds the power in health care decision-making. As medical experts, clinicians are traditionally thought to know what is best for the patient. However, following the civil rights movement in the U.S. and the evolution of medical jurisprudence, the latter half of the twentieth century saw the rise of a consumer choice model of the therapeutic relationship. This trend privileged individual autonomy, choice, and voluntary informed consent. The concept of "patient-centered care" emerged as the desired ethical ideal, and patients became viewed as active participants in the direction of their care rather than passive recipients [47]. Since parent- and family-involvement in treatment decision-making was also historically rare, the concept of "family-centered care" emerged from the pediatric context, where both the child's and family's information and perspectives came to be considered necessary to construct appropriate care plans [5].

The nature of the patient-family-clinician triad involves disparities in knowledge, vulnerabilities, and power. Pediatric patients, by virtue of their age, limited life experience, and in some cases decisional capacity, are in an inherently disadvantaged position, and are made additionally vulnerable by critical illness. Clinicians are in positions of power as they are sought based on their expert knowledge, skills, and ability to address the specific health concern [48]. Clinicians have obtained qualifications giving them socially sanctioned expert status, and expert status implies technical competence [49]. Since clinicians are expected to abide by certain professional norms and ethical codes of conduct, expert status also implies *prima facie*, professional trustworthiness. Most pediatric patients and families do not have the necessary knowledge to assess the technical competence of clinicians, or have the ability to judge the various medical instruments, medications, diagnostic tests, or interventions recommended by clinicians [48]. With various norms and codes of conduct in place, patients and families are likely to believe what their clinician says because they trust the source of information [48, 50].

In therapeutic relationships, patients and families may unreflectively place their trust within the discretionary power of the clinician [42]. In pediatrics, this typically occurs because the expert clinician is caring for the child who is sick. While there is usually shared identification of the good between the clinician, patient, and the family—that is, the pediatric patient's health and quality of life—it is the relational asymmetries between clinician and patient and family that can make the patient and family vulnerable to the discretionary power of the clinician [42, 48].

The empirical literature suggests that the benefits of patient-clinician interactions are greater when patient and clinician trust each other [51]. Trust is central to an effective therapeutic relationship [52], and is an important quality indicator of the therapeutic relationship [53]. Trust correlates with patients' satisfaction in clinicians [54], adherence to treatment, loyalty and continuity with the clinician [55], and it increases the likelihood of accessing health care, disclosing information, and overall health and well-being [56]. Indeed, therapeutic trust tends to be stronger in longer-term clinical relationships than shorter ones [56, 57].

From the parent's perspective in pediatric settings, trust is often defined as a parent's belief that the clinician, or healthcare team, is acting in the best interest of the child—that is, they act competently with good will in order to benefit the child. Although empirical research focusing on trust in pediatric settings is limited [58], the available evidence indicates that parents' trust of clinicians is associated with continuity of care, engendered by a partnership-building communication style [59, 60]. In a study of non-adherence in children and adolescents with end-stage renal disease, one-third of participants attributed their non-adherence to a loss of trust in their physician [61]. A study by Bricher [62] found that nurses use a range of strategies in order to develop trusting relationships with their pediatric patients, but found trusting relationships are difficult to maintain as nurses are often involved in administering procedures that the pediatric patients may consider painful or frightening.

With the lack of certainty that is inherent in end-stage cardiac disease, patients and families who have less medical knowledge than professionals may have or feel the need to adopt higher levels of trust with clinicians [56]. Similarly, in situations where quick and critical decisions have to be made, parents may feel the need to place high levels of trust in the healthcare team as the opportunity to seek second opinions or education about the course of treatment is minimal [63]. Nonetheless, if a patient or family places trust in a clinician, without some degree of caution or assurance of competence or good will of the clinician, the patient and perhaps also family well-being can be threatened [56]. The ways in which patients and families place trust in clinicians might affect their assessment of the possibilities of medical error or the benefits, risks, and limitations of modern medical science and technology, and render them more or less vulnerable to the possibility of iatrogenic harm [64].

14.6 Opportunities to Cultivate and Maintain Relations of Trust

Pediatric patients may be initially distrusting of others when someone unfamiliar, like a physician or nurse, wants to do something unfamiliar to them, which past experience suggest might be painful or uncomfortable [65]. The same thinking can be applied to families and their role in treatment decision-making for the child. This initial lack of trust may impact treatment adherence and health outcomes.

We argue that healthcare teams can work with patients and families and manage situations characterized by high-risk and uncertainty by striving toward a disposition of epistemic humility. In healthcare, epistemic humility is an approach that calls for partnership and dialogue between pediatric patient, family, and clinician in a deliberative decision-making process [48, 50]. Being epistemically humble means recognizing that medical judgments are always accompanied by uncertainty [66]. Appreciation of such uncertainty requires clinicians to adopt a disposition, or an attitude, that allows room for balancing clinical information, professional judgment, and pediatric patient and family perspectives, including those about the child's best interest. Epistemic humility is intentionally collaborative, as "both the HCP [healthcare provider] and patient are counting on each other in investigating a full picture of the patient's experience and determining the most appropriate management strategies" [48, p. 117]. This, in turn, may cultivate trust and support family-centered decision-making. We argue that the responsibility for epistemic humility and building trust falls more heavily on the clinician than on the pediatric patient and family given the inherent power hierarchy in the therapeutic relationship. However, we also contend that pediatric patients and families, as participants in a relationship, also have responsibilities to the extent feasible, as in post-operative treatment adherence. Family members who aim for a disposition of epistemic humility have a responsibility to demonstrate respect for the clinician's knowledge, virtue, and moral agency, and establish that they trust the clinician's intent to act in the best interest of the child. However, a full account of patient and family responsibilities is beyond the scope of this chapter.

14.6.1 Cultivating and Maintaining Trust Through Epistemic Humility

The literature suggests that parents tend to prefer collaborative or shared decision-making, because they trust and have confidence in the clinician with whom a therapeutic relationship is established [67]. The point about shared accountability for extremely difficult decisions is illustrated by this quote from a mother in a study about parental role in pediatric heart transplant decision-making:

> I need advice. I have talked with so many people I'm so confused; I don't know whom to trust. The only person I really trust is my original cardiologist at the pediatric cardiology center. Could he help me make this decision? *Can't we make the decision together?* I am so afraid if I don't make the right choice and the baby dies that I will have murdered him [68, p. 335].

While clinicians have vast medical knowledge, they do not have direct access to their patients' and families' everyday experiences—especially their lived experiences of risk and uncertainty. In search of the most appropriate clinical approach and to support family-centered decision-making, epistemic humility requires an inquiry into the patient and family experience. For example, the clinician can explore the extent to which a family member—namely, the substitute decision-maker—wants to have primary decision-making responsibilities. Then the clinician can provide the medical information, and invite the patient and family to tell their experiences of the illness, describe their personal values relevant to the decision, and discuss their perspective of what is in the best interest of the child. The child and family may also provide information that helps the clinician understand how risks associated with the transplant are contextualized *for them.* The child—if capable and to the extent possible—and family are then invited to ask questions. This may help resolve any outstanding uncertainties associated with the decision, or provide space to openly discuss them. The clinician will then make her recommendation of evidence-based options after synthesizing all of the relevant information. The clinician, by doing so, demonstrates a willingness to consider patient and family values, preferences, and discussion about best interests into her recommendations [69]. This may help demonstrate trustworthiness, build, and maintain trust relations. In addition, it provides a foundation for future decision-making as clinical scenarios requiring high risk heart transplantation rarely remain stable or improve; the general trajectory is deterioration, often leading to situations requiring further difficult decision-making. An example would be deterioration whilst awaiting a donor organ; knowing the values relevant to the prior decision-making is key to guiding subsequent decision-making when clinical changes occur. For example, if the key value is survival, then likely there will not be a change in decision in the absence of a truly catastrophic clinical change; but if the key value is quality of life and independence, in the setting of a major neurologic setback a family may choose to pursue compassionate end-of-life care.

What if the values of the patient and the family diverge? The most common situation for this to arise is in the adolescent age group wherein the patient has attained a level of understanding and involvement in their own self-care and decision-making that they are able to advocate for their own preferences. The goal remains collaborative or shared decision-making with an added layer of complexity which can be highly emotionally charged for all parties involved when there is discord in opinions within the family circle itself. A discussion around the options open to care teams to facilitate and/or manage these circumstances is beyond the scope of this chapter, but often require the involvement of additional support and resources for all parties involved—especially the adolescent patient facing a life-altering decision.

Epistemic humility does not require a rejection of clinicians' clinical expertise. Rather, it recognizes the limits of clinical knowledge and expertise in determining the most appropriate care plans, especially where there are high levels of risk, uncertainty, lack of professional agreement, or incomplete knowledge about the pediatric patient and their family's social situation. In a study of physician communication with parents of children with hypoplastic left heart syndrome, Kon et al. [63] suggest that physicians should consider the limits of their expertise in a manner consistent with epistemic humility. They argue that if there are genuine disagreements among specialists and a lack of professional consensus surrounding outcome data, this should be disclosed to parents in the spirit of informed consent. Epistemic humility is a commitment to continuous responsiveness to the patient and family's experience and recognition of the uncertainties and limitations of applying clinical knowledge and expertise to decision-making [48, 70].

Medical knowledge is a socially privileged kind of knowledge. While medical information today has proliferated on the Internet, and increased families' awareness of health issues and treatment options, most patients and families do not have the specialized clinical knowledge and skills to discern between good and bad evidence, or apply volumes of online information to their specific situation. The necessity of such expertise and skill places clinicians in an epistemically privileged position to decide how care ought to proceed [48, 50]. Nonetheless, pediatric patients and their families have lived with the risks and uncertainty and everyday challenges of the child's health condition for variable lengths of time—sometimes for years—and may have extensive experiences with multiple therapies, interventions, and have developed coping strategies. An approach grounded in epistemic humility can encourage mutual respect, but also generate mechanisms that align clinical expertise with the child's best interests and so foster, and sustain, trust.

14.7 Conclusion

Pediatric heart transplantation occurs in a context characterized by high levels of risk and uncertainty, which necessitates a strong family-centered trust relationship with clinicians. Pediatric patients and their families are burdened with multiple difficult decisions, both medical and moral, and require support as they go through the process of decision-making. The challenge for all parties is to determine what is best from a medical, physical, emotional, developmental, and psychosocial perspective, in order to provide a child who is a high-risk transplant candidate not only with an option for survival, but also with an option for an acceptable quality of life within their life circumstances, with or without transplantation. This can be achieved through a commitment to epistemic humility and a strong, family-centered trust relationship.

References

1. Hippocrates. In: Adam F, editor. The genuine works of Hippocrates. London: The Syndenham Society; 1869.
2. Wittgenstein L. On certainty. Anscombe GEM and von Wright, GH, editors. London: Blackwell; 1969.
3. Kopelman LM. The best-interests standard as threshold, ideal, and standard of reasonableness. J Med Philos. 1997;22:271–89.
4. Lindemann H. Why families matter. Pediatrics. 2014;134(Suppl 2):S97–103.
5. American Academy of Pediatrics. Patient- and family-centered care and the pediatrician's role. Pediatrics. 2012;129(2):394–404.
6. Pager CK. Dying of a broken heart: ethics and law in a case of hypoplastic left heart syndrome. J Perinatol. 2000;20:535–9.
7. Gold LM, Kirkpatrick BS, Fricker FJ, Zitelli BJ. Psychosocial issues in pediatric organ transplantation: the parents' perspective. Pediatrics. 1986;77(5):738–44.
8. Dellon EP, Shores MD, Nelson KI, Wolfe J, Noah TL, Hanson LC. Caregivers' perspectives on decision making about lung transplantation in cystic fibrosis. Prog Transplant. 2009;19 (4):318–25.
9. Lang L, Duff AJA, Brownlee KH. Introducing the need for lung transplantation in children with cystic fibrosis: parental experiences. J Cyst Fibros. 2005;4:259–62.
10. Rodrigue JR, MacNaughton K, Hoffmann RG III, Graham-Pole J, Andres JM, Novak DA, Fennell RS. Transplantation in children: a longitudinal assessment of mothers' stress, coping, and perceptions of family functioning. Psychosomatics. 1997;38(5):478–86.
11. Dipchand AI. Decision-making in the face of end-stage organ failure: high-risk transplantation and end-of-life care. Curr Opin Organ Tran. 2012;17:520–4.
12. Oxford English Dictionary. [Internet]. Uncertain. Oxford: Oxford University Press; 2014 [cited 2014 September 28]. Available from: http://www.oxforddictionaries.com/definition/english/uncertain#uncertain__4.
13. Bean RB, Bean WB. Sir William Osler: aphorisms from his bedside teachings and writings. New York: H. Schuman; 1950.
14. Fox RC. Medical uncertainty revisited. In: Scrimshaw SC, Albrecht GL, Fitzpatrick R, editors. The handbook of social studies in health and medicine. London: Sage Publications; 2000. p. 409–25.
15. Buetow S. The virtue of uncertainty in health care. J Eval Clin Pract. 2011;17(5):873–6.
16. Mishel MH. Reconceptualization of the uncertainty in illness theory. J Nurs Sch. 1990;22:256–62.
17. Stone AM, Scott AM, Martin SC, Brashers DE. Using information to manage uncertainty during organ transplantation. Qual Commun Res. 2013;2(1):42–60.
18. Martin SC, Stone AM, Scott AM, Brashers DE. Medical, personal, and social forms of uncertainty across the transplantation trajectory. Qual Health Res. 2010;20:182–96.
19. Brown J, Sorrell JH, McClaren J, Creswell JW. Waiting for a liver transplant. Qual Health Res. 2006;16:119–36.

20. Scott AM, Martin SC, Stone AM, Brashers DE. Managing multiple goals in supportive interactions: using a normative theoretical approach to explain social support as uncertainty management for organ transplant patients. Health Commun. 2011;26(5):393–403.
21. Tong A, Lowe A, Sainsbury P, Craig JC. Experiences of parents who have children with chronic kidney disease: a systematic review of qualitative studies. Pediatrics. 2008;121 (2):349–60.
22. Anthony SJ, Annunziato RA, Fairey E, Kelly VL, So S, Wray J. Waiting for transplant: physical, psychosocial, and nutritional status considerations for pediatric candidates and implications for care. Pediatr Transplant; in press.
23. Stubblefield C, Murray RLE. Parents' perceptions of their children's lung transplant experiences. J Fam Nurs. 1998;4(4):367–86.
24. Lerret SM, Weiss ME. How ready are they? Parents of pediatric solid organ transplant recipients and the transition from hospital to home following transplant. Pediatr Transplant. 2011;15(6):606–16.
25. Stubblefield C, Murrary RLE. Making the transition: pediatric lung transplantation. J Pediatr Health Car. 2000;14(6):280–7.
26. LaRosa C, Glah C, Baluarte HJ, Meyers KE. Solid-organ transplantation in childhood: transitioning to adult health care. Pediatrics. 2011;127(4):742–53.
27. Bora A. Risk, risk society, risk behaviour, and social problems. In: Ritzer G, editor. The Blackwell encyclopedia of sociology, vol. VII. Oxford: Blackwell; 2007. p. 3926–32.
28. Altham JEJ. Ethics of risk. P Aristotelian Soc. 1983; 84: 15–29.
29. Menon G, Raghubir P, Agrawal P. Health risk perceptions and consumer psychology. In: Haugtvedt C, Herr P, Kardes F, editors. The handbook of consumer psychology. New York: Psychology Press; 2008. p. 981–1010.
30. Tannert C, Elvers H-D, Jandrig B. The ethics of uncertainty: in light of possible dangers, research becomes a moral duty. EMBO Rep. 2007;8(10):892–6.
31. Edwards A, Elwyn G, Mulley A. Explaining risks: turning numerical data into meaningful pictures. Brit Med J. 2002;324:827–30.
32. Garcia-Retamero R, Galesic M. How to reduce the effect of framing on messages about health. J Gen Intern Med. 2010;25:1323–9.
33. Dipchand AI, Kirk R, Mahle WT, Tresler MA, Naftel DC, Pahl E, et al. Ten years of pediatric heart transplantation: a report from the pediatric heart transplant study. Pediatr Transplant. 2013;17(2):99–111.
34. Jeewa A, Manlhiot C, Kantor PF, Mital S, McCrindle BW, Dipchand AI. Risk factors for mortality or delisting of patients from the pediatric heart transplant waitlist. J Thorac Cardiov Sur. 2014;147(1):462–8.
35. Benden C, Edwards LB, Kucheryavaya AY, Christie JD, Dipchand AI, Dobbels F, Kirk R, Lund L, Rahmel AO, Yusen RD, Stehlik J. The registry of the international society for heart and lung transplantation: sixteenth official pediatric lung and heart-lung transplantation report—2013; Focus theme: age. J Heart Lung Transplant. 2013;32(10):989–97.
36. Dipchand AI, Kirk R, Edwards LB, Kucheryavaya AY, Benden C, Christie JD. For the international society for heart and lung transplantation. The registry of the international society for heart and lung transplantation: sixteenth official pediatric heart transplantation report—2013; Focus theme: age. J Heart Lung Transpl. 2013; 32(10): 979–988.
37. Allen UD. Minimizing infection risk after pediatric organ transplants: advice for practitioners. Paediatr Child Health. 2013;18(3):143–8.
38. Lunde DT. Psychiatric complications of heart transplants. Am J Psychiatr. 1969;126(3):363–73.
39. Lurie S, Shemesh E, Sheiner PA, Emre S, Tindle HL, Melchionna L, Shneider BL. Non-adherence in pediatric liver transplant receipients—an assessment of risk factors and natural history. Pediatr Transplant. 2000;4(3):200–6.
40. Baier A. Trust and antitrust. Ethics. 1986;96(2):231–60.

41. Jones K. Trust as an affective attitude. Ethics. 1996;107:4–25.
42. Rogers W, Ballantyne A. Gender and trust in medicine: vulnerabilities, abuses, and remedies. Intl J Fem Approaches Bioeth. 2009;1(1):48–66.
43. Ward P, Meyer S. Trust, social quality and wellbeing: a sociological exegesis. Dev Soc. 2009;38(2):339–63.
44. Hardin RL. Distrust: manifestations and management. In: Hardin RL, editor. Distrust. New York: Russell Sage Foundation; 2004. p. 3–33.
45. De Vries RG, Kim SY. Bioethics and the sociology of trust: introduction to the theme. Med Health Care Philos. 2008;11(4):377–9.
46. O'Neill O. Autonomy and trust in bioethics. Cambridge: Cambridge University Press; 2002.
47. Jonsen AR. The birth of bioethics. New York: Oxford University Press; 1998.
48. Ho A. Trusting experts and epistemic humility in disability. Intl J Fem Approaches Bioeth. 2011;4(2):102–23.
49. Goldman AI. Experts: which ones should you trust? Philos Phenom Res. 2001;63(1):85–110.
50. Ho A. "They just don't get it!" when family disagrees with expert opinion. J Med Ethics. 2009;35(8):497–501.
51. Ridd M, Shaw A, Lewis G, Salisbury C. The patient-doctor relationship: a synthesis of the qualitative literature on patients' perspectives. Brit J Gen Pract. 2009;59(561):116–33.
52. Dibben M, Lena M. Achieving compliance in chronic illness management: illustrations of trust relationships between physicians and nutrition clinic patients. Health Risk Soc. 2003;5:241–59.
53. Safran DG, Taira DA, Rogers WH, Kosinski M, Ware JE Jr, Tarlov AR. Linking primary care performance to outcomes of care. J Fam Practice. 1998;47:213–20.
54. Thom D, Ribisi K. Further validation and reliability testing of the trust in physician scale. Med Care. 1999;37:510–7.
55. Roberge D, Beaulieu M-D, Haddad S, Lebeau R, Pineault R. Loyalty to the regular care provider: patients' and physicians' views. Fam Pract. 2001;18(1):53–9.
56. Calnan M, Rowe R. Trust and health care. Soc Compass. 2007;1(1):283–308.
57. Kao AC, Green DC, Davis NA, Koplan JP, Cleary PD. Patients' trust in their physicians: effects of choice, continuity, and payment method. J Gen Intern Med. 1998;13:681–6.
58. Moseley KL, Clark SJ, Gebremariam A, et al. Parents' trust in their child's physician: using an adapted Trust in Physician Scale. Ambul Pediatr. 2006;6:58–61.
59. DeLemos D, Chen M, Romer A, Brydon K, Kastner K, Anthony B, Hoehn KS. Building trust through communication in the intensive care unit: HICCC. Pediatr Crit Care Med. 2010;11 (3):378–84.
60. Horn IB, Mitchell SJ, Wang J, Joseph JG, Wissow LS. African-American parents' trust in their child's primary care provider. Acad Pediatr. 2012;12(5):399–404.
61. Griffin KJ, Elkin TD. Non-adherence in pediatric transplantation: a review of the existing literature. Pediatr Transplant. 2001;5(4):246–9.
62. Bricher G. Paediatric nurses, children and the development of trust. J Clin Nurse. 1999;8:451–8.
63. Kon AA, Ackerson L, Lo B. How pediatricians counsel parents when no "best-choice" management exists: lessons to be learned from hypoplastic left heart syndrome. Archives Pediatr Adolesc Med. 2004;158:436–41.
64. Entwistle VA, Quick O. Trust in the context of patient safety problems. J Health Organ Manag. 2005;20L:397–416.
65. Peterson MR. At personal risk: boundary violations in professional-client relationships. New York: Norton; 1992.
66. Schwab AP. Epistemic humility and medical practice: translating epistemic categories into ethical obligations. J Med Philos. 2012;37:28–48.
67. Lipstein EA, Brinkman WB, Britto MT. What is known about parents' treatment decisions? A narrative review of pediatric decision making. Med Decis Making. 2012;32(2):246–58.

68. Higgens SS. Parental role in decision making about pediatric cardiac transplantation: familial and ethical considerations. J Pediatr Nurs. 2001;16(5):332–7.
69. Atkins K. Autonomy and the subjective character of experience. J Appl Philos. 2000;17 (1):71–9.
70. Upshur REG, Colak E. Argumentation and evidence. Theor Med Bioeth. 2003;24(4):283–99.

Chapter 15
Ethical Considerations in the Psychosocial Evaluation of Pediatric Organ Transplant Candidates, Recipients and Their Families

Sabrina F. Derrington, Aviva M. Goldberg and Joel E. Frader

Abstract This chapter will consider psychosocial aspects of pediatric organ transplantation for transplant candidates (those being considered for listing), organ recipients, and their families—including living related donors. We will outline an ethical approach to considering several questions: How ought mental health issues (for patient and caregiver), substance use, and assessment of available psychosocial support structures play into decisions about organ transplant candidacy in pediatrics? How should medical teams address problems with a parent's psychological function or coping skills when the child's transplant candidacy is threatened? What are the psychological sequelae of transplantation for children and adolescents and how do they affect medical decision-making?

Keywords Transplant · Child · Adolescent · Psychosocial · Nonadherence · Support · Risk factors · Substance use · Re-transplantation

S.F. Derrington (✉)
Department of Pediatrics, Division of Critical Care and Program in Palliative Care, Northwestern University Feinberg School of Medicine, Ann and Robert H. Lurie Children's Hospital, Chicago, USA
e-mail: sderrington@luriechildrens.org

A.M. Goldberg
Department of Pediatrics and Child Health, Faculty of Health Sciences, College of Medicine, University of Manitoba, Winnipeg, Canada
e-mail: agoldberg@hsc.mb.ca

J.E. Frader
Department of Pediatrics, Ann and Robert H. Lurie Children's Hospital, Division of Academic General Pediatrics and Primary Care/Program in Palliative Care, Chicago, USA
e-mail: jefrader@luriechildrens.org

R.A. Greenberg et al. (eds.), *Ethical Issues in Pediatric Organ Transplantation*, International Library of Ethics, Law, and the New Medicine 66,
DOI 10.1007/978-3-319-29185-7_15

15.1 Introduction

In the last 50 years, pediatric solid organ transplantation has rapidly matured and now constitutes the standard of care for many types of organ failure. Post-transplant outcomes have significantly improved as organ procurement strategies, surgical techniques, and immunosuppressive regimens have advanced. These developments have made the negative impact of poor adherence to post-transplant regimens even more obvious, since other modifiable causes of graft failure are decreasing in frequency. Efforts to reduce non-adherence have highlighted the importance of psychosocial factors in both the pre- and post-transplant periods [1, 2]. Recognition of the importance of mental and emotional health, social support, education, financial resources, adherence history, and substance abuse—for both patients and primary caregivers—has recently prompted increased attention to the pre-transplant psychosocial evaluation. In the United States, standards set by both the Organ Procurement and Transplantation Network (OPTN) and the United Network for Organ Sharing (UNOS) require such an evaluation but offer little guidance as to how psychosocial risk factors ought to impact patient candidacy for transplantation. While multiple or severe psychosocial concerns in adult transplant candidates sometimes lead to refusal to list, the sparse literature in pediatrics suggests that children, because of their vulnerability and developmental considerations, are and should be treated differently. This topic remains controversial and is plagued by lack of published data.

Life-threatening organ failure, waiting for and then eventually receiving a transplanted organ can have a profound, lasting psychological and emotional impact on a child or adolescent and his or her parent(s) and family [3]. Appreciation of these sequelae is necessary to inform analysis of the ethical dilemmas that may arise in the context of pre- or post-transplant care.

This chapter addresses the range of psychosocial complexities in pediatric solid organ transplant and considers the ethical implications, particularly for pediatric transplant teams. We review recently published recommendations for pre-transplant psychosocial evaluations in children and their families and examine lingering questions about how best to use these evaluations ethically. We discuss the psychological and emotional sequelae of organ transplantation, emphasizing the implications for medical decision-making in both primary and re-transplant scenarios. In addition, we highlight special considerations for psychosocial professionals and medical teams in dealing with transplant patients and families.

15.2 The Pre-transplant Psychosocial Evaluation

Why is evaluation of a patient's psychosocial status so important in the context of organ transplantation? Primarily, this question relates to the ongoing discrepancy between supply and demand for organs, which makes them a societal resource for which allocation must be carefully scrutinized. There is immense pressure to put available organs to the best use possible. While the precise definition of "best use" can be debated, modern allocation strategies attempt to balance equity (individuals with equivalent need for a transplanted organ should have an equal opportunity to receive one) and utility (available organs should be allocated in such a way as to maximize health and survival) [4, 5]. For each patient the goals include optimizing graft survival and function as well as providing individual benefit to the patient's quantity and quality of life. A patient's psychological health and social support system are key to optimizing quality of life and maintaining medical compliance after transplant, which subsequently impacts graft survival and patient morbidity/mortality [1, 6]. Therefore, early identification of psychosocial risk factors and intervention to optimize psychological health post-transplant are important aspects of good stewardship of this limited resource.

For pediatric patients, psychosocial evaluation must involve assessment of both the patient and the parent(s) or other primary caregiver(s). Particularly for younger patients, parents are primarily responsible for treatment adherence post-transplant; the well-being of the child is also heavily dependent on having a supportive family structure [7].

Appendix D of the OPTN Bylaws requires that transplant team personnel include psychosocial "professionals who are designated members of the transplant team and whose primary responsibility is coordinating the psychosocial needs of transplant candidates, recipients, living donors, and their families" [8, p. 69]. The Bylaws describe the responsibilities of these professionals to include conducting a psychosocial evaluation of patients being considered for transplant, but are silent regarding the content of this evaluation and what actions ought to follow from the results. There is likewise no publically accessible data on how individual transplant teams take psychosocial risk factors into account when deciding whether or not to list pediatric patients for transplant. With adult patients, significant variation exists between and within transplant centers as to which psychosocial factors are considered important and what ought to prevent listing for transplant [9, 10]. Levenson found that ongoing drug or alcohol abuse were the only issues widely agreed upon as absolute contraindications to transplant for adult patients, with wide variability for other issues such as social support structure and mental health [10]. Concern has been raised that this variability creates the potential for additional inequity in a system already wrought with ethical challenges [11]. There is also evidence that variability in the way psychosocial vulnerability is assessed may contribute to racial and ethnic disparities in access to organ transplantation [12–14]. Furth and colleagues asked a large sample of adult and pediatric nephrologists to decide whether or not to recommend patients for transplant based on a series of clinical vignettes. They found that the impact of parental education and compliance with therapy on recommendation for transplant varied depending on whether the child in the vignette was black or white. Non-compliance had a stronger negative effect on white patients and education had a stronger positive effect for black patients. Yet even for compliant patients with college-educated parents, white patients were twice as likely to be recommended for transplant than black patients (adjusted OR 2.06) [13].

In a recent review article, Lefkowitz et al. [15] defined the goals of pre-transplant pediatric psychosocial assessment: to identify patient and family strengths and risk factors that could affect post-transplant outcomes, and to use that information to tailor interventions and target psychosocial services both before and after transplant. Other authors have delineated similar goals, with the additional aim of informing and preparing the potential recipient and family for the transition from "a fatal illness [to] a life-long, chronic condition" [16]. Both articles suggest that psychosocial risk factors are *not* commonly used in pediatrics to decide eligibility for transplant, partly due to the lack of empirical data in children relating pre-transplant psychosocial functioning to post-transplant outcomes.

There is little question that non-adherence to immunosuppressive therapy and lack of close medical follow-up after transplant predicts both morbidity and mortality in pediatric transplant recipients [17–19]. But whether post-transplant adherence rates can be predicted by pre-transplant psychosocial functioning is less well established. Evidence from case reports demonstrate a range of post-transplant outcomes, both positive and negative, for adults with pre-existing psychopathology or compliance issues; results of empiric studies vary widely depending on the variables evaluated [6]. A recent meta-analysis of 61 studies of non-adherence in pediatric transplantation identified several risk factors for post-transplant non-adherence, including poorer family functioning (increased parental distress, lower family cohesion), poorer child psychological functioning (behavioral problems, increased child distress) and older age at the time of transplant (≥10 years). However the psychosocial characteristics were evaluated *after* transplant in a cross-sectional manner—thus the predictive direction of the effect and correlation with pre-transplant functioning, if any, cannot be known [1].

In a single-center study of pediatric heart transplant recipients, pre-transplant parental psychosocial assessment could predict post-transplant morbidity [20]. In this study, a social worker and a pediatric psychiatrist, blinded to patient identifiers, race, and medical outcome, reviewed charting from the pre-transplant psychosocial evaluation. The evaluation included assessment of parental psychiatric health, history of substance abuse or domestic violence, means of transportation, marital status, support systems, age and education, and categorized parents into high- and low-risk categories. Children of parents identified as high-risk were three times more likely to have documented sub-therapeutic levels of immunosuppressive agents, episodes of rejection, and hospitalization. As this was a retrospective study, it is unknown whether mental health treatment or other social support interventions could have substantively improved the situation and the outcomes.

Lefkowitz and Annunziato [15, 16], in two separate papers, recommend a pediatric psychosocial evaluation that includes:

1. Comprehension, expectations and outlook
2. Self-management, knowledge, and allocation of responsibility
3. Adherence and barriers to medical management

4. Behavioral health
5. Mental health screening
6. Intellectual performance and neurocognitive ability
7. Family functioning
8. Social support
9. Patient and parent adjustment

Evaluation of both the child and their parent and family is necessary, since children have limited ability to maintain and direct their own medical care and lifestyle, especially in the pre-adolescent years. This heavy reliance on parents to provide care means that young children should not be held responsible for poor adherence to medical treatment, dietary indiscretion, or their family's lack of social support or overall functioning. Reluctant to ever deny a child a life-saving organ because of parental inadequacies or a dysfunctional social structure, pediatric transplant teams are motivated to identify problems early on in the transplant evaluation process, and make every effort to address deficient resources and facilitate healthy change. This places immense pressure on the social workers, psychologists, and case managers who work with these patients and families. These professionals must also deal with the disappointing truth that, despite their ongoing efforts, the outcome may still be poor and they may not successfully affect all potentially modifiable risk factors.

15.3 Ethical Use of the Psychosocial Evaluation

The tension between just allocation of finite resources and the pressing needs of individual patients is always present in health care, but is rarely as acute as in transplant medicine. The transplant physician acts both as an advocate for an individual patient and a steward of scarce resources. While policies can guide the choice of transplant candidates and govern the fair allocation of organs once a candidate is activated on the waiting list, the ultimate decision about whether or not to activate a patient is usually that of an individual clinician or small team. Since there are no universally agreed upon criteria for psychosocial readiness, there is a risk that clinicians over- or under-value psychological issues and social/behavioral problems, with harmful results. Being too dismissive of psychosocial issues in one patient not only deprives that patient and family of the opportunity for potentially life-changing intervention, but does a disservice to other patients waiting for an organ, should the graft be lost due to non-adherence. Alternatively, a clinician who denies or delays transplant based on a negative psychosocial assessment prioritizes justice concerns over his or her obligations (beneficence and nonmaleficence) to the individual patient. Paradoxically, such decisions are vulnerable to subtle, even subconscious prejudices and may unjustly disadvantage certain patients. As in all pediatric medicine, the superseding fiduciary obligation is to the welfare and best interests of the child, but the responsibility in transplant medicine to temper that advocacy with stewardship creates ethical dilemmas for treating clinicians.

Thinking about "best interests" provides a framework for considering those difficult cases where refractory psychosocial problems threaten the possibility of a good outcome post-transplant. Analyzing the course of action that best supports the interests of a child involves an appraisal of the potential benefits and risks or burdens of each option. While bridging therapies like dialysis for kidney failure and ventricular assist devices (VADs) for heart failure can extend the lives of children with end organ disease, transplant is typically considered life-saving; those who do not receive a transplant can expect to die of their disease. When death is the alternative, most people (physicians, ethicists, or parents) accept a fairly high level of burden (e.g., post-transplant medication side-effects, risk of surgical complications, financial strain, etc.) as a reasonable trade-off for the benefit of survival, and may not even perceive there to be a real decision to make [21]. Transplant teams are likewise loathe to deny a child an organ, even when the likelihood of poor adherence, rejection, and morbidity from complications is high. Additionally, there is always the hope that adherence post-transplant will improve, and it seems both cruel and unwise to deny a child the chance at life based on predictions with limited reliability regarding post-transplant behavior. Pediatric transplant teams evaluate each case individually, functioning within professional guidelines, and go to great lengths to address psychosocial problems during the pre-transplant evaluation and subsequent waiting period. Another chapter in this book, by Goldberg and Fine, addresses the issue of non-adherence specifically in adolescent transplant candidates.

When the concern lies primarily in the psychosocial readiness of the child's caregivers, clinicians worry if it is ever appropriate to remove the child from his parent's custody in order to pursue transplant. If the parents are abusive, the choice is clear, and in most jurisdictions the health care team is obligated to report such cases to the appropriate child welfare authorities. But when loving parents seem incapable of providing the care required post-transplant, whether because of substance abuse, mental illness, cognitive disability, financial and social instability, or a history of poor adherence, should the transplant team involve child protective services to seek medical foster placement? This can sometimes be done with parental consent, but realistically often involves involuntary removal of children from their homes. The psychological impact on the child, already vulnerable due to her medical illness, may be profound. The involvement of child protective services has the potential to break any trust previously established with a family, which may be difficult to rebuild post-transplant. A complete discussion about utilizing child protective services to ensure a child's medical therapy is beyond the scope of this chapter, but there are serious concerns with how and when this step should be taken [22]. And although the likelihood of death without transplant may be certain, transplant itself is not a guarantee of good outcome. Many adult patients with end organ disease decide against transplantation, as do some children and parents. Therefore, it would be difficult to claim, in many situations, that the medical benefit to the child is so strong that taking custody is the only reasonable choice. In these rare situations the transplant team should request clinical ethics and legal consultation to ensure a balanced appraisal of the benefit-burden ratio for the child.

Removal of children from their parents should truly be a last resort, and only after every effort to improve the situation has been exhausted. In cases where abrogation of parental decision-making authority is ultimately deemed to be appropriate, attempts should be made to continue the parents' involvement in the child's life to whatever degree possible, including eventual reunification. The transplant team should continue to engage the family in the child's care throughout the transplant process, as appropriate; maintain the supportive elements of parental presence; and provide education and promote acquisition of caregiving skills.

Though likely rare, and not described in the available literature, there may yet be circumstances when the sum of psychosocial risk factors is so great that the child should not be transplanted, or should have transplant delayed, particularly if alternative life-prolonging therapy exists (e.g., dialysis, VADs). In certain medically complex cases the likelihood of achieving significant benefit from a new organ is substantially reduced. In situations with little chance of medical benefit, the burdens of pursuing transplant may exceed the benefits, and the team may be justified in recommending against transplantation. The ethical justification for this is typically based on the obligation to avoid doing harm. Harm is done when the intrinsic burdens of intensive care before and after high-risk transplant procedures are not balanced by a probability of benefit to the patient. The stress of repeated or extended hospitalization for the child in need of a transplant, the multiple unavoidable invasive procedures and interventions, and a complicated post-transplant course that will likely include further invasive and painful procedures are all harmful and unjustified if the graft (and therefore the patient) are very unlikely to survive. Patients awaiting transplant may also be denied the full range of palliative care that would otherwise be provided a child with a terminal condition [23]. Such cases must be thoughtfully considered from multiple perspectives in order to achieve a balanced ethical appraisal. The transplant team should request evaluation from ethics consultants (preferably a multidisciplinary committee), palliative care, and, when appropriate, other specialties such as psychiatry and child life. A decision not to offer transplantation should be communicated to the child's parents in a compassionate and supportive manner, and special care should be taken to provide age- and developmentally-appropriate information to the child or adolescent.

15.4 Living Donor Parents

Another unique issue for pediatrics is the fact that, for organs that have a living donor option, the child's parents may be evaluated simultaneously for both their psychosocial readiness to care for their child post-transplant and their readiness to serve as living donors. Guidelines on living donor evaluation stress the importance of separating the donor and recipient evaluation, both to protect confidentiality and to promote voluntariness of the donor [24]. While transplant decision-making duties for the child could, theoretically, be assigned to the non-donating parent, this is likely impractical and ignores the relational, collaborative model of decision-making that is natural to many families. Therefore, there will likely be some overlap for parents in their decision-making for themselves as donors and for their children.

Despite the duality of this parent-donor role, most parent donors and their partners report positive outcomes with their donation experience and do not feel it impaired their ability to consent. In a survey study of parents who donated kidneys to their children, almost all reported satisfaction with the decision, and claimed to have a more positive relationship with their children and partners after the transplant [25]. Donating parents seemed less worried about the medical risks of transplant than their non-donating partners, and felt that their decision-making process was free of coercion from the other partner, other relatives, and the medical staff. A Swedish interview study of parents post-transplant found similar reports of positive outcomes for many parents, though some admitted feeling pressure to be a living donor [26]. Narrative accounts of the parent donor experience speak to the naturalness of this decision, and its consistency with other parenting roles, which also require a degree of self-sacrifice [27]. Another questionnaire study found that parent donors had more psychological benefit from the transplant than did non-donating parents, and none regretted the decision [28]. Health care providers, too, recognize that the choice to donate carries some risk, but that it may be natural to parents since, as one respondent to an interview study stated "For a parent to child, I'm sure a parent would be prepared for risk of death even, if it was to save a child's life" [29, p. 160].

While most parent donors may feel that their choice is easy or natural, we must be careful not to stigmatize as unloving those parents who choose not to donate, or question their motivations. The studies on parent donors are encouraging, but responses are limited to families in which one parent ended up donating. We know much less about families in which neither parent donated and the reasons for their decision. As well, responses to surveys and interviews may be affected by a parent's desire, either conscious or unconscious, to be positive about their choice, as evidenced by psychology literature revealing that we sublimate negative emotions after major decisions in order to avoid post decisional regret [30]. As with other potential living donors, we must respect the ultimate choice of a parent and avoid letting our perceptions of that choice's validity affect the care we provide to their child.

Additionally, knowing the strong pull parents feel to donate for their child, and acknowledging that even when all efforts are made to eliminate coercion a parent may feel pressure from other family members or even from other members of the medical team, the transplant team has a responsibility to weigh the risk of morbidity/mortality to the donor as well as the likelihood of benefit for the child in deciding whether to make living related donor transplantation an option.

15.5 Challenges for the Psychosocial Team

Addressing the complex psychosocial needs of sick children and their families is a difficult task under any circumstances. For the social workers and psychologists working with pediatric transplant patients there is a great deal of pressure to "fix" the psychosocial issues that jeopardize the likelihood of a good outcome [31]. Challenges include scarce resources for mental health care, limited availability of need-based support services, and the difficulty of influencing set patterns of human behavior.

One article has described the use of a psychodynamic approach to address the psychological needs of pediatric liver transplant patients and their parents [31]. The techniques described rely on building a therapeutic alliance with a patient and family who have not requested (and often do not want) psychological intervention, which is incredibly time-intensive. This type of work also requires a skilled practitioner, whom many transplant teams may not have. The philosophy behind the psychodynamic approach is useful, however, as it begins with the recognition that parental traumatization plays a part in development of dysfunctional behaviors for both parent and child, and works to "identify risk and resilience in the face of life-endangering illness" [31, p. 137].

Ideally, the pre-transplant psychosocial evaluation identifies areas of strength as well as areas of concern, and allows the psychosocial team to target their interventions. These may include multimodal educational strategies for parents and child, cognitive-behavioral therapy to address maladaptive behaviors, and psychiatric treatment for Axis I disorders (clinical psychiatric issues, like depression or anxiety). Another important function is in clarifying expectations—both for patients and their parents, and for the transplant team [15]. Cognitive-behavioral preparation pre-transplant may increase autonomy, self-efficacy, and self-regulation, all of which have been associated with improved adherence in adult transplant recipients [16]. These skills are relevant both to adolescent patients and to parents, especially parents of young children, who need the emotional and psychological resources to help their children thrive both during and after transplant. For child recipients and their parents, developmentally-appropriate psychological and cognitive preparation may reduce acute anxiety and stress, decrease the likelihood of post-traumatic stress disorder, and promote long-term coping and adjustment to future medical challenges [15].

In working intensely with pediatric patients and their parents pre-transplant, the psychosocial team members may find an ethical challenge in moderating the emotional complexities of the team's relationships with patients and their families. Emotion plays a role in the team's desire to rescue the patient, particularly for physicians [32]. While the rescue impulse is natural and understandable, it may at times cloud the team's judgment and affect patient care (for example, re-transplanting a second or third time for a patient with significantly diminished chances of survival, or avoiding discussions about palliative care). A separate issue is the emotional attachment and even dependency that understandably often develops for parents who are in an incredibly vulnerable and terrifying position, and whose child is "saved" by the transplant team's intervention. They may spend significant amounts of time talking with the transplant social worker, nurse coordinator, and physicians, and feel that these people understand what they are going through more than close friends and family. The transplant team, especially the social worker and psychologist, may also become heavily emotionally invested, or they may remain detached, feeling that the relationship is simply a part of their job. The support provided by the transplant team is important both for functioning of the parent and well-being of the child, and care must be taken to maintain a focus on empowering the parent and providing them with tools for independent coping rather than fostering dependency. The team must also be wary of providing lower levels of support for families with whom it may be more difficult to develop synergy, especially with those who come from very different cultural or ethnic backgrounds, or for whom a language barrier exists.

15.6 Marijuana Use in Pediatric Transplant Candidates

It can be difficult to determine which patient characteristics should play a role in transplant candidacy and whether they require modification prior to transplant. One of the most controversial is whether substance use should play a role in listing decisions. Since adolescence is a time of experimentation during which substance use is common, policies that restrict candidacy based on substance use can have a negative impact on eligibility in this age group.

Alcohol abuse has long been considered a contraindication to liver transplant in adults, based on the concern that those with alcohol-related liver disease will return to use after their transplants and endanger their new grafts. In other solid organ recipients, alcohol dependence is associated with higher rates of infection and rejection [33]. Tobacco use, too, has documented negative effects on transplant and patient survival, with some calling for a six-month period of abstinence in lung recipients and cessation before other solid organs [34]. There appears to be consensus in the transplant community that the abuse or dependence on these substances should be of major concern during the transplant assessment, but a full discussion of the evidence for this consensus is beyond the scope of this chapter. There is much less consensus, however, regarding the use and abuse of marijuana in the adolescent transplant candidate, so we will give this area some focus.

Marijuana is commonly used by teenagers, with 45 % of twelfth graders in the U.S. admitting to cannabis use, and a similar percentage in a New Zealand cohort [35, 36]. A much smaller percentage (4.5 % in the New Zealand group) develop a cannabis use disorder, defined in the DSM-V as use which results in at least 2 of 12 symptoms that indicate "clinically significant impairment or distress," like inability to cut down and impairments in home, work or school functioning due to use [37]. While teenagers with organ failure may use cannabis recreationally, like their healthy peers, the drug may also alleviate some of the symptoms of organ failure, like nausea and poor appetite, and medical marijuana or THC pills are sometimes prescribed to patients with organ failure for exactly this reason [38].

Transplant centers are concerned about marijuana use, with some listing it as a relative or absolute contraindication to transplant. In a study of American liver transplant programs, occasional non-medical use of marijuana was considered a contraindication to liver transplant by 20.5 % of respondents, while daily non-medical use was considered an absolute contraindication in 46 % of respondents and a relative contraindication in another 43 % [39]. Marijuana consumption was found to be a contraindication to liver transplant by about 20 % of liver transplant programs in Italy [40]. A study of Canadian adult liver transplant centers found marijuana use an absolute contraindication for 17 % and a relative contraindication for another 17 % of respondents [41].

What might be the reason for this concern? There may be medical apprehension, as studies have found cancer rates to be higher in marijuana smokers, so there is a theoretical risk that marijuana use combined with immunosuppression could lead to higher rates of malignancy [42]. There is a case report of a kidney transplant patient developing aspergillosis after smoking marijuana [43] and of a heart transplant patient developing ventricular tachycardia, [44] but there are no large studies showing the negative effects of cannabis post-transplant.

There may also be psychological and legal concerns. Marijuana use remains illegal in most countries, and is restricted to adults in most countries that have legalized it, so the teenager using cannabis is almost always going to be breaking the law, unless it has been legally prescribed. This illegal behavior may lead the transplant team to conclude that the teenager who feels free to break the law may be less likely to take the "laws" of transplant adherence seriously. There may also be concerns that a teen who is using cannabis frequently may not be able to manage his transplant medications properly due to the effects of the drug on brain function, since high potency cannabis use can be associated with changes in executive function [45]. Indeed, Coffman reports on anecdotal evidence from transplant psychiatrists who see marijuana use as a predictor of post-transplant nonadherence [46].

Despite the concerns of the transplant community about marijuana use, only one study has specifically evaluated outcomes in adult liver transplant recipients [47]. In this retrospective study, transplant candidates who admitted to using marijuana were less likely to receive a transplant than non-users, but of those who did receive a transplant, graft outcomes were similar. There appear to be no similar studies of transplant recipients of other organ groups, and none in adolescent transplant candidates. As Weinreib and Lucey [48] have argued, the paucity of evidence for harm makes recreational marijuana use an overly strict criterion on which to deny transplant, unless the marijuana use is a sign of more serious addiction or psychiatric issues. Stolick and Carter have argued that the use of cannabis may have potentially beneficial effects, by decreasing nausea; improving appetite; and having potential, if still unproven, immunomodulatory effects [38]. While it is unlikely that marijuana use in adolescent transplant candidates will ever be wholeheartedly embraced, especially as it remains illegal in most jurisdictions, recreational or medicinal use does not, on its own, seem like a sufficiently justifiable reason to deny or delay transplant.

15.7 Psychological Sequelae of Transplantation

15.7.1 Psychological and Developmental Sequelae for Patients

Though solid organ transplantation for children has grown substantially since the 1980s, systematic data about the long-term psychosocial outcomes remain sparse, making it difficult to assess the overall balance of benefits to and burdens on patients. While in most situations the value of transplantation to save a child's life seems unassailable, transplant proponents have an obligation to assess the effects of their interventions on patients and their families. This data is necessary in order to provide adequate information for parents making a decision about transplantation, and to identify areas to improve post-transplantation quality of life and achievement of patients' developmental potential.

Recent reviews have suggested that while patients who had received kidneys as children assessed their quality of life as high, they had deficits in motor function, delayed educational achievement, and higher unemployment as adults. Among childhood liver recipients, more than a quarter had intelligence scores greater than two standard deviations below the mean for the tests used [49]. In one study, adult survivors of childhood heart transplantation surveyed at least ten years after initial transplant had quality of life scores for physical and mental health comparable to the general US population [50]. Another review of quality of life after pediatric solid organ transplantation noted frequent somatic complaints, physical problems, such as sleep disturbances and weight gain related to immunosuppressive regimens, high rates of psychiatric disorders, and problems with relationships among childhood kidney transplant recipients. Liver recipients reported views of their lives as similar to those of other children with chronic illnesses, with concerns about health, relationships, and their futures. Heart transplant patients reported improved function and outlook after transplantation, though a minority had serious psychological difficulties, such as depression and learning difficulties [51].

A 2012 report of young adults two decades after liver transplantation as children concluded that the subjects "have achieved a high level of education and employment despite suffering from the side effects of chronic immunosuppression and having lower (health-related quality of life) HRQOL and heath utility than their healthy peers" [52]. Another review about quality of life following organ transplantation noted on-going psychiatric morbidity in kidney and heart transplant populations, problems with school performance and overall development among some liver transplant patients, and general concerns regarding growth and final height, employment, and independent living among the recipients, though better quality of life than those with some other chronic conditions, such as rheumatoid arthritis, chronic lung disease, and inflammatory bowel disease [53]. A 2014 study examined pre-transplant intellectual and academic performance among children being evaluated for organ (kidney, liver, and heart) replacement. The research found that while the patients' functioning was within the average range their scores were "significantly lower than the normal population across organ groups" [54, p. 5]. The authors noted that pre-transplant deficits complicate interpretation of post-transplantation evaluations of quality of life and achievement.

A 2014 qualitative study of adolescents receiving heart transplants noted the patients' transformation from before surgery, when they had restricted activities and a high degree of dependence on others for their survival, to after transplantation, when the teens developed a sense of liberation, with growing autonomy and freedom. A follow-up study of adult survivors of pediatric heart transplant reported good quality of life and success in educational, employment, and financial achievement among those whose transplants had occurred at least ten years earlier [55].

Alonso et al. [56] also noted the difficulty of interpreting post-transplant results given the amount of pre-transplant morbidity, number and kind of pre-transplant interventions, and variations in age at the time of transplantation. The authors concluded that kidney and liver recipients have a relatively high need for special education, while kidney and heart transplant patients have post-transplant cognitive deficits. Patients with cyanotic heart disease had worse cognitive functioning post-transplantation, compared to cardiomyopathy patients. For intestinal transplantation, studies have shown a high prevalence of cognitive delay after transplant. The authors concluded that while many transplant recipients have normal cognitive function and academic achievement, "a disproportionate number have below average intelligence, learning difficulties, and problems meeting expectations in the classroom" [56, c119 p. 8].

Taken together, these findings suggest that pre-transplant factors and possibly the effects of the transplant process itself (surgery, anesthesia, and medications) contribute to a relatively high frequency of developmental deficits and long-term psychological morbidity in survivors. This information points to obligations of clinicians to minimize malnutrition, psychosocial deprivation, and other factors that may limit good outcomes, and to adequately inform patients able to appreciate these factors and their families of the potential, if not likelihood, that transplant will not solve all of the problems their children may have. At least in theory, such information might lead some families to judge that the benefits of transplantation may not outweigh the additional burdens and thus decide against transplantation.

15.7.2 Psychosocial Sequelae for Parents and Families

While pediatric transplant patients typically rate their quality of life as good, the proxy ratings by their parents are consistently lower across the cited studies [49–56].

These findings are summarized by the following: "Parent ratings [of the experiences of their children] were lower across all domains especially on emotional, social, school and physical summary scales" [51. p. 563]. This speaks not only to the resilience and adaptability of children, but also to the distress parents experience during and after transplant. It is well-established that parents and families of children undergoing solid organ transplant experience significant emotional and psychological stress, both in the pre-transplant phase, as they anxiously await availability of a life-saving organ, and post-transplant, as they cope with complicated immunosuppressive regimens and new worries about rejection. In fact, several authors have conceptualized the transplant experience as a traumatic event for parents. In one 2003 study, 50 % of parents of solid organ recipients reported at least moderately severe Post-traumatic Stress Disorder (PTSD) symptoms as measured by the Post-traumatic Stress Diagnostic Scale, with 45 % experiencing moderate to severe impairment of functioning as a result of these symptoms [57]. In children who have undergone liver transplantation, most studies of the effects on family members "have shown a considerable negative influence on parental emotional state and family life" [58, p. 551]. In one study, a majority of parents of children who received liver and/or intestinal transplants reported elevated levels of psychological distress in both the pre- and early post-transplant periods, and normal physical functioning but lower vitality scores compared to the normative population [59].

In contrast, a German sample of living-related liver donor parents reported higher general quality of life and lower anxiety and depression levels compared to the healthy population. These parents had significant physical symptoms post-donation, with some residual symptoms one year out from donation, but did not consider their donation to have had as much lasting personal emotional impact. However, 48 % considered the burden placed on their family to be average, high or very high [60]. For children who have had heart transplants, family stress may increase after transplantation, especially as others' expectations about child and parent functioning normalize several months following placement of the graft [61]. Particularly for adolescents, parents feel considerable anxiety around their children having to assume responsibility for medication management [61]. In general, following transplantation, families may deal with disruption of normal social support structures, financial problems, and increased family conflict, all of which may worsen parental stress and psychological sequelae, and ultimately impact the well-being of the child [57, 62].

The empirical data about the psychological and developmental outcomes following solid organ transplantation raise many questions. We cannot now know how much of what happens to surviving children results from pre-transplant morbidity and treatment, how much constitutes impairment secondary to transplantation itself (anesthesia and surgical time, especially time on cardiopulmonary bypass, neurotoxic effects of immunosuppressive medications, etc.), and what arises from other factors. It seems likely that the psychological and developmental outcomes have improved over time, though transplanted children continue to have psychiatric, developmental, and relationship challenges that demand special post-surgical attention, including neuropsychological evaluations, special education, and mental health services.

From an ethics perspective, the results summarized here require attention in the pre-transplant evaluation and consent process as patients and families may have unrealistic expectations about what transplantation can and cannot accomplish. Transplant physicians and surgeons need to provide families with accurate information about both physical and psychosocial outcomes after transplant, including potential effects on family functioning and well-being. Families have often pinned their hopes on the promise of transplantation, and transplant teams must avoid giving the impression that "everything will be fine" after surgery [63]. In cases where clinically significant pre-transplant neurological and/or developmental damage already exists, transplant team members need to caution families that transplantation, per se, will not reverse existing cognitive or physical impairment of the central nervous system. Whether any given degree of cognitive or neurological impairment should constitute a criterion for exclusion from transplantation remains controversial and is beyond the scope of this discussion [64]. Another chapter in this book, by Wightman and colleagues, addresses this issue specifically.

Special care must be taken with education and counseling pre-transplant when parents or other family members are being considered for living donation. Recommendations for psychosocial evaluation of unrelated donors have been published [65] and emphasize autonomous decision-making including freedom from coercion, absence of unrealistic expectations and psychopathology-driven motives, and adequacy of information. However, expecting parents to maintain a completely rational thought process about organ donation for their child is unrealistic, and the transplant team ought also to keep in mind the concept of relational autonomy, which incorporates an understanding of the interrelatedness of human beings and validates the role that emotion plays in health care decisions [66]. The psychosocial team members can play a significant role in supporting the parents' decision-making as they struggle to balance wanting to rescue their sick child, the need to continue providing for other dependent family members, and concerns about their own welfare. Especially in cases of fulminant liver failure, when there is no time for the recommended "cooling off" period, parents need the care and support of the psychosocial team as they are making a decision about donation, and after the fact, as they process and come to terms with the outcome. Living related kidney and liver donors experience a quality of life that is the same or better than the general population; however, they are at risk for psychiatric sequelae when the organ recipient does poorly, when they have unrealistic expectations about the procedure, and when they have severe medical complications post-donation [67].

15.8 Psychological Issues Related to Re-transplantation in the Pediatric Recipient

In 1993 Ubel, Arnold, and Caplan published an influential paper noting that the outcomes following re-transplantation were not as good as those following initial transplants, raising questions about fairness in the allocation of newly available organs [68]. While results following re-transplantation have likely improved over twenty years, concerns about the justice of re-transplanting patients who have been non-adherent continue. Few studies have reported data about re-transplant outcomes in children; we review information published since 2010.

The data on re-transplantation is mixed. In a single center study of liver transplant recipients, one- and three-year graft survival was similar between primary transplant and re-transplant groups [69]. In children who had received lung transplants in the U.S. between 1998 and 2008, 10 % required re-transplantation and this group had lower graft survival than the primary transplant group [70]. However, when corrected for those whose initial grafts lasted a year or more, survival rates mirrored that of primary transplant recipients. A single-center German retrospective analysis concluded that re-transplantation offered "an acceptable long-term outcome comparable to primary pulmonary transplantation" [71]. The authors did not specify their criteria for judging the outcomes acceptable. Using UNOS data, Desai et al. found results of intestinal re-transplantation were poor in both adults and children, especially for children under 2 years of age, when compared to outcomes after primary transplants [72].

Concerned that "approximately 25 % of the current pediatric waiting list…and half or more of all pediatric kidney recipients can be expected to eventually undergo re-transplantation," Van Arendonk et al. [73, p. 1360] analyzed registry data about outcomes for pediatric kidney re-transplantation. The authors had a particular interest in whether factors contributing to first graft failure could predict successive graft losses, which, in their view, would influence allocation decisions for re-transplant candidates. The study showed that "repeat transplants are not doomed to early failure even if prior graft survival was quite poor" [73, p. 1366].

A key ethical question is whether non-adherence to treatment regimens, especially the taking of immunosuppressive medications, should be a contraindication to re-transplantation for children and adolescents. Understandably, transplant teams may feel reluctant to re-list patients who have lost their grafts due to non-adherence, worrying that these recipients will repeat previous behavior, leading to loss of a graft that might have gone to someone who would better protect the graft. Dobbels et al. [74] reviewed the available literature on adherence and clinical outcomes after re-transplantation in order to better inform professionals' decisions about re-listing and re-transplantation. The authors found only a single paper with systematic data bearing on their question [75]. Dunn et al. followed 35 patients who were re-transplanted after having lost their graft to non-adherence and compared their outcomes to those of 552 patients who were re-transplanted after graft loss for other reasons. They found no significant differences between groups in actuarial patient or graft survival rates, renal function, or the incidence of biopsy-proven chronic rejection. However, previously non-adherent patients were more likely to exhibit non-adherence after their second transplant. The authors concluded that prior non-adherence should not be an absolute contraindication to re-transplantation, but should prompt intensive intervention and careful patient selection. This one study, of course, cannot settle controversies regarding clinicians' obligations to rescue patients versus their duty to husband the scarce resource of available organs, but it does emphasize the importance of evaluating each patient and family on an individual basis. Previously non-adherent adolescent re-transplant candidates may mature during their wait for a new organ, in part due to their experience of illness and/or with dialysis as they wait for another kidney. Such psychological change could substantially improve adherence after receipt of a new graft. In addition, better "care models incorporating self-management support," [74. p. 9] along the lines of successful systems for improving adherence in other chronic health conditions, could improve outcomes after re-transplantation, arguing against "punishment" of those with non-adherence following a prior transplant.

None of the above points to obvious solutions to questions about re-transplantation. Improvements in surgical techniques and immunosuppressive regimens over time make repeat transplantation reasonable when patients' physical and psychological conditions have not deteriorated excessively. Unfortunately, no systems exist to identify points of no return, either for physiologic or emotional decline. In addition, a few highly publicized cases in the 1990s highlighted the fact that some pediatric transplant patients may refuse re-transplantation and accept death because the medications and activities required to maintain their grafts left them feeling incapacitated or perpetually uncomfortable. Current drug therapy may be better. Nevertheless, one cannot say conclusively either at what point clinicians and families may feel efforts to provide support and psychological/psychiatric treatment have been sufficient, or when continued efforts to encourage or enforce adherence simply cannot succeed. An adolescent's experience of and reaction to his experience living with a chronic condition, and its management, has to be balanced against concerns about the patient's immaturity or recognition of normal psychological individuation or even rebellion. Based on current knowledge, we have inadequate approaches to individual clinical decision-making, as well as a need for mechanisms to ensure greater fairness in organ allocation.

15.9 Conclusion

Discussion of the key ethical issues in pediatric solid organ transplantation must include consideration of the complex psychosocial needs of patients and families. Mental health, psychological functioning, behavioral health, and social support for both child and parent play a significant role in their abilities to cope with the stressors of the transplant process and the challenges of the post-transplant regimen. Careful psychosocial evaluation and supportive intervention is a necessary and expected component of transplant care in order to optimize outcomes and provide maximum benefit to the child. Transplant teams may face difficult decisions when psychosocial pathologies threaten a good outcome post-transplant and do not respond to intervention; such situations must be evaluated on an individual case basis using the best interests of the child as a guiding ethical principle.

In working with pediatric patients with organ failure and their parents, transplant teams must appreciate the traumatic nature of the experience and the profound psychological and emotional sequelae for survivors and their families. In light of the data on psychosocial outcomes post-transplant, parents and patients deserve careful counseling and education in the pre-transplant period, and will need ongoing support from the team in a manner that supports independent functioning.

References

1. Dew MA, Dabbs AD, Myaskovsky L, et al. Meta-analysis of medical regimen adherence outcomes in pediatric solid organ transplantation. Transplantation. 2009;88:736–46.
2. Shellmer D, Brosig C, Wray J. The start of the transplant journey: referral for pediatric solid organ transplantation. Pediatr Transplant. 2014;18:125–33.
3. Qvist E, Jalanko H, Holmberg C. Psychosocial adaptation after solid organ transplantation in children. Pediatr Clin North Am. 2003;50(6):1505–19.
4. OPTN Vision Statement. Available at: http://optn.transplant.hrsa.gov/governance/about-the-optn/vision-goals/. Accessed 8 Dec 2014.
5. Courtney AE, Maxwell AP. The challenge of doing what is right in renal transplantation: balancing equity and utility. Nephron Clin Pract. 2009;111:c62–8. doi:10.1159/000180121.
6. Dew MA, Switzer GE, DiMartini AF, Matukaitis J, Fitzgerald MG, Kormos RL. Psychosocial assessments and outcomes in organ transplantation. Prog Transplant. 2000;10(4):239–59.
7. Fung E, Shaw RJ. Pediatric transplant rating instrument—a scale for the pretransplant psychiatric evaluation of pediatric organ transplant recipients. Pediatr Transplant. 2008;12:57–66.
8. OPTN Bylaws. Available at: http://optn.transplant.hrsa.gov/ContentDocuments/OPTN_Bylaws.pdf. Updated 1 Sept 2014, accessed 7 Dec 2014.
9. Flamme NE, Terry CL, Helft PR. The influence of psychosocial evaluation on candidacy for liver transplantation. Prog Transplant. 2008;18(2):89–96.
10. Levenson JL, Olbrisch ME. Psychosocial evaluation of organ transplant candidates: a comparative survey of process, criteria and outcomes in heart, liver and kidney transplantation. Pschosomatics. 1993;34(4):314–23.
11. Orentlicher D. Psychosocial assessment of organ transplant candidates and the Americans with Disabilities Act. Gen Hosp Psychiatry. 1996;18(Suppl. 6):5S–12S.
12. Moseley KL, Kershaw DB. African American and white disparities in pediatric kidney transplantation in the United States: unfortunate or unjust? Camb Q Healthc Ethics. 2012;21:353–65.
13. Furth SL, Hwang W, Neu AM, Fivush BA, Powe NR. Effects of patient compliance, parental education and race on nephrologists' recommendations for kidney transplantation in children. Am J Transplant. 2003;3(1):28–34.

14. Navaneethan SD, Singh S. A systematic review of barriers in access to renal transplantation among African Americans in the United States. Clin Transplant. 2006;20(6):769–75.
15. Lefkowitz DS, Fitzgerald CJ, Zelikovsky N, Barlow K, Wray J. Best practices in the pediatric pretransplant psychosocial evaluation. Pediatr Transplant. 2014;18:327–35.
16. Annunziato RA, Fisher MK, Jerson B, Bochkanova A, Shaw RJ. Psychosocial assessment prior to pediatric transplantation: a review and summary of key considerations. Pediatr Transplant. 2010;14:565–74.
17. Lurie S, Shemesh E, Sheiner PA, et al. Non-adherence in pediatric liver transplant recipients—an assessment of risk factors and natural history. Pediatr Transplant. 2000;4:200–6.
18. Ringewald JM, Gidding SS, Crawford SE, Backer CL, Mavroudis C, Pahl E. Nonadherence is associated with late rejection in pediatric heart transplant recipients. J Pediatr. 2001;139:75–8.
19. Fine RN, Becker Y, De Geest S, Eisen H, Ettenger R, Evans R, Rudow DL, McKay D, Neu A, Nevins T, Reyes J, Wray J, Dobbels F. Nonadherence consensus conference summary report. Am J Transplant. 2009;9(1):35–41.
20. Stone D, Banerjee M, Dupuis J, Leleszi J, Allasio D, Singh TP. Association of parental pretransplant psychosocial assessment with post-transplant morbidity in pediatric heart transplant recipients. Pediatr Transplant. 2006;10:602–7.
21. Pentz RD, Pelletier W, Alderfer MA, Stegenga K, Fairslough DL, Hinds PS. Shared decision-making in pediatric allogeneic blood and marrow transplantation: what if there is no decision to make? Oncologist 2012;17:881–885. doi: 10.1634/theoncologist.2011-0446. Originally published online 21 May 2012.
22. Downie J. A choice for K'aila: child protection and First Nations children. Health Law J. 1994;2:99–120.
23. Dellon EP, Leigh MW, Yankaskas JR, Noah TL. Effects of lung transplantation on inpatient end of life care in cystic fibrosis. J Cyst Fibros. 2007;6:396–402. doi:10.1016/j.jcf.2007.03.005.
24. Abecassis M, Adams M, Adams P, Arnold RM, Atkins CR, Barr ML, Bennett WM, Bia M, Briscoe DM, Burdick J, Corry RJ, Davis J, Delmonico FL, Gaston RS, Harmon W, Jacobs CL, Kahn J, Leichtman A, Miller C, Moss D, Newmann JM, Rosen LS, Siminoff L, Spital A, Starnes VA, Thomas C, Tyler LS, Williams L, Wright FH, Youngner S; Live Organ Donor Consensus Group. Consensus statement on the live organ donor. JAMA. 2000 Dec 13;284(22):2919–26.
25. Neuhaus TJ, Wartmann M, Weber M, Landolt MA, Laube GF, Kemper MJ. Psychosocial impact of living-related kidney transplantation on donors and partners. Pediatr Nephrol. 2005;20(2):205–9.
26. Kärrfelt HM, Berg UB, Lindblad FI. Renal transplantation in children: psychological and donation-related aspects from the parental perspective. Pediatr Transplant. 2000;4(4):305–12.
27. Zeiler K, Guntram L, Lennerling A. Moral tales of parental living kidney donation: a parenthood moral imperative and its relevance for decision making. Med Health Care Philos. 2010;13(3):225–36.
28. Kärrfelt HM, Berg UB, Lindblad FI, Tydén GE. To be or not to be a living donor: questionnaire to parents of children who have undergone renal transplantation. Transplantation. 1998;65(7):915–8.
29. Thomas EH, Bramhall SR, Herington J, Draper H. Live liver donation, ethics and practitioners: 'I am between the two and if I do not feel comfortable about this situation I cannot proceed'. J Med Ethics. 2014;40(3):157–62.
30. Svensson O. Differentiation and consolidation theory of human decision making: a frame of reference for the study of pre- and post-decision processes. Acta Psychol. 1992;80(143):168.
31. Schwab A, Rusconi-Serpa S, Schechter DS. Psychodynamic approaches to medically ill children and their traumatically stressed parents. Child Adolesc Psychiatric Clin N Am. 2013;22:119–39.
32. Jecker NS. The problem with rescue medicine. J Med Philos. 2013;38:64–81.
33. Parker R, Armstrong MJ, Corbett C, Day EJ, Neuberger JM. Alcohol and substance abuse in solid-organ transplant recipients. Transplantation. 2013;96(12):1015–24.

34. Corbett C, Armstrong MJ, Neuberger J. Tobacco smoking and solid organ transplantation. Transplantation. 2012;94(10):979–87.
35. Johnston LD, O'Malley PM, Bachman JG, Schulenberg JE. Monitoring the future national survey results on drug use, 1975–2012: volume I, Secondary School Students. Institute for Social Research, University of Michigan: Ann Arbor; 2013.
36. Boden JM, Fergusson DM, Horwood LJ. Illicit drug use and dependence in a New Zealand birth cohort. Aust NZJ Psychiatry. 2006;40(2):156–63.
37. American Psychiatric Association. Diagnostic and statistical manual of mental disorders. 5th ed. Arlington, VA: Am Psychiatr Publishing; 2013.
38. Baumrucker S, Mingle P, Harrington D, Stolick M, Carter GT, Oertli KA. Medical marijuana and organ transplantation: drug of abuse, or medical necessity? Am J Hosp Palliat Care. 2011;28(2):130–4.
39. Secunda K, Gordon EJ, Sohn MW, Shinkunas LA, Kaldjian LC, Voigt MD, Levitsky J. National survey of provider opinions on controversial characteristics of liver transplant candidates. Liver Transpl. 2013;19(4):395–403.
40. Caccamo L, Antonelli B, Rossi G. Attitudes of Italian liver transplantation centers toward the eligibility of controversial candidates. Liver Transpl. 2014;20(4):497–8.
41. Kroeker KI, Bain VG, Shaw-Stiffel T, Fong TL, Yoshida EM. Adult liver transplant survey: policies towards eligibility criteria in Canada and the United States 2007. Liver Int. 2008;28 (9):1250–5.
42. Hashibe M, Ford DE, Zhang ZF. Marijuana smoking and head and neck cancer. J Clin Pharmacol. 2002;42(11 Suppl):103S–7S.
43. Marks WH, Florence L, Lieberman J, Chapman P, Howard D, Roberts P, Perkinson D. Successfully treated invasive pulmonary aspergillosis associated with smoking marijuana in a renal transplant recipient. Transplantation. 1996;61(12):1771–4.
44. Sánchez Lázaro IJ. Almenar Bonet L, Sancho-Tello MJ, Martínez-Dolz L. Ventricular tachycardia due to marijuana use in a heart transplant patient. Rev Esp Cardiol. 2009;62 (4):459–61.
45. Ramaekers JG, Kauert G, vanRuitenbeek P, Theunissen EL, et al. High-potency marijuana impairs executive function and inhibitory motor control. Neuropsychopharmacology. 2006;31:2296–303.
46. Coffman KL. The debate about marijuana usage in transplant candidates: recent medical evidence on marijuana health effects. Curr Opin Organ Transplant. 2008;13(2):189–95.
47. Ranney DN, Acker WB, Al-Holou SN, Ehrlichman L, Lee DS, Lewin SA, Nguyen C, Peterson SF, Sell K, Kubus J, Reid D, Englesbe MJ. Marijuana use in potential liver transplant candidates. Am J Transplant. 2009;9(2):280–5.
48. Weinrieb RM, Lucey MR. Treatment of addictive behaviors in liver transplant patients. Liver Transpl. 2007;13(11 Suppl 2):S79–82.
49. Alonso EA, Sorenson LG. Cognitive development following pediatric solid organ transplant. Curr Opin Organ Transplant. 2009;14:522–5.
50. Petroski RA, Grady KI, Rodgers S, Backer CL, Kulikowska A, Canter C, Pahl E. Quality of life in adult survivors greater than 10 years after pediatric heart transplantation. J Heart Lung Transplant. 2009;28:661–6.
51. Anthony SJ, BarZiv SP, Ng VL. Quality of life after pediatric solid organ transplantation. Pediatr Clin N Am. 2010;57:559–74.
52. Mohammad S, Hormaza L, Neighbors K, Boone P, Tierney M, Azzam RK, Butt Z, Alonso EA. Health status in young adults two decades after pediatric liver transplantation. Am J Transplant. 2012;12:1486–95.
53. Sarwal MM, Bagga A. Quality of life after organ transplantation in children. Curr Opin Organ Transplant. 2013;18:563–8.
54. Reed-Knight B, Lee J, Cousins LA, Mee LL. Intellectual and academic performance in children undergoing solid organ pretransplant evaluation. Pediatr Transplant published online 2014: doi:10.1111/petr.12389.

55. Hollander SA, Chen S, Luuikart H, Burge M, Hollander AM, Rosenthal DN, Maeda K, Hunt SA, Bernstein D. Quality of life and metrics of achievement in long-term adult survivors of pediatric heart transplant. Pediatr Transplant. 2014. Published online doi:10.12384.
56. Alonso EM, Mohammad S, Sorenson LG. Cognitive development and functional outcomes after organ transplantation. In: Kirk AD, Knechtle SJ, Larsen CP, Pearson TC, Madsen JC, editors. Textbook of Organ Transplantation. 2014, Hoboken: Wiley.
57. Young GS, Mintzer LL, Seacord D, Casteneda M, Mesrkhani V, Stuber ML. Symptoms of posttraumatic stress in parents of transplant recipients: incidence, severity, and related factors. Pediatrics. 2003;111(6):e725–31.
58. Mohammad S, Alonso EM. Approach to optimizing growth, rehabilitation, and neurodevelopmental outcomes in children after solid-organ transplantation. Pediatr Clin N Am. 2010;57:539–57.
59. Tarbell SE, Kosmach B. Parental psychosocial outcomes in pediatric liver and/or intestinal transplantation: pretransplantation and the early postoperative period. Liver Trans Surg. 1998;4(5):378–87.
60. Kroencke S, Wilms C, Broering D, Roglers X, Schulz KH. Family strain and its relation to psychosocial dysfunction in children and adolescents after liver transplantation. Pediatr Transplant. 2014;18:851–9. doi:10.1111/petr.12367.
61. Adams J, Evangeli M, Lunnon-Wood T, Burch M. Restriction and dependence to autonomy and freedom: transformation in adolescent heart transplant recipients. Pediatr Transplant. 2014;18:637–50.
62. Kaller T, Petersen L, Petermann F, Fischer L, Grabhorn E, Schulz KH. Family strain and its relation to psychosocial dysfunction in children and adolescents after liver transplantation. Pediatr Transplant. 2014;18:851–9. doi:10.1111/petr.12367.
63. Brosig C, Pai A, Fairey E, Krempien J, McBride M, Lefkowitz DS. Child and family adjustment following pediatric solid organ transplantation: factors to consider during the early post-transplant. Pediatr Transplant. 2014;18:559–67.
64. Dobells F. Intellectual disability in pediatric transplantation: pitfalls and opportunities. Pediatr Transplant. 2014;18:658–60.
65. Dew MA, Jacobs CL, Jowsey SG, Hanto R, Miller C, Delmonico FL, United Network for Organ Sharing (UNOS), American Society of Transplant Surgeons, American Society of Transplantation. Guidelines for the psychosocial evaluation of living unrelated kidney donors in the United States. Am J Transplant. 2007;7(5):1047–1054. Epub 2007 Mar 12.
66. Walter JK, Ross LF. Relational autonomy: moving beyond the limits of isolated individualism. Pediatrics. 2014;133(Suppl 1):S16–23. doi:10.1542/peds.2013-3608D.
67. Jowsey SG, Schneekloth TD. Psychosocial factors in living organ donation: clinical and ethical challenges. Transpl Rev. 2008;22:192–5. doi:10.1016/j.trre.2008.04.008.
68. Ubel PA, Arnold RA, Caplan AL. Rationing failure: the ethical lessons of the retransplantation of scarce vital organs. JAMA. 1993;270:2469–74.
69. Heffron TG, Pilen T, Smallwood G, Henry S, Sekar S, Solis D, Casper K, Fasola C, Romero R. Liver retransplantation in children: the Atlanta experience. Pediatr Transplant. 2010;14:417–25.
70. Scully BB, Zafar F, Schecter MG, Rossano JW, Mallory GB, Heinle JS, Morales DLS. Lung retransplantation in children: appropriate when selectively applied. Ann Thorac Surg. 2011;91:574–9.
71. Muller C, Gorter H, Ballman M, Gottlieb J, Simon AR, Struber M, Hansen G. Pulmonary retransplantation in paediatric patients: a justified therapeutic option? A single-centre experience. Eur J Cardio-Thoracic Surg. 2011;39:201–5.
72. Desai CS, Khan KM, Gruessner AC, Fishbein TM, Gruessner RWG. Intestinal retransplantation; analysis of organ procurement and transplantation network database. Transplantation. 2012;93:120–5.
73. Van Arendonk KJ, Wang JMG, Deshpande NA, James NT, Smith JM, Montgomery RA, Colombani PM, Segev DL. Practice patterns and outcomes in retransplantation among pediatric kidney transplant recipients. Transplantation. 2013;95:1360–8.

74. Dobbels F, Hames A, Aujoulat I, Heaton N, Samyn M. Should we retransplant a patient who is non-adherent? A literature review and critical reflection. Pediatr Transplant. 2012;16:4–11.
75. Dunn TB, Browne BJ, Gillingham KJ, Kandaswamy R, Humar A, Payne WD, Sutherland DE, Matas AJ. Selective retransplant after graft loss to nonadherence: success with a second chance. Am J Transplant. 2009;9(6):1337–46. doi:10.1111/j.1600-6143.2009.02625.x Epub 2009 May 13.

Chapter 16
Ethical Issues in the Care of the Adolescent Transplant Recipient

Aviva M. Goldberg and Richard Fine

Abstract This chapter will explore the ethical issues that arise in the care of the adolescent transplant recipient. We will explore the dilemmas that arise in four clinical situations: (1) the adolescent who refuses transplant, (2) the non-adherent teen who is being considered for transplant, (3) contraception for female transplant recipients and (4) transition from pediatric to adult care. Ethical issues will include (1) adolescent decision-making and developing capacity (2) measuring adherence and using it as a criterion for transplant eligibility (3) maternal-fetal conflict in the use of drugs with teratogenic potential and (4) ethical concerns that arise during transition from pediatric to adult care.

Keywords Transplant · Ethics · Fertility · Nonadherence · Adolescent decision-making · Transition

16.1 Introduction

Adolescents[1] are a distinct group in pediatric transplantation. While they face many of the same physical and psychosocial challenges as other transplant recipients, they also have some unique issues. They are developing their independence, becoming more responsible for their medical care, and testing boundaries in both the medical and non-medical areas of their lives. They are preparing to transition from child-

[1]We will use the words "adolescent", "teen" and "teenager" interchangeably through the chapter

A.M. Goldberg (✉)
University of Manitoba, Winnipeg, MB, Canada
e-mail: Agoldberg@hsc.mb.ca

R. Fine
Stony Brook Children's, Stony Brook, NY, USA
e-mail: Richard.Fine@stonybrookmedicine.edu

R.A. Greenberg et al. (eds.), *Ethical Issues in Pediatric Organ Transplantation*, International Library of Ethics, Law, and the New Medicine 66,
DOI 10.1007/978-3-319-29185-7_16

hood to adulthood, including a transition from pediatric to adult transplant centers. They are experiencing the normal challenges of adolescence, including their emerging sexuality and its consequences, in a decidedly abnormal context of transplant care. They have experienced more struggles and must manage greater pressures than many of their healthy peers. They may have been transplanted as babies or young children and may now be dealing with the consequences of those early surgeries, including the possibility of re-transplant if their original graft is failing.

This period of change and instability is reflected in the alarming statistics on graft survival in adolescents. Other than the first year post-transplant, when adolescents generally do very well, kidney graft survival in this group is worse than in both adults and younger children [1, 2]. Much of this graft failure can be directly linked to nonadherence with medications, and this nonadherence is so common that some suggest that "the adolescent's quest for autonomy and independence render terms like medication compliance an oxymoron" [3, p. 475]. Others have pointed out that nonadherence is prevalent, but not inevitable, suggesting that there is the potential to reduce or avoid it, even in this high risk population [4].

In addition to nonadherence, adolescence presents other unique challenges to the transplant patient and his or her caregivers, some of which this chapter will explore. We will discuss the dilemmas that arise in four clinical situations: (1) the adolescent who refuses transplant, (2) the non-adherent teen who is being considered for transplant, (3) contraception for female transplant recipients and (4) transition from pediatric to adult care. Ethical issues will include adolescent decision-making and emerging capacity, measuring adherence and using it as a criterion for transplant eligibility, maternal-fetal conflict in the use of drugs with teratogenic potential, and resource allocation concerns that arise during transition.

16.2 The Nonadherent Transplant Candidate

Psychosocial criteria are routinely used in the evaluation of potential transplant candidates both explicitly in candidacy guidelines and by individual practitioners in their decisions regarding transplant referral [5–7]. One of these criteria is a prediction of adherence, which is the patient's ability to comply with the rigors of post-transplant immunosuppression and monitoring [8]. A more complete discussion of psychosocial assessment can be found in the chapter by Derrington et al. elsewhere in this book. Without good adherence, the transplant recipient is in danger of rejection, which can result in graft loss and eventual death. While a second transplant is possible after graft loss due to rejection, the patient may have become sensitized, meaning that they develop antibodies against the human leukocyte antigen (HLA) of the first transplant donor. If sensitization develops, it can be extremely difficult to find a second donor who is a suitable match.

There is also a real pressure on the transplant community because of the lack of available organs to meet the demands for transplantation. Most deceased donor allocation systems attempt to reach some sort of balance between equity and utility: getting the most use out of donated organs while trying to give equitable access to all those in need [9, 10]. Patients who are not expected to benefit from transplant, or whose chance of benefit is less than those of other potential recipients, often receive lower priority for access. For those organs that have a potential living donor route (kidney, liver and lung), respect for the living donor requires that we not accept an organ if there is little anticipated benefit to the recipient [11].

If a recipient is suspected to be non-adherent, some might argue that she should not be listed for transplant, both because she may endanger her future transplant success and because the donated organ may be wasted if she does not adhere to post-transplant care. In this section, we will attempt to determine how suspected nonadherence in an adolescent should affect decisions of transplant eligibility for primary and repeat transplants.

16.2.1 Defining Understanding Nonadherence Pre- and Post-transplant

Adherence in the medical context is generally defined as "the extent to which a person's behavior (taking medications, following a recommended diet and/or executing life-style changes) corresponds with the agreed recommendations of a health care provider" [8]. Nonadherence, conversely, is behavior which does not correspond with those recommendations. In solid organ transplantation, nonadherence is seen when patients omit medication doses, take doses late, fail to adhere to fluid or nutritional goals or miss medical appointments and monitoring procedures.

Despite a relatively simple definition for nonadherence, the reported frequency of the problem varies, even in similar populations [12]. This discrepancy occurs because both the threshold for nonadherence and the manner of measuring it can vary. Clearly, perfect adherence (never missing a scheduled dose at the scheduled time) is a lofty and likely unattainable goal, but how much nonadherence is acceptable? There is no clear amount of adherence that renders a transplant "safe" from rejection. Over- identifying nonadherence can be dangerous since the patients truly in danger of graft loss are lost within a larger group, but under-identifying nonadherence can lead to missed opportunities for intervention. There are many methods for measuring adherence: electronic monitoring, blood drug levels, pill counts, self-recall, health care provider impression and medication refill rates have all been used, with differing frequencies of reported adherence depending on the means of measurement [13].

Despite the lack of consistency in definition and measurement technique, what is clear is that nonadherence is common in adolescent transplant recipients, and often greater than that of both older and younger recipient [14, 15]. While rates of nonadherence in the general adult population are generally reported around 25 %, high-quality pediatric trials have reported the rates in adolescents as higher than 50 % [14, 16]. In one study of kidney recipients, nonadherence rates in adolescents were reported at 57 %, compared to 17 % in pre-pubertal children [17]. In heart transplant patients, nonadherence was much more common in adolescents than in younger recipients, and in pediatric liver transplant patients, older adolescents were more likely to be nonadherent than younger teens [18, 19].

It is also clear that graft survival is adversely affected by nonadherence. In kidney transplant recipients, risk of graft loss increased seven-fold in those who were non-adherent [20]. Time to graft loss was significantly shorter in adolescent kidney transplant patients who were nonadherent to medical therapy [21]. In a systematic review by Dobbels et al. between 13 and 75 % of graft loss in pediatric kidney transplant recipients could be explained by nonadherence [12]. Nonadherence is associated with graft rejection and increased mortality in adolescent heart transplant patients [22], liver patients [23], and lung patients [24]. Even minor subclinical nonadherence can be associated with high risk of graft rejection, as noted in a study of heart transplant patients where the risk of rejection was almost 14 times higher in the group that was less than perfectly adherent [24].

Teenagers become nonadherent for different reasons. Greenstein and Siegel have described three types of nonadherence typical to the adolescent transplant recipient [25]:

1. Accidental—those who forget to take their medications
2. Invulnerable—those who believe that missing medications will not result in harm
3. Decisive—those who are aware of the consequences, but choose to stop their medications anyway

Co-morbid psychiatric conditions, like depression and post-traumatic stress disorder (PTSD), are also contributors to nonadherence in this population [26, 27]. In addition to patient factors, socioeconomic factors and factors related to the patient's underlying condition play a role in adherence. For example, in one study adolescents who had kidney disease for >5 years were more likely to be nonadherent than those more recently diagnosed [28]. Treatment factors can influence adherence, with adolescents reporting more difficulty with complicated regimens, those that require evening doses of medications and those that have unpleasant cosmetic side effects [28]. Factors, like race, income level, education level and proximity to the transplant center also play important roles in both adherence and the perception of nonadherent populations [29].

When interviewed, adolescents have many barriers to adherence, including:

1. Medication factors, including the taste of medications, the difficult schedules and cosmetic side effects [30]
2. Social and psychological factors including depression, home life difficulties and a feeling of being overwhelmed by their disease management
3. Benefits to deliberate non adherence including more attention from parents and perceived options other than transplant (like dialysis) [31]

16.2.2 *Arguments for Denying Transplant on the Basis of Known or Suspected Nonadherence*

Though literature suggests that only about 5 % of transplant candidates are rejected based on nonadherence, this statistic likely underestimates the total number of patients for whom adherence played a role in getting listed, since patients deemed nonadherent by their treating physician may not be referred to the transplant center for assessment [32]. When interviewed, clinicians consider pre-transplant nonadherence to be a significant factor in deciding whether to refer someone for transplantation and a prediction of adherence is a part of most recommendations for recipient candidacy, but admit that this often a difficult behaviour to assess [33]. As previously mentioned, nonadherence puts the patient at risk, since a failed transplant can leave the patient more sick and more vulnerable than he[2] may have been if transplant was delayed.

There is some evidence that pre-transplant nonadherence can predict post-transplant nonadherence. A prospective study of 141 heart, liver and lung transplant patients found that pre-transplant nonadherence to medication was the only predictor of post-transplant graft loss in a multivariate analysis [34]. Adherence in this study was defined quite strictly—subjects were asked the question "During the last 14 days, how often did you not take your medication on time?" Any answer other than "never" labeled the patient as non-adherent. Despite this strict definition, which placed many of the patients in the nonadherent group, the relationship between pre-transplant and post-transplant nonadherence was relatively strong, with an odds ratio of 4.37 and a 95 % confidence interval of 1.18–16.16. In a study of heart transplant patients, univariate analysis showed prior noncompliance with medications and appointments was associated with noncompliance to cyclosporine dosing [24].

Other major reason to delay or forgo transplant in the nonadherent teen are justice and utility concerns. Rhodes has argued that justice concerns would favor non-transplantation in a nonadherent teen since the nonadherent recipient is less likely to derive benefit than is another candidate [35]. Transplant clinicians cannot only be concerned with the needs of the patient in front of them, but must be cognizant of the other patients on the waiting list. This is truly a lifeboat ethics scenario: two or more individuals are in direct competition for the same limited resource. Since the demand for organs exceeds the supply, some patients will not be transplanted and will die on the waiting list: 2–3 % of people active on the kidney transplant waiting list will die every year, and 10–20 % of would-be candidates for heart, liver and lung transplant die while waiting [36]. Even more concerning is the recognition that the pediatric priority that is a part of most deceased donor allocation systems means that a teen, even one who is nonadherent, who is approved for a transplant waiting list is likely to receive his transplant much sooner than would

[2]We will alternate the use of male and female pronouns throughout the chapter to refer to adolescents, except for the section on teratogenicity, for obvious reasons.

an adult candidate, regardless of the adult's adherence status. If the graft later fails due to nonadherence, neither the prioritized pediatric recipient nor the adult lingering on the waiting list will have achieved sustained benefit.

Clinicians recognize their dual responsibility in patient advocacy and responsible stewardship, and agree that transplant in a patient thought to be nonadherent cannot always be supported. A poll of the audience at a symposium on congenital heart disease, published by Kaufman and Jessup, supported a second heart transplant for a 13-year-old with medical complications (91 % in favor) but did not support a third transplant for that same patient after repeated nonadherence (14 % in favor) [37]. Surveys of transplant professionals show that they often list nonadherence as an absolute or relative contraindication to transplant [38].

16.2.3 Arguments for Transplantation When Nonadherence Is Known or Suspected

If it could be proven that adherence pre-transplant was a perfect predictor of post-transplant adherence, a compelling case could be made for denying or delaying transplant in this population. However, predictors of adherence are only suggestive, not definitive, and can sometimes be influenced by assumptions and biases of the transplant team. Since we cannot accurately predict non-adherence, it makes little sense to make what is, quite possibly, a life-and-death decision based on this flawed assessment.

There is no strong evidence that pre-transplant adherence to non-medical therapies, like dialysis attendance, can predict with certainty post-transplant medication adherence, though it makes intuitive sense that this might be the case [33]. The tasks that we ask a patient to perform during their pre-transplant care are significantly different from post-transplant care. For example, a teen who is late to dialysis, does not stick to his fluid restriction (often one litre per day or less) and misses his phosphate binding medication (which are taken with meals) is likely to be labeled as nonadherent. Post-transplant, a teen is expected to take medications with completely different side effects, drink more fluids (often greater than 3 litres per day), and to time medications to a specific time of day, not to meal intake. This pre-transplant patient may also have factors contributing to nonadherence that are alleviated by transplant, including less nausea, clearer cognition, improved mood and greater optimism. As respondents to an interview study on this topic stated, "I am totally unable to predict who may be compliant with [post-transplant] medication and who won't be. I've had extraordinary lack of success with predicting that" [33 online].

Even for teens with admitted nonadherence post-transplant, there is little guarantee that their nonadherence will continue with a second transplant. In a study from Minnesota, patients who received a second transplant after losing their first to non-adherence were compared to those who lost their first graft for other reasons.

While the loss of a second graft to non-adherence was higher in the group who were previously non-adherent (14 vs. 2 %) the majority of patients did well and overall graft and patient survival were similar between the groups [39]. There is, therefore, hope in that non-adherent patients can change their behavior for the better, even when they admit to being nonadherent in the past.

A significant concern with using nonadherence as a justification to deny transplant to a teen is the chance that this determination could be used to further marginalize patients or groups who already have diminished access, since the definition of adherence is somewhat subjective and therefore open to bias. Mosley and Kershaw have postulated that this may be part of the reason that African American (AA) adolescents have less access to transplant than their Caucasian peers [29] and similar biases have been noted in other minority populations, like the Aborigine population of Australia and Aboriginal populations in Canada [40].

A final caution of using nonadherence is quite practical—our assessments rely on honesty from the transplant candidate, which may well be compromised if the patient knows that she could be denied transplant based on her answers. Many clinical assessments of nonadherence are based on patient self-report. If we use self-reported nonadherence as a reason to not list a patient for transplant, patients may feel motivated to withhold information or provide answers that they think the team is expecting. This can lead to grave consequences down the road, since attempts at reducing the risk for nonadherence cannot be made if the team remains unaware of the problem.

16.2.4 Dealing with Nonadherence Suspicions Ethically

From the arguments above, it seems clear that resource allocation and concerns for post-transplant outcomes cannot, on their own, justify the denial of transplant to a teen at risk for nonadherence: the assumptions underlying this assessment are too great, and the consequences too dire. On the other hand, as responsible stewards of the transplant system with a responsibility to donors, donors' families and other wait list candidates, transplant professionals must do all we can to mitigate the factors that can lead to post-transplant nonadherence.

First, we must get better at defining, predicting and preventing nonadherence and its specific challenges in the adolescent age group. We must be careful not to oversimplify both the causes of and solutions to nonadherence. As Frader has noted, in cases of nonadherence there is an impulse to label the teen as the person with the problem, instead of looking deeply at the complex medical, psychological and social contributors to the teen's situation [35]. Dobbels has suggests that, instead of blaming teens for their nonadherent behavior, we get better at identifying risk factors early and training transplant teams to communicate effectively with these teens [12].

We can employ practical solutions that make it easier to adhere to our medical plans. Immunosuppressants that can be given as a single daily dose, instead of twice daily dosing, can improve adherence, as can avoiding medications with unpleasant cosmetic side effects [41]. Future strategies, like monthly injectable medications that can induce tolerance would obviate the need for adherence to a complicated immunosuppressive regimen, but these strategies are not yet ready for clinical use [42].

Adherence is not an issue that can be solved by patients and heath care providers alone. We must recognize the role that socioeconomic factors, like poor education and poverty, play into the ability of patients to adhere to the complexities of post-transplant care. In our role as advocates, we must work towards solutions that reduce these inequities and equip our patients with the tools they need to achieve better adherence. As well, the imbalance between supply and demand puts providers into the difficult situation of triaging patients. If this shortage can be ameliorated, the focus could return to the needs of the individual patients and avoid the risk of bedside rationing.

As Warady and Lantos suggest, the most balanced and humane approach to the nonadherent adolescent seeking transplant seems to be "an individualized assessment of barriers to adherence, likelihood of nonadherence and the potential benefits of interventions that might improve adherence" [43, p. 193]. Rhodes acknowledges that while justice concerns may favor refusing transplant to nonadherent teens, these concerns should be tempered with "caring concerns," which would require "a sincere effort to meet their patient's immediate needs and to work with her so that sometime in the future she might become a transplant candidate" [35, p. 380].

16.3 The Adolescent Who Refuses Transplant

The medical model of health care supports seeking a transplant for an adolescent with end stage solid organ failure in most circumstances. Kidney transplantation offers both survival and quality of life advantages over dialysis [44]. Liver transplantation and heart transplantation can save children from fatal end organ damage [45, 46]. Lung transplantation and small bowel transplantation remain more controversial, but there are at least some children for whom the benefits of transplant are felt to outweigh the risks [47, 48]. Of course, some children will have medical comorbidities that make transplant unsafe or unwise, so it may not be possible to transplant every child with end organ failure. Overall, however, transplantation is often the default choice for children with end organ disease; usually, when a medical team offers transplant listing, parents and patients accept the offer.

Occasionally, a medical team will offer transplantation but the adolescent patient will refuse that offer, sometimes with their parents' support and sometimes against their parents' wishes. In several well publicized cases, which will be discussed later in the chapter, teens have refused transplant even in life-threatening situations, and courts have sometimes intervened to force the child into transplant. How should the transplant team approach an adolescent who is deemed medically acceptable for transplantation, but who refuses to be listed? What ethical issues are involved?

16.3.1 Arguments Against Accepting an Adolescent's Transplant Refusal

While adolescents may have the same cognitive abilities as adults, there are psychological differences which affect their decision-making, including a tendency to be more greatly influenced by peers, higher tolerance for risk, and focus on short-term decision-making over long-term consequences [49]. Some argue that this renders adolescents incapable of making their own health care decisions, especially in high stakes, life-threatening situations. If an adolescent's decisions regarding transplant are based on immature reasoning, it makes sense that this refusal should not be accepted by the teen's parents or medical team, and that she should be treated more like a child, whose decision-making is routinely assumed by parents or legal guardians.

Beauchamp and Childress argue that capacity is a sliding scale, with more capacity required to make high stakes decisions [50]. Transplantation is definitely a high stakes decision, for while life with a transplant may be difficult—with its attendant risks and requirements for lifelong immunosuppression—a refusal to transplant is often fatal. For these high stakes decisions, many argue that the average adolescent lacks the emotional maturity and capability to refuse medically beneficial therapy. Ross has argued that children need the time to develop "enabling virtues" like self-control, in order to advance their lifetime autonomy [51]. In Ross's argument, allowing teenagers to dissent to life-saving medical procedures robs them of this time, and sacrifices the autonomy that they will develop later in life for present day choices that may be foolish or poorly grounded. Gilmore and Herring argue that adolescents may be able to consent to some types of therapy, but should not be similarly allowed to refuse that same therapy, especially in serious or life-threatening situations [52]. They reference Coggan [53], who argues that there are different types of autonomy:

1. "ideal desire" autonomy, that reflects what a person should want
2. "best desire" autonomy, that reflects what a person overall desires dictate, and
3. "current desire" autonomy, that reflects a person's immediate inclinations

Overriding adolescent decision-making could be seen as respect for best desire autonomy, even if it contravenes current desire autonomy, because the adolescent may well realize, in adulthood, that the transplant was beneficial and lifesaving. A good example of this may be the case of Hannah Jones.

Adolescents may have changing values over a relatively short period of time, which can make honouring their decisions in life threatening situations more problematic. For example, Hannah Jones was a 13-year-old whose heart had been weakened by chemotherapy, and who was offered transplant but refused. At the time, she stated that she understood that refusal could result in her death [54]. Child protection authorities met with Hannah and did not want to force her into treatment, and the British High Court upheld her choice to refuse the transplant [55]. Within a year, however, Hannah changed her mind, consented to listing and received a transplant. Interviewed about her choice at age 16, she stated “When I turned down the chance for a transplant, I didn’t leave hospital thinking I was going to die. In my mind, I thought I was going home to get better. I tried not to think about the alternative” [56 online]. For teens like Hannah Jones, the initial refusal might be short-sighted, since they may well change their minds; these teens will possibly be thankful if their earlier choice is not respected and they are compelled into the transplant.

Silber has argued that paternalism may be justified in adolescent health care, especially in cases with a mental health component, since protection of life outweighs respect for autonomy in children, and because adolescents may have impaired capacity incomplete brain development and/or comorbid mental health concerns [57]. Adolescents who are transplant candidates are, necessarily, sick, and their illness may prevent them from being fully capable. For example, a teen who, with the intention of killing herself, purposely ingests acetaminophen and develops fulminant hepatic failure would presumably have some mental health concerns [58]. A refusal to accept transplant in this case might be consistent with the adolescent’s previous suicidal intent, but it would be potentially reasonable to override the adolescent’s dissent, in this situation, under the assumption that her current statements are not consistent with those she would have once her mental illness was successfully treated.

This paternalistic perspective to override an adolescent’s decision-making authority regarding transplant acceptance may make some intuitive sense. How many of us can look back on our adolescence and state that we always made rational, well deliberated decisions? Statistics on teen suicide, accidental deaths, unintended pregnancy and drug and alcohol abuse point to the self-harming and dangerous behaviors that are relatively common in this age group [59–61]. We recognize that adolescents make mistakes, and feel a responsibility to protect them from bad choices that may rob them of healthy adult lives. Legislation that holds adolescent drivers to a zero blood alcohol level or restricts their night driving are examples are examples of justified paternalism for adolescents, over and above that which is imposed on adults, in an attempt to save them from the harm that their decisions could cause [62, 63].

16.3.2 Arguments for Aquiescing to a Transplant Refusal

While some jurisdictions place legal limits on the ability of adolescents to consent to, or refuse, treatment, law recognizes generally that teenagers need to be involved, at least to some degree, in their medical decision-making. Major medical groups, like the Canadian Pediatric Society, American Academy of Pediatrics and the British Medical Association advise that a minor should be given more medical decision-making authority as he or she grows [64–67]. Most societies and authors argue that adolescents who meet criteria for providing consent have a right to be involved in and, in some cases, direct, their own medical treatment. Those criteria differ slightly among authors and legal jurisdiction, but generally they are the same requirements that we would apply to medical decision-making in adults, namely that the decision-maker can display [68, 69]:

1. **understanding** of the information required to make a decision, including information about the alternatives
2. **appreciation** of the nature and significance of the decision
3. the ability to **reason** regarding the decision and its consequences in order to make a choice
4. the ability to express a **choice**
5. a set of consistent and stable **values**

Many adolescents, particularly those in their older adolescent years, would be able to demonstrate these abilities to the satisfaction of their parents, medical caregivers, and/or the legal system. Research shows that children with serious and/or chronic illnesses may even develop the insight to make these decisions earlier than their healthy peers [70]. This could be especially important for teens considering a second transplant, since their previous lived experience would be directly relevant to their post-transplant expectations.

Courts have indeed granted adolescents the right to make their own major health care decisions when they have shown an ability to provide fully informed consent. In Britain, the Gillick criteria for competence were developed after a mother argued that physicians should not be allowed to provide contraceptive advice to her daughter, who was less than 16 years old at the time [71]. The court disagreed, stating:

> …whether or not a child is capable of giving the necessary consent will depend on the child's maturity and understanding and the nature of the consent required. The child must be capable of making a reasonable assessment of the advantages and disadvantages of the treatment proposed, so the consent, if given, can be properly and fairly described as true consent.

In a Canadian case, the Supreme Court of Canada ruled that adolescents less than 16 years old could have their choices regarding medical treatment respected when:

> after a careful analysis of the young person's ability to exercise mature and independent judgment, the court is persuaded that the necessary level of maturity exists [72]

From a consequentialist perspective, forcing an adolescent to accept a transplant that he or she does not want may not extend or improve his life. Successful transplantation requires the lifelong "buy in" of the adolescent, in terms of taking medication, attending medical appointments and adhering to other medical plans. The dismal adherence rates of adolescents (specifics of which are discussed in the previous section) speak to the difficulty in achieving adherence even in the motivated adolescent. How poorly, then, can we expect adolescents to do post-transplant if forced into the surgery?

Of course, forcing transplant on a reluctant teen may be logistically impossible, even if it is legally or ethically permissible. How could one actually "force" a refusing teen into surgery or treatment? Would the team be able or willing to use mechanical or chemical restraints to get the patient into the operating room? After transplant, would the team be willing or even able to compel the teen to take his medication and comply with other follow-up procedures, like biopsies? While there is no data to support this, anecdotal accounts suggest that there are cases where teens have been coerced into transplant or have grudgingly acquiesced, but that true forced transplant is unlikely to occur.

We must also recognize that donor organs remain a scarce resource, so a transplant for the reluctant adolescent may mean that another person on the waiting list dies without a graft. This is not a theoretical concern—over 6900 people died in the U.S. in 2013 while on an organ waiting list [73]. One could argue, then, that this scarce resource, often called the "gift of life," should be given to a person who truly wants it, since they are more likely to take better care of it and to get longer graft survival from it. If an adolescent displays an unwillingness to accept that "gift," perhaps we should respect that choice and give it to someone else.

While a transplant may be considered beneficial by the medical team proposing it, the patient's calculus of risk over benefit may be markedly different, even when presented with the same medical facts. Some adolescents may have comorbidities that limit the improvements they can expect post-transplant. Transplantation comes with some very real risks, including infection and malignancy, which may be in and of themselves life-threatening [74]. Some types of transplantation, like lung or intestinal transplant, may not always offer survival advantage over conservative management [75]. Even in heart disease, refusal of transplant may not be immediately life threatening. While dialysis is generally inferior to kidney transplantation —in both survival and quality of life—only a minority of adult dialysis patients in one study chose to be listed for transplant [76]. Reasons for refusal included the side effects of immunosuppressant therapy, unpredictability of transplant outcomes, and unfavorable outcomes in fellow patients. In the U.S., courts have found that competent adults, or their substitute decision-makers, may refuse transplant even in life-threatening situations [77]. Some potential recipients may have lifelong coping issues related to their childhood transplant and would not characterize transplant as an "unmitigated good" [78]. While one might argue that transplant is usually in a patient's best interests, we must recognize the situations which may be exceptions to that general rule.

16.3.3 Dealing with a Transplant Refusal

Medical professionals need to understand their own biases in these situations. We are much more likely to be troubled by an adolescent's refusal of recommended treatment than we would be with their under-informed assent. Transplant professionals manage complex situations and offer transplant after a careful weighing of the benefits and risks to particular patients. While we recognize that transplant may not be the best choice in all situations, our identities as transplant professionals are based on the belief that transplant is the right choice for many. It is only natural for the transplant team to be distressed, disappointed or even insulted when an adolescent refuses what we believe to be the best option in a life-threatening situation. An honest reflection on our feelings about transplant refusal can help us better understand why and how we might be pushing a reluctant teen towards accepting a transplant that they say they do not want.

While an adolescent's refusal to undergo transplant can be troubling, it is not without potential solution. Sinclair recommends a shared decision-making in these difficult cases, that would involve an open and honest discussion of the health care team's concerns with the transplant refusal and a search for the reasons behind the initial refusal, in an attempt to better understand the teen's perspective [79]. If any of the adolescent's identified concerns are modifiable they should be addressed. Patients may have unvoiced fears about surgery, pain or post-surgical recovery that can be addressed with adequate discussion, education and involvement of colleagues like clinical health psychology, pain services, child life, etc. Undiagnosed mental health issues, like depression or post traumatic stress disorder, can be diagnosed and treated. Concerns about cosmetic side effects can be discussed and options with effects can be explored (e.g. steroid free or steroid minimization protocols). Myths about transplant can be dispelled and stories of success post-transplant can be shared. Linking the potential recipient to a peer mentor who has been successfully transplanted may help the adolescent share their hopes and fears with a peer in a similar situation [80–82].

We should also be mindful of how urgently the decision to transplant needs to be made, and consider, in less urgent cases, whether further discussion can persuade a reluctant adolescent to agree to transplant. Two British legal cases highlight the difference in approach. In one, a 12-year-old girl refused heart transplant and lived seven more years with slowly deteriorating health after the courts decided not to force her into transplant [79]. In the other, a court ordered a teen with rapidly progressive heart failure (who was expected to die within a week) to undergo a transplant she initially refused [83]. A publication ban prevents the public from knowing more about the court decisions that led to the different outcomes, but it seems likely that part of the decision had to do with how quickly the decision needed to be made. It seems less defensible to force an adolescent into transplant when the decision is not immediately life-threatening.

It is also prudent to listen carefully to the adolescent professing a refusal, to determine whether they are saying "not ever" or merely "not now." With a "not ever" refusal, a teen is refusing transplant now and at any point in the future. If the proposed transplant is truly life-saving, then the refusing adolescent will never become an adult, and his decisions today will rob him of that long-term autonomy discussed earlier. In these cases, especially if there is time pressure to make the decision, there may be a valid case for pushing the teen harder towards transplant, since there is no chance for the adolescent to contemplate, consult or change his mind. This does not relieve the transplant team of the responsibility to explore the reasons behind the transplant refusal, but shifts it to the post-surgical period, which will hopefully be less time-pressured. Of course, the adolescent in this scenario may well feel that these discussions are coming too late in the transplant course, since the surgery and its attendant responsibilities for post-transplant care are now a *fait accompli*.

In the "not now" scenario, an adolescent refuses transplant in the short-term, but is not completely ruling it out down the road. This is a more common scenario for not only kidney transplantation, in which dialysis offers an alternative to immediate transplant, but now, possibly, for heart transplant candidates, what with the increased use of ventricular assist devices (VADs). While quality of life on dialysis is generally lower than that after transplant, it is not of such terrible quality that a hasty decision needs to be made, in most cases. In the case of a thoughtful adolescent, "not now" may be a very mature choice—the teen recognizes that her current struggles or limitations make transplant less likely to be successful, and is choosing to wait until those struggles can be better managed. If there is an alternative bridging strategy, imperfect though it may be, a "not now" refusal may be the beginning a path towards eventual, successful transplant.

Importantly, a transplant refusal may not be the same as a refusal of all medical care. If a transplant refusal is respected, the medical team must ensure that other medical care that could improve the child's primary disease or increase the child's comfort continues. Palliative care colleagues can be especially helpful in the transition from a cure-seeking path to a comfort-seeking path [84].

16.3.4 Conclusion

Adolescents do not always make decisions that are perceived to be in their best interests by parents or medical teams, nor that are consistent with the best available medical evidence or advice. Medical professionals are obligated to offer care that is in a teen's best interests, but also to respect the emerging capabilities of adolescents in charting their own medical course. An adolescent who refuses transplant may have modifiable reasons for refusal, and those should be explored and ameliorated to the greatest degree possible. In some situations, especially when the chances of transplant success is low and/or when the patient's refusal is deemed to be mature, informed and consistent, it may be reasonable to accept a refusal, even if it means that the adolescent to die. The more capable the adolescent and the more consistent her wishes over time, the more seriously the refusal needs to be considered and respected. In very rare cases, where the refusal is felt to be imminently life-threatening and uninformed, involuntary or inconsistent, clinicians may need to resort to child protective services or the courts to compel an adolescent into accepting a transplant. The legal route, however, should be the last resort since it will, almost certainly, erode the adolescent's trust in the health care system and in his transplant team and is no guarantee that the adolescent will adhere to the post-transplant care necessary to maintain the graft. In general, we advocate an approach that balances respect for the adolescent's emerging autonomy with recognition that the decisions involve maturity and wisdom that may simply be beyond the adolescent's current capabilities. This recognition can lead to the development of support structures necessary to help the adolescent on a path to success, however that may be defined in each particular case.

16.4 Fertility and Contraception for Post-transplant Adolescent Women

While both male and female transplant recipients may have fertility concerns post-transplant, only female recipients need to worry about the potential teratogenicity of their medications and the impact of pregnancy on their own transplant function. The teratogenic effects of immunosuppressant medications are still not completely understood, but there is good evidence that many of the medications that are best for preventing graft rejection and reducing side effects are, at least potentially, harmful to a developing fetus. Adolescent girls post-transplant may become sexually active, and may become pregnant, either intentionally or unintentionally. How should providers deal with contraceptive counseling and teratogenic medications in this population? Do patients have similar rights to other adolescents seeking contraceptive counseling or do the complications of post-transplant care impose ethically justifiable restrictions on access to confidential care?

16.4.1 Pregnancy Post-transplant and Risks of Unplanned Pregnancy

Current consensus guidelines recommend delaying pregnancy post-transplant until the recipient has stable graft function and is at low risk for opportunistic infection [85]. Some women experience difficulties with fertility, due to their underlying disease, medications used pre-transplant and their general post-transplant health, but all post-menarchal females should be considered potentially fertile after solid organ transplant [86]. Pregnancies post-transplant are considered high-risk, with rates of eclampsia, prematurity, low birth weight and perinatal death that are higher than the general population [87]. Live birth in women with a functioning kidney graft does not seem to negatively impact maternal or graft survival, but liver and heart-lung patients likely face more risk than kidney transplant patients, and less is known about outcomes due to their relative rarity [88–90]. Pregnancy complications can lead to neonatal issues that can affect a child's long term physical and neurocognitive development. Information has generally been gleaned from voluntary pregnancy registries, and the risks of long-term outcomes, like developmental delay and autoimmune disease in offspring, are not yet completely understood. This lack of clear data has led the transplant community to be cautious in its recommendations and to careful, educated planning of pregnancy for all pre-menopausal female transplant recipients [85].

In addition to the risks of pregnancy on a recipient's graft function and on fetal viability, many of the medications used after solid organ transplantation have known or suspected teratogenicity. The data is somewhat limited because there are few prospective studies on these drugs in pregnant women, so they cannot generally be classed as Federal Drug Administration (FDA) Category A, the safest category.[3] Currently used immunosuppressant medication range from prednisone, which is FDA pregnancy Category B, to mycophenolate mofetil and rapamycin, which are pregnancy Category D. Many transplant patients may be on other medications with teratogenic potential, like ACE inhibitors (pregnancy Category C in the first trimester, Category D thereafter) and statins (pregnancy Category X).

The Centers for Disease Control (CDC) categorizes all current contraceptive methods as category 2 for uncomplicated solid organ transplant patients, meaning that the advantages of using the method generally outweigh the theoretical or proven risks [91]. Despite the fact that contraception is generally considered safe, solid organ transplant recipients do not always receive appropriate contraceptive counseling. In a Brazilian study, only 49 % of adult women post kidney transplantation had received contraceptive counseling [92]. While 72 % were using some kind of contraception, 93 % of pregnancies in this group of women were unplanned.

The data on sexual activity in youth who have received transplants is mixed. An American study of over 32,000 youths between the seventh and twelfth grades found that those with chronic disease had very similar patterns of sexual activity to those without chronic disease, including proportion ever having intercourse, age of sexual debut, pregnancy involvement or patterns of contraceptive use and sexual orientation [93]. Conversely, a French study of adolescents of the same ages as the American study found that 38 % of girls with chronic illness had sexual intercourse, compared with 28 % of healthy controls [94]. A study of kidney transplant recipients found a lower rate of romantic relationships and sexual intercourse than healthy peers [95] and one that compared stable kidney transplant patients to those with youth onset diabetes showed lower rates of romantic relationship and higher rates of perceived sexual dysfunction in the transplant group [96].

---

[3]The FDA Categories of Drug Safety During Pregnancy are as follows:

A—Controlled human studies show no fetal risks; these drugs are the safest.

B—Animal studies show no risk to the fetus but no controlled human studies have been conducted, or animal studies show a risk to the fetus but well-controlled human studies do not.

C—No adequate animal or human studies have been conducted, or adverse fetal effects have been shown in animals but no human data are available.

D—Evidence of human fetal risk exists, but benefits may outweigh risks in certain situations (e.g., life-threatening disorders, serious disorders for which safer drugs cannot be used or are ineffective).

X—Proven fetal risks outweigh any possible benefit.

There is very little published data on contraceptive practices in adolescent transplant recipients. One study in adolescent liver transplant recipients who had been transplanted in childhood found them to be much more likely than their non-transplanted peers to use condom-based contraception (58 vs. 12 %) [97]. Unfortunately, the typical-use failure rate for condoms is 18 % and may be even higher in adolescents, so condom based contraception is not the best pregnancy prevention strategy for this group [98].

This data suggests that sexual activity in transplant recipients is not uncommon, but that contraceptive counseling is likely inadequate to prevent unwanted pregnancy. Combined with the known and unknown teratogenic effect of common post-transplant medications, transplant professionals need to be aware that the sexual health needs of their teenage patients may not be met by current practice. So, how should a transplant practitioner ethically counsel an adolescent recipient about her risks and responsibilities regarding sexual heath, and is it ever fair to alter a teen's immunosuppressant strategy because of fears of unintended pregnancy?

16.4.2 Balancing Respect for Autonomy and Non-maleficence in Adolescent Transplant Recipients

Two conflicting responsibilities of the transplant team in the care of adolescent recipients are respect for the recipient's autonomy and confidentiality and the responsibility to avoid harm to the recipient and, possibly, her potential offspring.

Contraception counseling in adolescents is one area where the minor's right to confidentiality is usually respected. The American Academy of Pediatrics writes that "policies supporting adolescent consent and protecting adolescent confidentiality are in the best interests of adolescents" [60 online]. The laws and professional standards in many jurisdictions require health professionals to provide confidential contraceptive counseling to adolescents [99]. The motivation for this is partly to respect the emerging autonomy of the adolescent patient, but also to encourage adolescents to seek out health care when they might otherwise not: if a patient worries that her parents might have access to information on her sexual behavior she might be less willing to share that information with a health care provider. Yet the confidentiality is not absolute: in many jurisdictions providers are required to report to child welfare authorities if an adolescent discloses a sexual relationship with a much older partner, or one that is in a position of power [100].

Health care providers also have an obligation to protect patients from foreseeable harm and to inform them of the side effects of their medications, and this obligation may be regulated and monitored. In the United States, the FDA mandates a Risk Evaluation and Mitigation Strategy (REMS) for mycophenolate mofetil (MMF), a commonly used immunosuppressant in solid organ transplant. The program aims "to tell doctors, nurses, pharmacists and patients about the risks of taking mycophenolate during pregnancy" [101]. While it does not restrict access of MMF to women of childbearing age or require negative pregnancy test prior to prescribing the medication, it does urge the use of two methods of contraception (one hormonal and one barrier or two different types of barrier). As well, it requires completion of a patient-prescriber acknowledgement for prescribers, patients and, in the case of minors, their parent or guardian. It does not tell prescribers what to do if patients choose not to complete the form or specifically address adolescent issues.

Many of the goals of the REMS program are laudable. Clearly, providers need to educate young women about the teratogenic risks of their medications so that they can make informed and responsible choices. The REMS program does not restrict young women from accessing MMF, even in pregnancy, as long as the prescriber engages in an evolution of the risks and benefits to the patient. The program also recognizes that minors may be sexually active, and includes pediatric providers in the program.

Providers should be careful, however, not to use the REMS program in a manner that undermines the adolescent's ability to access confidential care. REMS requires parents to sign the patient-provider form for minors, acknowledging the teratogenic risks of the medication. It does not require the adolescent to divulge her contraceptive plans to her parents. The complexity of post-transplant care often requires the involvement of parents to help administer and track medication doses, even in teens. The safety of contraception in transplant patients is similar to that of the general population and choice of contraceptive should not affect the efficacy of immunosuppressant medication. Therefore, it seems reasonable that an adolescent girl could make contraceptive choices separate from her overall transplant care and so that care should remain confidential.

Providers should also be aware that risk management programs are not a substitute for open and honest communication about the teratogenicity of medications on an ongoing basis. Analysis of American pregnancy prevention programs for patients on isoretinoin, a highly teratogenic acne medication, has shown that fetal exposure was still a major problem even with the most comprehensive program (iPLEDGE) [102]. Women taking Category X medications have alarmingly high rates of nonadherence to oral contraception, suggesting that many are insufficiently protected from pregnancy [103].

It seems clear that direct communication with a responsible teen make sense, even with the added complexity of transplant care. What, then, should be the approach to the "irresponsible" teen, who is presented with the facts on pregnancy and chooses to continue to have intercourse with suboptimal contraception? Whitney and McCullough have argued that physicians may have a moral duty to withhold isoretinoin, another highly teratogenic medication, when a "young woman is so irresponsible and impulsive that the physician judges that she is not currently capable of undertaking reliable contraception and that a conversation about this aspect of her life would not educate or inform her" [104, p. 35]. But, unlike other teratogenic medications, which may have reasonable non-teratogenic alternatives, transplant medications are carefully chosen with the wellbeing of the transplant recipient the primary concern. There are alternatives to rapamycin and MMF for women, which themselves have side effects—for example, azathioprine is sometimes used instead of MMF when gastrointestinal side effects are severe. This seems justifiable because the risks and benefits of these medication changes accrue to the same person—the transplant recipient. What does not seem justifiable is to unilaterally restrict teratogenic medications on the basis of an individual's contraceptive choices, because the benefits in that case accrue to a potential fetus (who has not yet even been conceived) and the risks accrue to the transplant recipient. In general, women with reproductive potential are allowed to make choices while pregnant about their own bodies, even if those choices could harm a potential fetus. As well, critics of Whitney and McCullogh's approach have pointed out that physicians are not infallible judges of moral character, and that a decision to withhold potentially helpful medications, without discussion, on the basis of "responsibility" is excessively paternalistic [105]. There is also the chance that withholding medications will erode trust in the physician-patient relationship, and lead to issues with adherence to other medications down the road. As discussed in the section on adherence, this is an outcome best avoided.

16.4.3 Conclusion

More important than the legal documentation on teratogenic medications is the recognition from transplant providers that they have a special onus to explain the teratogenicity of these medications and establish a therapeutic relationship that allows teens to make informed choices. Promotion of healthy sexuality, effective contraception and clear communication will go farther, in the long run, than will requiring signatures on a form or mandating disclosure to a parent or guardian. A program like REMS, if used as part of an open and shared decision-making model, can demonstrate respect for an adolescent and his or her choices regarding sexuality.

16.5 Ethical Issues in the Transition from Pediatric to Adult Care

If transplantation of solid organs is to be successful then all pediatric recipients will grow into adulthood and ultimately require transition from the pediatric health care provider/clinic to an adult program. This would seem to be a simple process of referral from a pediatric care giver to an adult medicine colleague. However, in 2000 Watson and colleagues [106] identified some disastrous consequences following the transfer of 20 pediatric renal allograft recipients to adult facilities. The increase in non-adherence and graft failure shortly after transfer to adult nephrology care was alarming. This report stimulated intense interest in the concept of transition and led in 2011 to the publication of a consensus statement jointly developed by the International Society of Nephrology (ISN) and the International Pediatric Nephrology Association (IPNA). The transition process can raise complex ethical questions for the pediatric provider; in this section, we will explore some of these issues, including the transition of responsibility from pediatrician to internist, the patient's evolving maturity and responsibility and the pressure to transplant patients before they lose their pediatric priority.

16.5.1 The Importance of Successful Transition in Pediatric Transplantation

Transition is a multifaceted, active process that attends to the medical, psychosocial, and educational need of adolescents and young adults as they move from child (pediatric) to adult centered care [107]. The goal of successful transition is to provide high quality, coordinated, uninterrupted healthcare which is patient centered, age and developmentally appropriate, culturally competent, flexible, responsive and comprehensive. In addition, the transition process must provide support to the parent(s)/guardian of the adolescent/young adult during the transition process [108].

Done well, transition can be a developmentally appropriate process that allows a teen to continue the good management of his transplant that began in the pediatric years. The danger, however, is that transition can disrupt a fragile situation, leading to complications and potential graft loss. The ethical onus is on the pediatric provider to anticipate and plan for transition with the knowledge that this can be a risky time. The 2007 Organ Procurment Transplant Network(OPTN)/Scientific Registry for Transplant Recipients(SRTR) annual report indicated that the five year renal graft survival rates for those transplanted at age 11–17 was substantially less than the 1–5 and 6–10 year old age groups and was comparable to the >65 year old age group for both live-related and deceased donor recipients. The etiology of the decreased graft survival rates was not specifically delineated but non-adherence with medications was a prime suspect. Ethical concerns are therefore raised when financial or institutional factors encourage transition precisely at the age that it riskiest for graft loss.

16.5.2 Practicalities and Challenges in Transition

There are many models of transition. In the past the most commonly used was direct transition which involved the single step model of transition from pediatric to adult care. Subsequently, intermediary step models have been developed: transition clinic, adolescent clinic and young adult clinic. There are no long-term data to identify the optimal approach to successful transitioning, since many of these models have only been recently developed.

The ISN/IPNA consensus statement delineated a transition process which focused on the need for transition champions in both the pediatric and adult units who would be responsible for coordinating the transition process. Certainly such advocates are necessary for the process to be successful. The consensus statement also details 11 competencies that the adolescent/young adult should be expected to exhibit prior to being transferred to the adult unit. Both these concepts should form the framework of a transition policy to be developed by both the transferring pediatric and accepting adult unit. Ideally at least one pediatric health care professional (champion) and one adult health care professional (champion) should be involved in the care of the adolescent/young adult prior to, during and following the transitioning process: continuity is exceedingly important.

There are four main barriers to successful transition: (1) reluctance of older adolescents and young adults to leave the security of a familiar environment of pediatric care; (2) reluctance of parent(s)/guardian to reduce their oversight of the healthcare of the adolescent/young adult; (3) reluctance of the pediatric healthcare professionals to relinquish their responsibility for care of the adolescent/young adult; (4) the availability, familiarity and willingness of adult healthcare professionals to assume responsibility for care of the adolescent/young adult [109].

As with other adolescent care, transition requires a transfer of decision making authority and responsibility from the parent/guardian to the teen. However, it is ethically difficult to advocate transfer of responsibility, despite its desirability, if the teen is not capable of accepting that responsibility. While some jurisdictions have a legal bright line for consent to medical treatment, it is well recognized that this transition is best seen as a gradual evolution, rather than an abrupt shift at the age of majority. The ISN/IPNA consensus statement advocates the introduction of the concept of transition should be introduced in early adolescence (12–14 years of age) [110]. Later steps in transition, including formal transfer to the care of an adult provider, will occur many years later, and their timing is determined by many factors, both related to the particular patient and external to her. Transitioning should not occur at a time of crisis. If graft function is deteriorating then transitioning should be delayed until the next step in clinical management has been accomplished. Patients with cognitive delay, common in solid organ transplant recipients, will likely need more individualized plans. These patients should have an assessment of their capacity to undertake the functions of daily living such as decision making, vocational/educational objectives, peer relationships and sexuality. Continued or even permanent involvement of the parent(s)/guardian may be required and the length of time of the transitioning process may need to be significantly increased in such individuals [111]. Likewise, a major issue in the transitioning of a cognitively delayed individual is the capability of adult care professionals to undertake the care of such patients, especially if they have medical comorbidities or diseases that previously limited life expectancy to the pediatric years. An ethical approach to care requires advocating for this careful transition, even if local policies, regulations or insurance concerns create difficulties.

Health care coverage and other benefits can be a major factor in transition. Willoughby et al. [112] reviewed 1001 primary pediatric transplant recipients between 1995 and 2001 in whom the US. Medicare coverage was lost in 51.6 %, and the graft loss rate doubled at 3 years post-transplant in those who lost coverage ($p < 0.001$) compared to those who retained eligibility. Failure rates in the 4th year after cessation of coverage were 6.5 % with no coverage compared to 3.7 % with coverage and the duration of graft function was 13.7 % years with coverage compared to 7.7 years without coverage. The authors projected the long term graft survival was 25 years with coverage and 15 years without. Even socialized health care systems can create insurance problems if prescription drugs are not universally covered, as is the case in Canada. Consequently, prior to transitioning to adult care it is imperative that insurance coverage issue be resolved so that the transitioned adolescent/young adult is not confronted with financial burdens during the transition process. This may raise ethical concerns during the transplant eligibility discussion, since it makes little sense to proceed with transplantation if funding for organ sustaining immunosuppressive drugs and continued funding for medical care, cannot be assured for life of the graft.

While the need for transition is clear, its implementation is far from universal, which can lead to inequities in the management of patients who are not exposed to a robust transition process. Despite the development of consensus guidelines to assist in the development of a comprehensive transition process a survey [113] of European Pediatric Nephrology units two years after publication of the guidelines indicated that only one third of the units integrated the guidelines into their transition practice. One major problem was patient volume: 47 % of the units transferred <5 patients annually with chronic disease stage 5 (primarily post-transplant patients) to a median of 5 adult units. It is incumbent on the transitioning team, then, to educate themselves and their adult colleagues and advocate for appropriate resources, even in smaller programs.

Transition can be a difficult time for parents, who must step back from the care of a child who they saw through life threatening illness. While it is natural for both the parents and medical team to have difficulty with this necessary step, it is clear that respect for autonomy, beneficence/nonmaleficence and justice are all satisfied by a robust and successful transition process, since the well transitioned teen has a higher likelihood of graft survival and less need for re-transplant. As with other difficult transitions from adolescence to adulthood, support for the entire family will aid in this recognition.

16.5.3 Pressure to Transplant Before Transition

In many allocation systems, pediatric patients are given priority in access to deceased donor organs, even when size matching is not a major concern, as in kidney transplantation. This is justified because there are time limited growth and developmental opportunities in childhood that are stunted by organ failure. Priority can also possibly be justified by fair innings, which would allow priority on the basis of children with organ failure having had less healthy years than those who develop their disease later in life. As Ross and Thistlewaite argue, however, fair innings on its own is not sufficient justification as it should ideally supplement equal opportunity, which recognizes that all wait list candidates are of equal worth regardless of their age [114]. In most systems, pediatric priority is lost at some time in later adolescence, and teens who reach this age while still waiting then need to wait with other adults, though in other systems priority is maintained as long as the patient started dialysis in childhood [115, 116]. This loss of priority can place providers in the difficult position of rushing a workup or listing before that priority is lost. While the motivation to assist the teen is well-meaning, a teen who is not ready for transplant, medically or psychosocially, may not achieve the full benefit of that priority if he becomes nonadherent or has difficulty with transitioning. As well, there is a valid case to be made that transplanting a teen too early is a violation of utility, since the teen is less likely to maintain his graft if inadequately prepared. These difficult cases require a frank discussion with the teen and her family about the team's concerns, and a recognition that waiting for an organ such as a kidney while undergoing dialysis, even for a long time, might be better for the teen than would be a transplant at the wrong time. Maintaining pediatric priority until the time of transplant is obviously attractive from a transition perspective, but is open to criticism because it could be seen to disadvantage adults, especially young adult who developed their disease just after the pediatric cutoff.

16.6 Conclusion

Adolescents, at the threshold between dependence and independence, have unique challenges in the transplant process. This chapter has highlighted the areas where ethical concerns can arise, including evaluation of the reluctant transplant candidate, approaches to evaluating and ameliorating nonadherence, contraceptive counseling in the female transplant recipient, and transition from pediatric to adult centres. While there is still much research to be done in these areas, acknowledgement and reinforcement of the ethical principles that gird transplant assessment, allocation and clinical management in general can guide the transplant professional towards ethical care for these patients and their families.

References

1. Andreoni KA, Forbes R, Andreoni RM, Phillips G, Stewart H, Ferris M. Age-related kidney transplant outcomes: health disparities amplified in adolescence. JAMA Intern Med. 2013;173(16):1524–32.
2. Magee JC, Bucuvalas JC, Farmer DG, Harmon WE, Hulbert-Shearon TE, Mendeloff EN. Pediatric transplantation. Am J Transplant. 2004;4(Suppl 9):54–71.

3. Nevins TE. Non-compliance and its management in teenagers. Pediatr Transpl. 2002;6 (6):475–9.
4. Feinstein S, Keich R, Becker-Cohen R, Rinat C, Schwartz SB, Frishberg Y. Is noncompliance among adolescent renal transplant recipients inevitable? Pediatrics. 2005;115(4):969–73.
5. Cass A, Cunningham J, Anderson K, Snelling P, Colman S, Devitt J, Preece C, Eris J. Decision-making about suitability for kidney transplantation: Results of a national survey of Australian nephrologists. Nephrology (Carlton). 2007;12(3):299–304.
6. Knoll G, Cockfield S, Blydt-Hansen T, Baran D, Kiberd B, Landsberg D, Rush D, Cole E. Kidney transplant working group of the canadian society of transplantation. Canadian society of transplantation consensus guidelines on eligibility for kidney transplantation. CMAJ. 2005;173(10):1181–4. Erratum in: CMAJ. 2005;173(12):1490.
7. Batabyal P, Chapman JR, Wong G, Craig JC, Tong A. Clinical practice guidelines on wait-listing for kidney transplantation: consistent and equitable? Transplantation. 2012;94 (7):703–13.
8. Sabaté E. Adherence to long-term therapies: evidence for action. Geneva: World Health Organisation, Geneva; 2003.
9. Council on Ethical and Judicial Affairs, American Medical Association. Ethical considerations in the allocation of organs and other scarce medical resources among patients. Arch Intern Med. 1995;155:29.
10. Geddes CC, Rodger RS, Smith C, Ganai A. Allocation of deceased donor kidneys for transplantation: opinions of patients with CKD. Am J Kidney Dis. 2005;46(5):949–56.
11. Abecassis M, Adams M, Adams P, Arnold RM, Atkins CR, Barr ML, Bennett WM, Bia M, Briscoe DM, Burdick J, Corry RJ, Davis J, Delmonico FL, Gaston RS, Harmon W, Jacobs CL, Kahn J, Leichtman A, Miller C, Moss D, Newmann JM, Rosen LS, Siminoff L, Spital A, Starnes VA, Thomas C, Tyler LS, Williams L, Wright FH, Youngner S. Live organ donor consensus group. consensus statement on the live organ donor. JAMA. 2000;284 (22):2919–26.
12. Dobbels F, Hames A, Aujoulat I, Heaton N, Samyn M. Should we retransplant a patient who is non-adherent? A literature review and critical reflection. Pediatr Transpl. 2012;16(1):4–11.
13. Shemesh E. Measuring adherence to medications: are complex methods superior to simple ones? Pediatr Transpl. 2012;16(4):315–7.
14. Dobbels F, Ruppar T, de Geest S, Decorte A, van Damme-Lombaerts R, Fine RN. Adherence to the immunosuppressive regimen in pediatric kidney transplant recipients: a systematic review. Pediatr Transplant. 2010;14:603–13.
15. Pinsky BW, Takemoto SK, Lentine KL, Burroughs TE, Schnitzler MA, Salvalaggio PR. Transplant outcomes and economic costs associated with patient noncompliance to immunosuppression. Am J Transpl. 2009;9(11):2597–606.
16. Dobbels F, Van Damme-Lombaert R, Vanhaecke J, De Geest S. Growing pains: nonadherence with the immunosuppressive regimen in adolescent transplant recipients. Pediatr Transpl. 2005;9(3):381–90.
17. Ettenger RB, Rosenthal JT, Marik JL, Malekzadeh M, Forsythe SB, Kamil ES, Salusky IB, Fine RN. Improved cadaveric renal transplant outcome in children. Pediatr Nephrol. 1991;5 (1):137–42.
18. Oliva M, Singh TP, Gauvreau K, Vanderpluym CJ, Bastardi HJ, Almond CS. Impact of medication nonadherence on survival after pediatric heart transplantation in the U.S.A. J Heart Lung Transpl. 2013;32(9):881–8.
19. Berquist RK, Berquist WE, Esquivel CO, Cox KL, Wayman KI, Litt IF. Nonadherence to post-transplant care: prevalence, risk factors and outcomes in adolescent liver transplant recipients. Pediatr Transpl. 2008;12(2):194–200.
20. Butler JA, Roderick P, Mullee M, Mason JC, Peveler RC. Frequency and impact of nonadherence to immunosuppressants after renal transplantation: a systematic review. Transplantation. 2004;77(5):769–76.

21. Chisholm-Burns MA, Spivey CA, Rehfeld R, Zawaideh M, Roe DJ, Gruessner R. Immunosuppressant therapy adherence and graft failure among pediatric renal transplant recipients. Am J Transplant. 2009;9(11):2497–504.
22. Ringewald JM, Gidding SS, Crawford SE, Backer CL, Mavroudis C, Pahl E. Nonadherence is associated with late rejection in pediatric heart transplant recipients. J Pediatr. 2001;139 (1):75–8.
23. Stuber ML, Shemesh E, Seacord D, Washington J 3rd, Hellemann G, McDiarmid S. Evaluating nonadherence to immunosuppressant medications in pediatric liver transplant recipients. Pediatr Transpl. 2008;12(3):284–8.
24. De Geest S, Abraham I, Moons P, Vandeputte M, Van Cleemput J, Evers G, Daenen W, Vanhaecke J. Late acute rejection and subclinical noncompliance with cyclosporine therapy in heart transplant recipients. J Heart Lung Transpl. 1998;17(9):854–63.
25. Greenstein S, Siegal B. Compliance and noncompliance in patients with a functioning renal transplant: a multicenter study. Transplantation. 1998;66:1718–26.
26. Maikranz JM, Steele RG, Dreyer ML, et al. The relationship of hope and illness-related uncertainty to emotional adjustment and adherence among pediatric renal and liver transplant recipients. J Pediatr Psychol. 2007;32:571–81.
27. Wolff G, Strecker K, Vester U, Latta K, Ehrich JH. Non-compliance following renal transplantation in children and adolescents. Pediatr Nephrol. 1998;12:703–8.
28. Blowey DL, Hebert D, Arbus GS, et al. Compliance with cyclosporine in adolescent renal transplant recipients. Pediatr Nephrol. 1997;11:547–51.
29. Moseley KL, Kershaw DB. African American and white disparities in pediatric kidney transplantation in the United States—unfortunate or unjust? Camb Q Healthc Ethics. 2012;21 (3):353–65.
30. McAllister S, Buckner EB, White-Williams C. Medication adherence after heart transplantation: adolescents and their issues. Prog Transpl. 2006;16(4):317–23.
31. Bullington P, Pawola L, Walker R, Valenta A, Briars L, John E. Identification of medication nonadherence factors in adolescent transplant patients: the patient's viewpoint. Pediatr Transpl. 2007;11(8):914–21.
32. Levenson JL, Olbrisch ME. Psychosocial evaluation of organ transplant candidates. A comparative survey of process, criteria, and outcomes in heart, liver, and kidney transplantation. Psychosomatics. 1993;34:314.
33. Anderson K, Devitt J, Cunningham J, Preece C, Jardine M, Cass A. If you can't comply with dialysis, how do you expect me to trust you with transplantation? Australian nephrologists' views on indigenous Australians' 'non-compliance' and their suitability for kidney transplantation. Int J Equity Health. 2012;11:21-9276-11-21.
34. Dobbels F, Vanhaecke J, Dupont L, Nevens F, Verleden G, Pirenne J, De Geest S. Pretransplant predictors of posttransplant adherence and clinical outcome: an evidence base for pretransplant psychosocial screening. Transplantation. 2009;87(10):1497–504.
35. Seiden DJ, Frader J, Gatter R, Rhodes R. Retransplantation and the "noncompliant" patient. Camb Q Healthc Ethics. 1999 Summer;8(3):375–81.
36. Canadian Institute for Health Information. Canadian organ replacement register annual report: treatment of end-stage organ failure in Canada, 2003 to 2012 (Internet). (Feb 1 2014; cited Dec 1 2014). Available from: https://secure.cihi.ca/free_products/2014_CORR_Annual_Report_EN.pdf.
37. Kaufman BD, Jessup M. Adult and pediatric perspectives on heart retransplant. World J Pediatr Congenit Heart Surg. 2013;4(1):75–9.
38. Tong A, Jan S, Wong G, Craig JC, Irving M, Chadban S, et al. Rationing scarce organs for transplantation: health care provider perspectives on wait-listing and organ allocation. Clin Transpl. 2013;27(1):60–71.
39. Dunn TB, Browne BJ, Gillingham KJ, et al. Selective retransplant after graft loss to nonadherence: success with a second chance. Am J Transpl. 2009;9:1337–46.

40. Cass A, Devitt J, Preece C, Cunningham J, Anderson K, Snelling P, Eris J, Ayanian J. Barriers to access by Indigenous Australians to kidney transplantation: the IMPAKT study. Nephrology (Carlton). 2004;9(Suppl 4):S144–6.
41. Kuypers DR, Peeters PC, Sennesael JJ, Kianda MN, Vrijens B, Kristanto P, Dobbels F, Vanrenterghem Y, Kanaan N. ADMIRAD study team. Improved adherence to tacrolimus once-daily formulation in renal recipients: a randomized controlled trial using electronic monitoring. Transplantation. 2013;95(2):333–40.
42. Dhanireddy KK, Maniscalco J, Kirk AD. Is tolerance induction the answer to adolescent non-adherence? Pediatr Transpl. 2005;9(3):357–63.
43. Lantos JD, Warady BA. Should a nonadherent adolescent receive a second kidney? Virtual Mentor. 2012;14(3):190–3.
44. Rianthavorn P, Ettenger RB. Medication nonadherence in the adolescent renal transplant recipient: a clinician's viewpoint. Pediatr Transpl. 2005;9(3):398–407.
45. DeMaso DR, Douglas Kelley S, Bastardi H, O'Brien P, Blume ED. The longitudinal impact of psychological functioning, medical severity, and family functioning in pediatric heart transplantation. J Heart Lung Transpl. 2004;23(4):473–80.
46. Limbers CA, Neighbors K, Martz K, Bucuvalas JC, Webb T, Varni JW, Alonso EM. Studies of pediatric liver transplantation functional outcomes group. Health-related quality of life in pediatric liver transplant recipients compared with other chronic disease groups. Pediatr Transpl. 2011;15(3):245–53.
47. deSante J, Caplan A, Hippen B, Testa G, Lantos JD. Was Sarah Murnaghan treated justly? Pediatrics. 2014;134(1):155–62.
48. Cummings CL, Diefenbach KA, Mercurio MR. Counselling variation among physicians regarding intestinal transplant for short bowel syndrome. J Med Ethics. 2014;40(10):665–70.
49. Scott ES, Reppucci D, Woolard JL. Evaluating adolescent decision making in legal contexts. Law Hum Behav. 1995;19(3):221–44.
50. Beauchamp TL, Childress JF. Principles of biomedical ethics. 5th ed. New York, NY: Oxford University Press, Inc; 2001.
51. Ross LF. Health care decision-making by children: is it in their best interest? Hastings Cent Rep. 1997;27:41–5.
52. Gilmore S, Herring J. 'No' is the hardest word: consent and children's autonomy. Child Family Law Q. 2011;23(1):3–25.
53. Coggon J. Varied and principled understandings of autonomy in English law: justifiable inconsistency or blinkered moralism? Health Care Anal. 2007;15(3):235–55.
54. Barkham P. Hannah's choice. The guardian (internet) 2008 November 12. (cited 2014 November 11). Available from: Patrick Barkham, The Guardian, November 12 2008.
55. Dyer C. Trust decides against legal action to force girl to receive heart transplant. BMJ. 2008;12:337.
56. Stocks J. Mail online July 30 2011.
57. Silber TJ. Justified paternalism in adolescent health care. Cases of anorexia nervosa and substance abuse. J Adolesc Health Care. 1989;10:449–53.
58. Rhodes R, Aggarwal S, Schiano TD. Overdose with suicidal intent: ethical considerations for liver transplant programs. Liver Transpl. 2011;17(9):1111–6.
59. Statistics Canada (Internet) (last update Jul 25 2012; cited Dec 1 2014). Available from: http://www.statcan.gc.ca/pub/84-215-x/2012001/tbl/t003-eng.htm.
60. American Academy of Pediatrics Committee on Adolescence. Pediatrics. 2014;134(4): e1244–56.
61. Kann L, Kinchen S, Shanklin SL, Flint KH, Kawkins J, Harris WA, Lowry R, Olsen EO, McManus T, Chyen D, Whittle L, Taylor E, Demissie Z, Brener N, Thornton J, Moore J, Zaza S. Centers for disease control and prevention (CDC). Youth risk behavior surveillance–United States, 2013. MMWR Surveill Summ. 2014;13:63.
62. C.C.S.M. c. D104 The drivers and vehicles act manitoba 2005.
63. Bill 126—Zero Blood alcohol concentration (BAC) for young drivers and escalating sanctions for novice drivers Ontario (Canada) 2010.

64. American Academy of Pediatrics, Committee on Bioethics. Informed consent, parental permission, and assent in pediatric practice. Pediatrics. 1995;95:314–7.
65. Frader JA, Flanagan E. Minors as decision makers (internet). American Academy of pediatrics bioethics case-based teaching guide; 2011 (cited 2014 November 11). Available online at: http://www2.aap.org/sections/bioethics/PDFs/Curriculum_Session5.pdf.
66. Canadian Pediatric Society Bioethics Committee. Treatment decisions regarding infants, children and adolescents. Paediatr Child Health. 2004;9(2):99–114.
67. British Medical Association. Consent, rights and choices in health care for children and young people. London: BMJ Books; 2000.
68. Buchanan AE, Brock DW. Deciding for others: the ethics of surrogate decision making. Cambridge: Cambridge University Press; 1989.
69. Appelbaum PS, Grisso T. Assessing patients' capacities to consent to treatment. N Engl J Med. 1988;319(25):1635–8. Erratum in: N Engl J Med 1989;320(11):748.
70. Weir RF, Peters C. Affirming the decisions adolescents make about life and death. Hastings Cent Rep. 1997;27(6):29–40.
71. Gillick v West Norfolk & Wisbech area health authority, UKHL 7. 17 October 1985.
72. A.C. v. Manitoba (Director of child and family services), Supreme Court of Canada, June 26, 2009, Docket 31955.
73. United Network for organ sharing (internet) (Last updated Dec 1 2014; cited Dec 1 2014). Available from: http://www.unos.org.
74. Benden C, Goldfarb SB, Edwards LB, Kucheryavaya AY, Christie JD, Dipchand AI, Dobbels F, Levvey BJ, Lund LH, Meiser B, Yusen RD, Stehlik J. International society for heart and lung transplantation. The registry of the international society for heart and lung transplantation: seventeenth official pediatric lung and heart-lung transplantation report–2014; focus theme: retransplantation. J Heart Lung Transpl. 2014;33(10):1025–33.
75. Singh TP, Milliren CE, Almond CS, Graham D. Survival benefit from transplantation in patients listed for heart transplantation in the United States. J Am Coll Cardiol. 2014;63 (12):1169–78.
76. Nizič-Kos T, Ponikvar A, Buturović-Ponikvar J. Reasons for refusing kidney transplantation among chronic dialysis patients. Ther Apher Dial. 2013;17(4):419–24.
77. Pennsylvania. Superior court. In re Duran. West's Atl Rep. 2001;769:497–508.
78. Orfali K, Anderson-Shaw L. When medical cure is not an unmitigated good. Perspect Biol Med 2005;48(2):282–292.
79. Sinclair SJ. Involvement of adolescents in decision making for heart transplants. MCN Am J Matern Child Nurs. 2009;34(5):276–81.
80. Jerson B, D'Urso C, Arnon R, Miloh T, Iyer K, Kerkar N, Annunziato RA. Adolescent transplant recipients as peer mentors: a program to improve self-management and health-related quality of life. Pediatr Transpl. 2013;17(7):612–20.
81. Wright L, Pennington JJ, Abbey S, Young E, Haines J, Ross H. Evaluation of a mentorship program for heart transplant patients. J Heart Lung Transpl. 2001;20(9):1030–3.
82. Zelikovsky N, Petrongolo J. Utilizing peer mentors for adolescents with chronic health conditions: potential benefits and complications. Pediatr Transpl. 2013;17(7):589–91.
83. Dyer C. English teenager given heart transplant against her will. BMJ. 1999;319(7204):209.
84. Fowler A, Freiberger D, Moonan M. Palliative and end-of-life care in pediatric solid organ transplantation. Pediatr Transpl. 2014 Nov 25. (epub ahead of print).
85. McKay DB, Josephson MA, Armenti VT, August P, Coscia LA, Davis CL, Davison JM, Easterling T, Friedman JE, Hou S, Karlix J, Lake KD, Lindheimer M, Matas AJ, Moritz MJ, Riely CA, Ross LF, Scott JR, Wagoner LE, Wrenshall L, Adams PL, Bumgardner GL, Fine RN, Goral S, Krams SM, Martinez OM, Tolkoff Rubin N, Pavlakis M, Scantlebury V, Women's Health Committee of the American Society of Transplantation. Reproduction and transplantation: report on the AST consensus conference on reproductive issues and transplantation. Am J Transpl. 2005;5(7):1592–9.
86. Sucato GS, Murray PJ. Developmental and reproductive health issues in adolescent solid organ transplant recipients. Semin Pediatr Surg. 2006;15(3):170–8.

87. Wyld ML, Clayton PA, Jesudason S, Chadban SJ, Alexander SI. Pregnancy outcomes for kidney transplant recipients. Am J Transpl. 2013;13(12):3173–82.
88. Levidiotis V, Chang S, McDonald S. Pregnancy and maternal outcomes among kidney transplant recipients. J Am Soc Nephrol. 2009;20(11):2433–40.
89. Ville Y, Fernandez H, Samuel D, Bismuth H, Frydman R. Pregnancy in liver transplant recipients: course and outcome in 19 cases. Am J Obstet Gynecol. 1993;168(3 Pt 1):896–902.
90. Wagoner LE, Taylor DO, Olsen SL, Price GD, Rasmussen LG, Larsen CB, et al. Immunosuppressive therapy, management, and outcome of heart transplant recipients during pregnancy. J Heart Lung Transpl. 1994;13:993–1000.
91. Centers for Disease Control and Prevention. US medical eligibility criteria for contraceptive use, 2010. MMWR Recomm Rep. 2010;59 (RR-4):1–6. (Jun 22 2012; cited Dec 1 2014). Available from: www.cdc.gov/mmwr/preview/mmwrhtml/rr5904a1.htm?s_cid=rr5904a1_e.
92. Guazzelli CA, Torloni MR, Sanches TF, Barbieri M, Pestana JO. Contraceptive counseling and use among 197 female kidney transplant recipients. Transplantation. 2008;86(5):669–72.
93. Surís JC, Resnick MD, Cassuto N, Blum RW. Sexual behavior of adolescents with chronic disease and disability. J Adolesc Health. 1996;19(2):124–31.
94. Choquet M, Du Pasquier Fediaevsky L, Manfredi R. Sexual behavior among adolescents reporting chronic conditions: a French national survey. J Adolesc Health. 1997;20(1):62–7.
95. Melzer SM, Leadbeater B, Reisman L, Jaffe LR, Lieberman KV. Characteristics of social networks in adolescents with end-stage renal disease treated with renal transplantation. J Adolesc Health Care. 1989;10(4):308–12.
96. Henning P, Tomlinson L, Rigden SP, Haycock GB, Chantler C. Long term outcome of treatment of end stage renal failure. Arch Dis Child. 1988;63(1):35–40.
97. Kosola S, Lampela H, Lauronen J, Mäkisalo H, Jalanko H, Qvist E, Pakarinen MP. General health, health-related quality of life and sexual health after pediatric liver transplantation: a nationwide study. Am J Transpl. 2012;12(2):420–7.
98. Centers for Disease Control and Prevention. Contraception (cited Dec Nov 17 2014). Available at: http://www.cdc.gov/reproductivehealth/unintendedpregnancy/contraception.htm#44.
99. Guttmacher Institute. An overview of minors' consent law. State policies in brief as] of June 1, 2014 (June 1 2014; cited Nov 14 2014). Available from: www.guttmacher.org/statecenter/spibs/spib_MACS.pdf.
100. Tackling Violent Crime Act S.C. 2008, c. 6 (Canada).
101. Mycophenolate REMS (cited December 1 2014). Available from: https://www.mycophenolaterems.com/.
102. Choi JS, Koren G, Nulman I. Pregnancy and isotretinoin therapy. CMAJ. 2013;185(5).
103. Steinkellner A, Chen W, Denison SE. Adherence to oral contraception in women on category X medications. Am J Med. 2010;123(10):929–34.
104. Whitney SN, McCullough LB. Physicians' silent decisions: because patient autonomy does not always come first. Am J Bioeth. 2007;7(7):33–8.
105. Kon AA. Silent decisions or veiled paternalism? Physicians are not experts in judging character. Am J Bioeth. 2007;7(7):40–2.
106. Watson AR. Non-compliance and transfer from pediatric to adult transplant unit. Pediatr Nephrol. 2000;14:469–72.
107. Blum RW, Garell D, Hodgman CH, et al. Transition from child-centered to adult health-care systems with chronic conditions. A position paper of the Society for Adolescent Medicine. J Adolesc Health. 1993;14:570–6.
108. McDonagh JE. Growing up and moving on: transition to adult care. Pediatr Transpl 2004;9:364–372.
109. Bell LE, Bartosh SM, Davis CL, et al. Am J Transpl 2008;8:2230–2242.
110. Watson AR, Harden PN, Ferris ME, et al. Kidney Int. 2011;80:704–707.
111. Kaufman M Pediatr Transpl. 2006;10:413–417.
112. Willoughby l, Fukami S, Bunnapradist S, et al. Pediatr Transpl. 2006;11:127–131.

113. Forbes TA, Watson AR, Zurowska A, et al. Pediatr Nephrol. 2014;29:1617–1624.
114. Ross LF, Thistlethwaite JR Jr. Age should not be considered in the allocation of deceased donor kidneys. Semin Dial. 2012;25(6):675–81.
115. Transplant Manitoba Gift of Life Program. Kidney Allocation. Available at: http://www.transplantmanitoba.ca/transplant-program/kidney-allocation. Accessed 9 Dec 2014.
116. Eurotransplant Kidney Allocation Systems Available at: http://www.eurotransplant.org/cms/mediaobject.php?file=Chapter4_thekidney7.pdf. Accessed 9 Dec 2014.

Afterword

We invited an international group of experts to write chapters which identified and explored complex ethical questions pertaining to pediatric organ transplantation. Our hope was that the book would provide a theoretical and practical overview of the specific ethical issues chosen as well as the legal issues they raise. The authors shared rich insights, analyses and reflections from their views as ethicists, physicians, nurses, lawyers, trainees, academics and philosophers. Pediatric transplant medicine involves unique circumstances that require a tailored ethical approach for each presenting situation. Correspondingly it was not our intention to provide black and white answers to the nuanced issues presented. Our goal was not to settle the discourse but rather to highlight the complexity of these challenges and provide considerations for practice and ethically defensible strategies and good practices.

With continuing medical advances ethical issues in the field of transplantation will continue to emerge for which clinicians, scholars and policy makers will grapple. In terms of next steps, as there is a paucity of books on transplantation ethics generally, and pediatrics more specifically, we hope this book will inspire continued dialogue and debate about the emerging issues pertaining to the intersection of ethics, pediatrics and transplantation.

It is our hope that this book will serve as a resource for pediatricians, transplant health care professionals, trainees, graduate students, scholars, practitioners of bioethics and health policy makers on a number of important issues. This collection of 16 essays provides an in-depth discussion and overview of the spectrum of ethical issues in the field. We hope that readers will come away with a deeper appreciation of the ethical issues and practical guidance.

R.A. Greenberg et al. (eds.), *Ethical Issues in Pediatric Organ Transplantation*,
International Library of Ethics, Law, and the New Medicine 66,
DOI 10.1007/978-3-319-29185-7

Index

R.A. Greenberg et al. (eds.), *Ethical Issues in Pediatric Organ Transplantation*, International Library of Ethics, Law, and the New Medicine 66,
DOI 10.1007/978-3-319-29185-7